# The TAB Electronics Guide to Understanding Electricity and Electronics

*G. Randy Slone*

**TAB Books**
Imprint of McGraw-Hill

New York  San Francisco  Washington, D.C.  Auckland  Bogotá  Caracas  Lisbon  London
Madrid  Mexico City  Milan  Montreal  New Delhi  San Juan  Singapore  Sydney  Tokyo  Toronto

# McGraw-Hill
*A Division of The **McGraw·Hill** Companies*

©1996 by The McGraw-Hill Companies, Inc.
Published by TAB Books, an Imprint of McGraw-Hill

Printed in the United States of America. All rights reserved. The publisher takes no responsibility for the use of any materials or methods described in this book, nor for the products thereof.

pbk     5 6 7 8 9    FGR/FGR    9 0 0 9 8
hc      3 4 5 6 7 8 9    FGR/FGR    9 0 0 9 8 7

Product or brand names used in this book may be trade names or trademarks. Where we believe that there may be proprietary claims to such trade names or trademarks, the name has been used with an initial capital or it has been capitalized in the style used by the name claimant. Regardless of the capitalization used, all such names have been used in an editorial manner without any intent to convey endorsement of or other affiliation with the name claimant. Neither the author nor the publisher intends to express any judgment as to the validity or legal status of any such proprietary claims.

**Library of Congress Cataloging-in-Publication Data**
Slone, G. Randy.
    The TAB electronics guide to understanding electricity and
 electronics / by G. Randy Slone.
      p.   cm.
    Includes bibliographical references and index.
    ISBN 0-07-058215-7 (h). — ISBN 0-07-058216-5 (p)
    1. Electronics.   2. Electricity.   I. Title.
TK7816.S57    1995
621.381—dc20                                                        95-22897
                                                                                          CIP

McGraw-Hill books are available at special quantity discounts to use as premiums and sales promotions, or for use in corporate training programs. For more information, please write to the Director of Special Sales, McGraw-Hill, 11 West 19th Street, New York, NY 10011. Or contact your local bookstore.

Acquisitions editor: Roland Phelps
Editorial team: Joanne Slike, Executive Editor
                      Andrew Yoder, Managing Editor
                      Bill Taylor, N3LRA, Book Editor
Production team: Katherine G. Brown, Director
                        Joann Woy, Indexer
Design team: Jaclyn J. Boone, Designer                                             0582165
                 Katherine Lukaszewicz, Associate Designer                    EL3

# Contents

**Preface   x**
**Introduction   xi**
**Dedication   xii**

**1 Getting started   1**
   Establishing reasonable goals   1
      What is unreasonable?   1
      What is reasonable?   2
   Obtaining the informational tools   4
      Text books   4
      Data books   4
      Periodicals   6
   Setting up a lab   7
      The workbench   8
      Hand tools   8
      Miscellaneous supplies   12
      Electrical lab power   12
   Basic test equipment   14
   Starting a parts and materials inventory   18
      Salvaging   18
      What to salvage   20
      Salvaging electronic components   21
      Buying from surplus dealers   23

**2 Basic electrical concepts   25**
   Electronic components   25
      Resistors   26
      Potentiometers, rheostats, and resistive devices   28

       Capacitors   30
       Inductors   32
       Diodes   33
       Transistors   34
       Integrated circuits   34
           A final note on parts identification   35
  Characteristics of electricity   35
       Voltage   38
       Current   39
       Resistance   41
       Alternating current (ac) and direct current (dc)   41
       Conductance   42
       Power   42
  Laws of electricity   42
       Ohm's law   43
       Parallel circuit analysis   46
       Series circuit analysis   50
       Series/parallel circuits   52
       Power   54
       Common electronic prefixes   56

## 3 The transformer and ac power   57
  Ac waveshapes   57
  Ac frequency   58
  Ac amplitude   59
  Ac calculations   61
  Inductance   63
       Dc resistance   69
  Transformers   70
  Soldering   72
       Soldering overview   73
       Soldering for the first time   74
       Soldering procedure   75
       Desoldering procedure   76
  Assembling and testing the first section of a lab power supply   76
       Safety is emphasized throughout this book   77
       Materials needed for the completion of this section   77
       Mounting the hardware   79
       Wiring and testing procedure   80
       Testing the first section of the lab power supply   83

## 4 Rectification   89
Introduction to solid-state devices   89
- Diode principles   91
- Referencing   101

Assembly and testing: second section of a lab power supply   101
- Testing bridge rectifier modules   102

## 5 Capacitance   107
Capacitor types and construction   107
Basic capacitor principles   110
- Filter capacitors   114

Designing raw dc power supplies   119
Assembling and testing: third section of a lab power supply   122
- Testing the power supply   123

Food for thought   125

## 6 Transistors   127
Preliminary definitions   127
Introduction to transistors   128
Transistor principles   129
Common transistor configurations   134
- Adding some new concepts   134
- The common-emitter configuration   135
- The common-collector configuration   138
- The common-base configuration   140
- Transistor amplifier comparisons   141

Impedance matching   141
Assembling and testing of the last section of a lab power supply   142
- Assembling the circuit board   144
- Mounting Q1 and Q2   146
- Circuit description   147
- Testing the power supply   148
- In case of difficulty   149
- Checking transistors with a DVM   149

## 7 Special-purpose diodes and optoelectronic devices   151
Zener diodes   151
- Designing simple zener-regulated power supplies   153

Varactor diodes   156
Schottky diodes   157
Tunnel diodes   157
Diacs   157
Fast recovery diodes   158
Noise and transient suppression diodes   158
A basic course in quantum physics   159
    LEDs (light emitting diodes)   160
    Optoisolators, optocouplers, and photoeyes   161
    Photodiodes, phototransistors, and photoresistive cells   162
    Laser diodes   162
    LCDs (liquid crystal displays)   163
    CCDs (charge coupled devices)   164
Circuit potpourri!   164
    Preliminary steps to project building   165
    Flashing lights, anyone?   165
    Three lights are better than two!   167
    A mouse in the house   169
    A sound improvement   170
    A delay is sometimes beneficial   171
    A long-running series   173
    Keep it steady   175
    Double your pleasure   175
    Show the blow   176

## 8 Audio amplification systems   179
Transistor biasing and load considerations   179
    Amplifier classes   181
    Additional amplifier classification   184
Audio amplifier output configurations   185
    Audio amplifier definitions   185
    Power amplifier operational basics   189
Building high-quality audio systems   193
    Building a professional-quality audio amplifier   198
    Building a printed circuit board   200
    Amplifier assembly   201
    Initial set-up   205
    Amplifier power supply   207
    Speaker protection circuit   207

## 9 Power control  211
Silicon-controlled rectifiers (SCRs)  211
The triac  214
UJTs, diacs, and neon tubes  215
Building a soldering iron controller  216
Circuit potpourri  219
- Watts an easier way?  219
- Isolation may be good for the soul  220
- Curiosity catcher  222
- See ya later, oscillator  223

## 10 Field-effect transistors  227
FET operational principles  228
- FET parameters  230
- FET biasing considerations  231
- Static electricity: an unseen danger  232

Building a high-quality MOSFET audio amplifier  233
Circuit potpourri  236
- Sounds like fun  236
- Emergency automobile flasher  238
- Home-made ac  239

## 11 Batteries  241
Battery types  241
- Battery ratings  242
- Battery care  243
- A few words of caution  243
- Recharging batteries  243

Building a general-purpose battery charger  244

## 12 Integrated circuits  247
Operational amplifiers  248
IC or hybrid audio amplifiers  251
IC voltage regulators  251
Special-purpose ICs  252
Improvement of lab-quality power supply  252
Building a quad power supply  254
Circuit potpourri  255

Noise hangs in the balance  255
Hum reducer  256
Let the band pass  257
High-versatility filters  257
How was your trip?  259
Watt an amplifier  260

## 13 Digital electronics  261

Logic gates  262
    Combining logic gates  265
Multivibrators  266
Digital clocks  271
Shift registers  271
Digital memory devices  272
Summary  272
Circuit potpourri  273
    How is your pulse?  273
    Improved digital pulser  274
    Digital logic probe  275
    Bounceless is better  276
    High-stability crystal timebase  276
    PWM motor control  277

## 14 Computers  279

How a computer works  279
    Hardware, software, and firmware  282
    Memory and data storage  283
    Input/output devices  284
    Programming computers  284
    Computer processing of analog signals  285

## 15 More about capacitors and inductors  287

Inductive reactance  287
Capacitive reactance  288
    Capacitive and inductive comparison  289
    Combining inductance and capacitance  290
Reflected impedance  291
Resonance  291

Passive filters 294
Circuit potpourri 298
    Proportional, integral, and differential action 298
    Square waves galore 300
    Basic tone generator 302
    Where's the fire? 302
    Building a lab function generator 303
    Building a percussion synthesizer 304

## 16 Radio and television 309
Modulation 309
Transmission and reception of RF signals 311
Radio receivers 314
AM and FM detectors 316
Television 317

## Appendix A Symbols and equations 321
## Appendix B Sources for electronic materials 333
## Index 337

# Preface

WHEN I WAS 10 YEARS OLD, I PURCHASED AN INTRODUCTORY electronics book at a local pharmacy. I cherished this book because I knew it contained marvelous and fascinating knowledge about electricity; a mysterious and awesome force unseen by the naked eye! However, after carrying it back and forth to school every day for months and spending countless study-hall hours trying to decipher the strange terms and symbolisms, I finally became discouraged and concluded that it was beyond my comprehension.

In reflecting on that experience, I now recognize that it would have been almost impossible for me to understand that book. As with so many other introductory-level electronic textbooks, too many assumptions were made regarding the reader's educational background, especially in regard to mathematics.

In part, because of my childhood experience, I became very excited (and grateful) when this project was offered to me by the people at TAB/McGraw-Hill. For many years, I have wanted to write the book that I should have bought in that small-town drugstore.

In a sense, you could say this text was written through the eyes of a 10-year-old boy; the awe, mystery, and excitement is still there!

A special thanks to my good friend John Adkins, who helped me with most of the photography, and who plays a *mean* five-string banjo.

I would also like to voice my appreciation to Roland Phelps and to the staff at TAB/McGraw-Hill for their help and sincere interest in this project.

# Introduction

THROUGHOUT MY MANY YEARS OF TEACHING EXPERIENCE, I have discovered two essential processes involved with every successful educational venture; you learn by doing, and you do it because you enjoy it. *Understanding Electricity and Electronics* has been designed from beginning to end with these two principles in mind. The many projects incorporated into the text are not boring, theory-oriented demonstration circuits. Quite to the contrary, I have spent considerable time and research to include the most practical, educational, and fun projects available for a general-interest audience. My philosophy is simple; if a reader is going to invest the time and money into understanding and building electronic projects, why not make them fun and impressive?

*Understanding Electricity and Electronics* has been designed to be the best "first book" for anyone interested in becoming proficient in the electrical/electronic fields. The beginning chapters guide you, via step-by-step methods, in establishing an inexpensive electronics workplace and in acquiring all of the informational and mechanical tools needed. The succeeding chapters combine electronic projects with the appropriate text, both of which advance at an easy-to-digest pace. Expensive electronic test equipment is not required for building, testing, or utilizing the projects; the only essential piece of equipment needed is a volt-ohm meter. The appendices provide a handy reference for commonly used equations, symbols, and supply sources. In short, *Understanding Electricity and Electronics* contains everything that the novice needs to know, to have, and to do to become a competent electronics hobbyist and experimenter.

To my Lord and Savior, Jesus Christ, from whom all good things originate, to the glory of our Heavenly Father.

# Getting started

AS THE OLD PROVERB STATES, "EVERY LONG JOURNEY starts with the first step." This chapter deals with the first steps in developing a new and fascinating activity in your life. My goal is to make the *journey* a comfortable, entertaining, and rewarding experience. If you will supply a little time, effort, and patience, you can accomplish your goal.

## Establishing reasonable goals

All successful individuals achieve their varying degrees of success by establishing and accomplishing goals. Many people establish goals without even realizing it. In many cases, these goals can be classified as dreams, aspirations, or concepts. Reflect back to the last time that you were really pleased with some accomplishment in your life. This accomplishment required personal motivation, planning, effort, success, and the satisfaction brought on by that success. The actual goal was determined during the planning stage, but it was not achieved until the success stage. Between these stages, occurred all of the work and effort.

Take these concepts one step further. Assume that an unobtainable goal has been established. An indefinite amount of work and effort might be invested in this elusive goal, but the end result will be discouragement and defeat. To make matters worse, the individual who established this unreasonable goal might become hesitant to set new goals, because of the fear of a similarly wasted effort and failure.

In my opinion, the establishment of unreasonable goals is the most significant obstacle that you must face in your journey toward becoming proficient in the field of electronics.

### What is unreasonable?

For the sake of illustration, assume for the moment that this book is entitled *Building Your Own Automobiles*. Few people would buy such a book; only those motivated by the desire to build their own car from scratch. Such an idea is ludicrous for several reasons. For one, it isn't practical. The cost of buying the individual pieces and parts to build a finished car would cost five to ten times as much as a new car that has been factory built and tested. The builder would have to be very knowledgeable, and experienced in a wide variety of the specialized skills existing within the automobile industry: such as, diagnostic alignment, automobile electrical systems, bodywork and finishing techniques, etc. In addition, it staggers the mind to imagine the vast array of the specialized shop tools that would be required for such a project!

There are similar circumstances in the electronics field. Even electronic geniuses don't build their own television sets from thousands of tiny parts that they pick up at their local electronics dealer. It is also helpful to understand that most electronic systems are not invented, designed, and built by single individuals. For example, the consumer electronic products that we all enjoy every day (TVs, radios, CD players, VCRs, etc.) are actually evolutionary products. They have been re-designed and improved over a period of many years, by many different design engineers.

Generally, goals involving the *from scratch* building of complex electronic systems are usually unreasonable. Also, from a conceptual point of view, the field of electronics is vast, and divided into many specialized fields. If your goal is to understand everything about electronics, I wish you a very long life. I believe you may need more than one lifetime to accomplish a goal of that magnitude!

## What is reasonable?

This is the exciting part in regard to the electrical/electronic fields; because the possibilities are limited only by your ingenuity and imagination. I have known many people who got started in electronics as a rewarding hobby, only to find themselves in a high-paying career before they knew it! For example, a good friend of mine became interested in home computers as a hobby. As he continued to expand his computer system, his personal financial situation forced him to locate the most inexpensive places to purchase the pieces for his system. As he began to impress his friends and relatives with his computer system, they decided to buy their own systems. One day, almost by accident, he discovered that he could supply them with systems identical to his own for substantially less money than the local computer store; even after he added in a healthy profit for himself. As a result, he changed his career in the midstream of life (he was an investment counselor), and opened a very successful computer store. I know this story very well, because I bought my first computer from him!

I have had many friends and acquaintances who began tinkering with electronics in their homes as a hobby. Eventually, they found themselves deluged with friends, neighbors, and relatives bringing them everything from portable television sets to computer monitors; all harping the same request, "When you get a few minutes, would you please take a look at this. I think its a simple problem because it worked just fine yesterday." This leaves the besieged electronics tinkerer with one of two choices: either begin charging for repair services, or become a candidate for the "Good Samaritan of the Year Award". Of those who began to charge for their services, many have found lucrative and rewarding careers.

Many people have the mistaken belief that a career in electronics is not possible without a formal degree from an accredited college or technical institution. A formal degree will certainly enhance and accelerate your career progress, but there are many career pathways for non-degreed individuals as well. For example, many electronic salespersons do not have a deep, intricate knowledge of the products they sell; a functional and applicative understanding is all that is required. The consumer electronics repair field is loaded with people who were called *tube jockeys* back in the 1950s. As the field of electronics evolved from vacuum tubes to solid-state technology, they advanced right along with it by reading the various electronic periodicals published within their field. Many younger people, who are successful in these same fields, received their education from correspondence or vocational schools.

Non-degreed electronic hobbyists often find careers within their hobby. For example, the hobbyist who collects a large parts inventory for personal use may begin to sell these parts at substantial profits. Many local electronic stores have had their beginnings in this manner; not to mention some large national parts distribution chains.

Your personal interests will play a big role in discovering and opening doors to possible career opportunities. A hobbyist who likes to tinker with automotive sound systems may start a part-time business installing these systems in the local community. A little effort, perseverance, and dedication to performing quality work can convert it to a full-time lucrative career.

Non-degreed electrical and/or electronic career opportunities are common in most industrial manufacturing facilities. The majority of electrical maintenance personnel I have trained over the years have had little or no formal classroom training in the electrical/electronic fields. Within the industrial manufacturing community, any prior experience in this area (even at the hobbyist level) is usually given weighty consideration in hiring and job promotions.

The list of possibilities goes on and on. Don't expect to go around designing sentient robots, or building laboratories that look like they're out of an old Boris Karloff movie. However, don't stumble over the diamonds, while you are looking for the gold! It is always reasonable to expect to go as far as your effort and perseverance will take you.

## Obtaining the informational tools

The informational tools you will need for a successful hobby or career in the various electrical/electronic fields can be broken down into four categories: text books, data books, periodicals, and catalogs.

### Text books

Text books are generally self-explanatory as to their usefulness to any specific individual. As your experience grows, you will probably collect a reasonable library, according to your needs and interests. For the novice, I would recommend an *electronics dictionary* and a beginning-level *electronics math* book. It would also be wise for you to enroll into an electronics book club, or a similar technical book service. The monthly catalogs from these organizations will help to keep you abreast of current releases that are of great benefit to your continued advancement. A list of these organizations is included in the appendix at the back of this book.

### Data books

The manufacturers and distributors of electronic components publish data books, containing cross-referencing information and individual component specifications. A few examples of such books are *NTE Semiconductors*, *The GE Semiconductor Replacement Guide*, and *SK Replacement Cross-Reference Directory*. As you can see, the data book titles are self-explanatory.

Your first project in the field of electronics is to obtain all of the electronic data books that you can "get your hands on". The reason, for having included this section at the beginning of this book, is to give you ample time to order and receive a fair quantity of data books before you start your first projects. They are *that* essential. Many electronic manufacturers will supply their data books free of charge, if you simply call and ask them. This is especially true if you have started a small part-time or full-time business. Try this approach first; then, if you are not successful, they can be purchased from many different electronic supply companies.

Manufacturers' data books can be general or specific in nature. Try to acquire the general or broad-based data books in the beginning. As your interests begin to lean toward certain specific areas, you can obtain what you need at a later date.

Data books provide needed information in two critical areas: cross-referencing and parts specifications. Back in the days of vacuum tubes, the tube manufacturers would identify their tubes with certain generic numbers. In other words, a 12AU7 tube would always be labeled *12AU7*, regardless of the manufacturer. Unfortunately, this tradition did not carry over into the solid-state field. Although there are generic numbers for solid-state components, they are only used occasionally. Instead, you must rely on the manufacturer's cross-references, which are supplied in their data books. For example, suppose you needed to replace a defective transistor labeled *NTE 130*. If you had some NTE 130s in your personal stock, or if your local electronics shop carried the NTE line of components, you would simply use another NTE 130 as a replacement. But, if you didn't have one and the local parts store only carried the SK line of components, you would have to cross-reference the NTE 130 to its SK equivalent. In this case, it would be an SK3027. When you consider that there are dozens of major parts manufacturers, each using their own unique part numbers, you begin to appreciate the usefulness of an exhaustive cross-reference library.

Cross-referencing will also play an important role in acquiring a respectable parts inventory. All of the large surplus and wholesale electronic houses offer many electronic components at buy-out or wholesale prices. In most cases, you will have to cross-reference these parts to know if you can use them. In addition, if you salvage parts from used equipment to place in an inventory, you will have to cross-reference the used parts to know what they are.

One final word on cross-referencing electronic parts; it is not as difficult as it may seem from just reading this book. Upon receipt of your first few data books, spend a few minutes scanning through the cross-reference section. You should easily recognize

*Obtaining the informational tools*

how the parts are arranged according to sequential numbers and letters.

Electronic data books also provide the detailed specifications for electronic parts. When you begin to build or design electronic projects, you will need a working knowledge of the specifics of the various parts you intend to use. For example, the device parameters will define the electrical conditions for reliable operation (breakdown voltage, power dissipation, maximum current, etc.), and the pictorial diagrams will provide the necessary mechanical information (case style, lead designation, pin definitions, etc.).

If you have not understood some of the terms I have used in describing and explaining data books, don't worry. When you have a chance to skim through one, much of what is written here will become clear. The rest will be understood as you begin to build a few of the projects covered in the following chapters. The important thing to do right now is to GET THEM! A list of the sources, from which to obtain electronic data books, is provided in the appendix of this book.

## Periodicals

Periodicals are very important to anyone involved in the electrical/electronic fields for a variety of reasons. First, they keep the electronics enthusiast current and up-to-date in the latest technology. It has been said that a college-degreed electrical engineer will become obsolete in five years without a strong personal effort to stay current with technological advances. I do not necessarily agree with that statement; I believe it may take less than five years! Consider this; the IRS allows a business owner to totally depreciate a computer system in three years. The advancements in digital technology are so rapid that a typical computer system could be outdated in only one or two years after it has been purchased. The majority of electronic text books are either revised, or taken off the market in 3 to 5 years. Though this may seem to imply that a proficient electronics person must become a chronic bookworm, you actually stay current by reading periodicals.

Most periodicals and electronics magazines are constructed in such a way that they are enjoyable and entertaining to read. The general-interest periodicals will always contain something of interest to almost everyone and they will motivate the reader with new ideas and perspectives. The newest innovations in the industry are covered, along with their practical aspects. In many cases, the reader is provided with home brew projects to utilize these innovations. Periodicals also make great wish books, with their large variety of product advertisements. The publishers of these magazines also recognize that all of their readers are not at the same technical level; thus, the home projects will vary from easy to difficult, and from practical to just plain fun. Without even realizing it, you can stay up-to-date and have a great time doing it.

I recommend that you subscribe to several of the general-interest electronic periodicals. By the time you finish this text book, you should be able to understand and build most of the projects provided in these magazines, especially those that are of special interest to you. In addition, you will pick up little bits of information here and there that will help your progress, and spur your interest. One word of caution: do not become discouraged if you experience some degree of confusion as you read through these magazines for the first time. Electricity and electronics are, generally, not that difficult to understand. Simply be content with what you do understand, and the rest will follow in time as you progress toward your goal. A list of several of the best general-interest electronic periodicals is included in the appendix at the back of this book.

## Setting up a lab

While you are waiting for your first few data books and electronic periodicals to arrive in the mail, it would be prudent to turn your attentions to setting up an electronics lab. The *lab* is the room or area in your home or business where you will build, test, or repair electronic equipment. It is also the place where you will probably spend considerable time studying, experimenting, sorting parts, jumping for joy, and stewing in frustration. The lab is a **dangerous** place for novices and children, an eyesore to visiting guests, a probable aggravation to your spouse, a collection area for a large volume of "yet to be salvaged" electronic junk, and a secure area for all of your expensive tools and test equipment. The environment in the lab should be quiet, comfortable, and well lighted. With all of these considerations in mind, it is wise to put a little forethought into the best location to "set up shop".

I highly recommend choosing a room with a door that can be locked to keep out children, pets, and the overly curious. In addition to the possibility of personal injury to the uninvited, you will probably be working with equipment or projects that are easily damaged or "tampered with." Frequently, you will want to leave your work "in progress" overnight, or even for days at a time, so it must not be disturbed.

Garages are usually not the best place for a lab because of the difficulty in keeping a controlled environment. In addition to your personal discomfort from temperature extremes, most electronic equipment is very sensitive to the moisture that will condense on it in a cold environment (especially if you are bringing it in from a warm home or car). A damp environment is hard on tools and test equipment because of corrosion problems. If you plan on setting up your lab in a basement, a dehumidifier would be a wise investment. A spare bedroom or den is an excellent choice for a lab. Keep in mind that a lab does not have to be an enormously large place. It need only contain a workbench (about 3' × 6'), a 3- or 4-level bookshelf, a bare wall or shelving unit for small parts cabinets, and a closet or floor area for storage. A nice luxury would be a desk for studying, reading, drawing schematics, and miscellaneous paperwork.

### The workbench

A good size for an electronics lab workbench is about 3 × 6 feet. Of course, this can greatly vary according to your needs and what you might already have available. In most instances, it needs to be sturdy enough to hold about 80 pounds, stable enough to not be easily shaken (very annoying when trying to solder small or intricate parts), void of any cracks (small parts have a way of finding them), and it should be the correct height for comfortable use.

Various community organizations and churches often sell heavy-duty, fold-up tables, used for group meetings and meals, at very affordable prices. These same tables can be purchased at office supply stores. They make excellent electronic workbenches, and I personally use two of them for my office and lab. The formica tops are very durable, resistant to heat, and aesthetically pleasing.

If you prefer, building a lab workbench is a simple project. Particle board or plywood makes a good top, and 2 × 4s are adequate for legs. An old hollow-core (or solid-core) door also makes an excellent top with 2 × 4s or saw horses for legs. Commercial electronic workbenches, providing many convenient features, are also available, but they are very expensive.

### Hand tools

If you are any kind of "do-it-yourselfer," you probably already have the majority of hand tools you will need for working in electronics. However, in the event that your forté has been working on diesel trucks, keep in mind that the majority of your work will be with small items. A good electronics tool box will consist of small-to-medium sizes of the following common tools; needle nose pliers, side cutters, wire strippers, screwdrivers (both flat and phillips), nut drivers, socket sets, wrenches, tack hammers, files, and hack saws. In regard to powered hand tools, a 1/4-inch drill and a scroll saw are a good beginning for most work. A high-speed, hand-held grinder, with a variety of attachments, (such as the *Dremel Moto-Tool*) will be extremely handy for any fabrication project. A *nibbling tool* does a great job of cutting printed circuit board material, chassis material, and various types of metal or plastic enclosure boxes. Of course, there are specialized tools intended exclusively for use in the electronics industry. Most electronic parts suppliers will carry a fair selection of these.

It is best not to go overboard; spending a small fortune on a great variety of hand-held tools in the beginning. The tools mentioned so far are only suggestions. You will probably save a great deal of money by adding tools only as you need them.

One of the most important tools to the electronics enthusiast is the soldering iron. Do not get a soldering iron confused with a soldering gun. Soldering guns have a pistol grip, and are intended primarily for heavy-duty soldering applications. Their usefulness in the electronics field is very limited because they can easily damage printed circuit boards and electronic components. A soldering iron is straight like a pencil, with a very small point or tip. A typical soldering iron with stand is shown in Fig. 1-1.

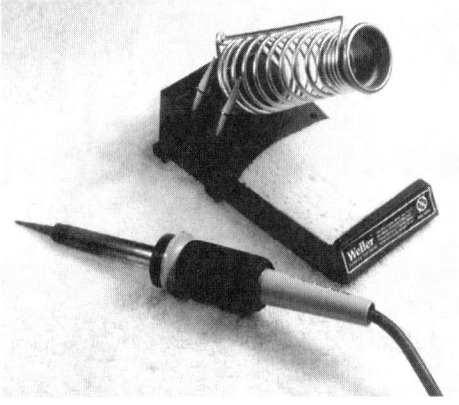

■ **1-1** *A typical soldering iron with stand and extra tips.*

The critical variable, with a soldering iron, is its tip temperature. If the tip temperature is too high, it could destroy heat-sensitive electronic components. If the tip temperature is too low, the solder might not flow or adhere properly to the joint and a poor electrical connection might result.

If the *load* on a soldering iron always remained constant, it would be fairly easy to maintain a constant tip temperature. But in actual use, the load will vary depending on the size of the joint and the amount of solder used. In other words, a large solder joint will conduct more heat away from the soldering iron tip, than will a smaller joint, causing the tip temperature to fall to a lower level. If the wattage (or heating power) to the tip were increased to compensate for this temperature drop, the tip might then become too hot when the iron is not in use, and potential destruction of components could result. Because of performance-versus-price reasons, there are four commonly available types of soldering irons on the marketplace:

1. *Nonadjustable*   Specified by wattage.
2. *Regulated*   Specified by wattage and tip temperature.
3. *Adjustable*   Wattage (power) control.
4. *Adjustable*   Temperature control.

The most common and least expensive type of soldering iron is the first type listed. It is nonadjustable, and is rated (or specified) by its wattage. This is an acceptable type with which to start out, or to throw into the tool box for emergency repairs when away from the lab. The disadvantage with this type of iron is the variation in tip temperature relative to the load. For general purpose electronic work, try to find one that is rated at about 30 watts.

The second type of soldering iron listed is the type that I use most of the time. The tip of this iron contains a special thermostatic switch that will maintain the tip temperature reasonably close to a specified point.

The third type of soldering iron incorporates a *light-dimmer* circuit to vary the amount of wattage the soldering iron is allowed to dissipate. For bigger jobs, the wattage can be increased to facilitate easier soldering. For smaller jobs, the wattage can be decreased for a lower tip temperature. The wattage control is manual with this type of iron, and the user must use caution not to get the tip temperature too hot when soldering smaller components.

The fourth type of soldering iron is the most versatile, and it is the most expensive. A heat sensing device is implanted close to the tip, so that the tip temperature can be continuously monitored and controlled by an electronic controller located in the holding stand for the soldering iron. These *soldering stations* have an adjustment control in their bases; thus, the exact tip temperature can be set and maintained, regardless of the load placed on the iron itself.

In addition to the soldering iron, you will need some accessories to perform quality soldering jobs. If your soldering iron doesn't come with a stand, buy one! I ended up burning my table, an expensive pair of pants, and my hand before I finally learned this lesson. Also, be sure the stand has a sponge holder. This essential tip-cleaning convenience is well worth the small additional expense. Pick up a few extra sponges, too.

A quality soldering iron will have a variety of different size tips available for it. If it doesn't, look for another soldering iron. You should choose a couple of small tips for intricate work, and some medium-size tips for general-purpose work.

When purchasing solder, buy only the *60/40 resin-core* type. Acid-core solder should never be used on electronic equipment.

If the need arises to remove a soldered-in component (or to correct a mistake), you will need a desoldering tool. Most desoldering tools consist of a spring-loaded plunger-in-a-tube housing with a hollow tip at one end. To remove unwanted solder: melt the solder with a soldering iron, place the desoldering tool tip close to the molten solder, and press the trigger. The trigger releases the spring-loaded plunger; thus creating a vacuum in the tube, and causing the molten solder to be sucked up into the tube. When purchasing the desoldering tool, be sure that replacement tips are available. The teflon tips will flare out with use, decreasing the effectiveness of the desoldering tool. A typical desoldering tool is shown in Fig. 1-2.

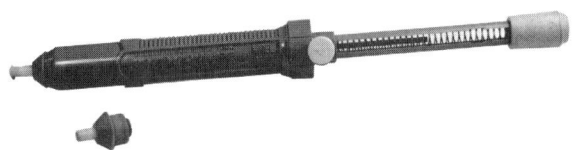

■ **1-2** *A common type of desoldering tool with an extra tip.*

Soldering is a learned skill, but certainly not a difficult one to master. A little practice, coupled with a conscientious attitude, is all that is required. The actual techniques to soldering are covered in chapter 3 of this book.

## Miscellaneous supplies

For the most part, the miscellaneous supplies that you will accumulate over a period of time will be a matter of common sense. This section lists a few items to help you get started.

For making temporary connections, and setting up certain tests, you will need a number of *alligator clip* leads. These consist of a short length of insulated wire (usually 10 to 15 inches) with a spring-loaded alligator clip at each end. I recommend purchasing (or fabricating) about a dozen.

A variety of sizes and colors of wire will be needed to build various projects. Some electronic supply stores offer a *variety pack*, which would be ideal for starting a new lab. The minimum you will probably need is a spool of medium hook-up wire (stranded, insulated, about 18 to 22 gauge), a spool of small hook-up wire (stranded, insulated, about 26 to 28 gauge), and a small spool of shielded, coaxial cable.

A few helpful cleaning supplies would be cotton swabs, isopropyl alcohol, some very fine sandpaper or emory paper, a can of flux remover, and a variety of small brushes (save your old toothbrushes for the lab).

## Electrical lab power

The area you have chosen for your lab will hopefully have a sufficient number of 120-Vac wall outlets to accommodate your needs. If, by any chance, you live in an older home that does not have grounded outlets (3 holes for each plug; hot, neutral, and ground), they should be upgraded to the grounded type in your lab. If you have any questions on how to do this, it would be wise to have a professional do it for you. Ungrounded or incorrectly wired outlets are dangerous!

You will find it very convenient to install an outlet strip (a rectangular enclosure with multiple outlets, as in Fig. 1-3) somewhere within easy access on your workbench. Try to find one with a lighted on-off switch, a plastic (or nonconductive) housing, and a 15-amp circuit breaker. It is best to find some way of mounting it securely to the workbench so that you can easily remove plugs with only one hand.

■ **1-3** *Multiple outlet ac power strip.*

Whenever the need arises to service or repair line-powered electronic equipment (the term *line-powered* means that the equipment must be plugged into a standard 120-Vac outlet), you will probably want to purchase an isolation transformer to power it (Fig. 1-4). This will become critically important if you are also using line-powered test equipment to perform tests and measurements. It will not be necessary to purchase an isolation transformer for any of the projects in this book, but you might want to keep your eyes open for a good bargain for future needs. The theory and purpose of isolation transformers are covered in a later chapter.

■ **1-4** *An isolation transformer specifically designed for an electronics test bench.*

If your interests lie in the industrial electronics field, you might find a need to provide 220-Vac power to your lab. This is easily accomplished with a step-up power transformer. As in the case of isolation transformers, an appropriate step-up transformer can be added on at any time in the future if the need arises.

## Basic test equipment

The most frequently used (and most important) piece of test equipment to the electrical or electronics enthusiast is commonly called the DVM (digital voltmeter). A *DVM* is used to measure and display voltage, current, and resistance. A typical DVM is shown in Fig. 1-5. In regards to function, a DVM is basically the same instrument as a VOM (volt-ohm-milliammeter), a DMM (digital multimeter), or a VTVM (vacuum tube voltmeter). The term *DVM* is most frequently used today, and it is the term I will use through the remainder of this book. When talking with experienced electrical or electronics personnel, they will probably refer to a DVM as a "voltmeter", or simply "meter."

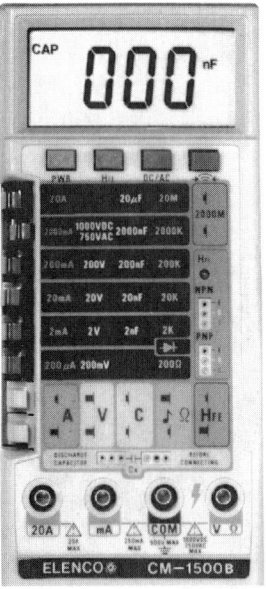

■ 1-5 *A versatile, multi-function digital voltmeter (DVM). Photograph courtesy of Elenco Electronics Inc., Wheeling IL.*

Modern DVMs can typically perform a variety of functions other than simple voltage, current, and resistance measurements. Some might come in very handy; others are somewhat *gimmicky*. The only important additional feature to look for, when shopping for a

DVM, is a "diode test" function. The purpose for this will be explained later in this book. The other functions are between you and your pocketbook.

A DVM is the only piece of test equipment you will need to accomplish the goals of this textbook. If you continue to pursue the electrical/electronic field, you will use this instrument for many years to come, so try to find one from a reputable company that is durable and well proven. A well established electronics dealership might provide some valuable guidance in this area. As you progress through this book, you will discover many uses for a DVM. But at this point, I offer the following words of caution: **Don't try to use a DVM without reading and understanding the operator's manual!** This caution also applies to other types of test equipment. Electricity is dangerous!

The remainder of this section covers the additional types of commonly used electronic test equipment. This is not a suggestion that you should go out and buy all of this equipment. Depending on your interests, you might never need some of these; some pieces you might decide to buy as you progress through this book, and fully understand their function and purpose. You might even decide to build some of it yourself for the fun, satisfaction, and great savings. In any event, read through the remainder of this section, get some basic concepts (don't get upset if you don't understand it all right now), and consider it for future reference as needed.

The *oscilloscope* (Fig. 1-6) can provide a visual representation of voltage and current variations (commonly called *waveforms*) within an operating circuit. In addition to displaying these waveforms, the oscilloscope can also be used to measure their amplitude and frequency.

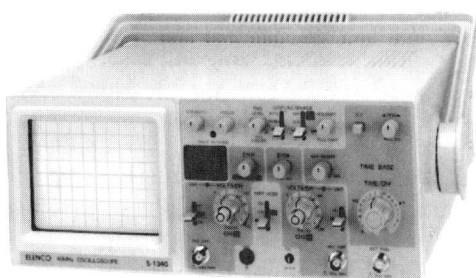

■ **1-6** *A modern dual-trace oscilloscope. Photograph courtesy of Elenco Electronics Inc., Wheeling, IL.*

In many ways, an oscilloscope is similar to a small television set. The cathode-ray tube (picture tube) is used for displaying the voltage or current waveforms. The waveform amplitude is calculated by measuring the vertical height of the waveform, and comparing it to the vertical sensitivity adjustment on the front panel of the oscilloscope.

The frequency of the waveform is calculated by measuring the horizontal length of one complete waveform (one complete cycle), and comparing it to the horizontal-sweep frequency adjustment on the front panel of the oscilloscope.

In addition to amplitude and frequency measurements, an oscilloscope is used for *waveform analysis*. Simply stated, this means the person performing a test should have a good idea of what the waveform should look like at the point being checked. If the waveform is not correct, the defect in the waveform can often identify the problem.

*Logic probes* and *logic pulsers* are used by personnel involved with digital electronics. Logic probes give a visual (and sometimes aural) indication of the logic state of the check-point in question; either high, low, or pulsing. A logic pulser "injects" a continuous train of highs and lows (called *pulses*) into a digital circuit, so that its operation might be observed.

*Power supplies* are used to externally power circuits that are being built, serviced, or tested. For example, if you wanted to functionally test an automobile radio, it would be necessary to connect it to a +12-Vdc power supply to simulate the automobile battery. Lab power supplies are usually line powered, and adjustable over a wide range of voltages and currents. Most electronic enthusiasts will collect a variety of power supplies (some purchased, some home-built, some salvaged from used equipment) for maximum lab versatility. Fig. 1-7 shows an example of power supplies.

*Frequency counters* are instruments used to count the frequency of any periodic waveform. They are more convenient, and more precise for measuring frequency, than an oscilloscope.

*Signal generators* produce a test signal to be injected into an electronic circuit for testing and design purposes (Fig. 1-8). Signal generators typically produce a selectable "sine" wave or "square" wave test signal with adjustable amplitude and frequency. Signal generators that produce a greater variety of test waveforms (such as triangular waves) are called *function generators*. A special type of signal generator, that automatically varies the output frequency within preselected limits, is called a *sweep generator*.

■ **1-7** *A triple-output lab power supply. Photograph courtesy of Interplex Electronics Inc., New Haven, CT.*

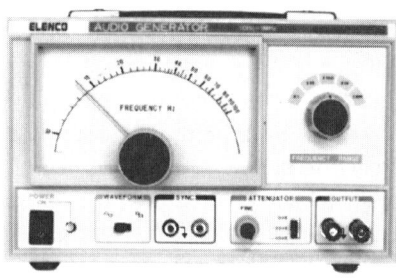

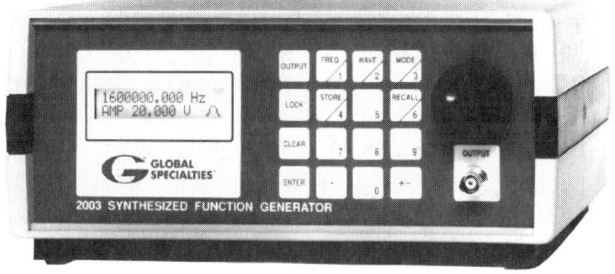

■ **1-8** *An audio frequency signal generator and function generator. Photographs courtesy of Elenco Electronics Inc., Wheeling, IL and Interplex Electronics Inc., New Haven, CT.*

*Current-transformer ammeters* (commonly known by a manufacturer's trade-name *Amprobe*) are most often used by electricians for measuring high values of ac current. (This type of ammeter will not measure dc currents.) This instrument measures the strength of the moving magnetic field created around any wire

*Basic test equipment*

through which ac current is flowing. The circuit does not have to be opened and no physical contact needs to be made to the wire. It then converts this field strength reading to a proportional current reading.

## Starting a parts and materials inventory

Collecting and organizing a good parts inventory is one of the more enjoyable and profitable aspects of the electrical/electronics field. But, like most other endeavors in life, there is a right way and a wrong way to do it. The wrong way will cost you plenty. The right way will open up a fascinating hobby-within-a-hobby (or career-within-a-career) that is both entertaining and educational. If you follow many of the suggestions I have outlined in this section, you can easily acquire a parts inventory that will rival most local electronic parts stores, at very little expense.

### Salvaging

The cheapest way to acquire electronic parts is to remove them from electronic "junk" that someone has thrown away, or given to you. The key to success here is to be able to differentiate the good junk from the bad junk. Unless your interest lies exclusively in the TV repair business, old defective television sets should probably remain in the bad junk category for several reasons. First, they're big, bulky, hard to move, and take up too much space. Second, they can be dangerous to tinker around with, unless you know what you're doing. The CRT (picture tube) is subject to implosion (the opposite of explosion, but with the same catastrophic results) if accidentally fractured. Also, the second anode of the CRT can retain a very nasty high voltage for months. Whereas the shock probably wouldn't do you any physical harm, the involuntary response from your muscles could! Third, the quantity of useful parts that can be removed from TV sets is usually small, unless you have a need for high-voltage components. Fourth, they're hard to get rid of after you're through with them.

The good junk category would include radios, VCRs, CD players, stereo systems, computers, tape players, electronic musical instruments, automobile stereo systems, and some types of commercial or industrial electronic equipment.

Good junk can be acquired in a variety of ways. Most electronic repair shops get stuck with a large volume of junk from customers who don't want to pick up their electronic equipment when they find out how expensive it will be to repair. Many shop owners will give it away for the asking. The local garbage dump is also a good place to find scrap electronics. Don't forget to spread the word around (to all of your friends, neighbors, and relatives) that you would like to have any of their electronic junk destined for the trash can. If you're really ambitious, contact the maintenance superintendents or plant engineers at the local manufacturing plants in your area. Manufacturers often upgrade to new electronic systems, and will discard their old systems for salvage prices. (I once purchased 3 truck loads of extremely valuable electronic equipment from a large manufacturer for only a penny a pound!)

Keep in mind that an electronic bargain is in the eye of the *knowledgeable* beholder. For illustration, lets assume your particular interest is in the audio electronics field (speaker systems, amplifiers, CD players, etc.). If you receive catalogs from the various electronic suppliers listed in the appendix of this book, you'll be amazed at the low prices on top quality internal components for audio systems. (These are the same companies that many of your local electronic dealerships probably buy from.) Unfortunately, the biggest problem you will run into is obtaining suitable cabinets, housings, or enclosures into which to mount your internal components for a finished product. High-quality speaker cabinets might cost $100 or more. Professional-quality *project enclosures* are very expensive, or unavailable in the size needed. This is one of the areas where the junk market really pays off. Junked CD players, VCRs, and computers often have beautiful cases that might require little or no modification to install your parts. If an expensive speaker system goes bad (usually from cone dry rot or misuse), most people simply throw them away; cabinets and all! The speakers can easily be replaced for a third (or less) the cost of a new speaker system of equal quality. Of course, these are just a couple of examples of the cost effectiveness of collecting the right kinds of electronic junk. Try to be somewhat selective, according to your interests, or you might wind up with more junk than living space in your home.

A good place to find materials for cabinet or enclosure fabrication is the local scrap metals yard. You will be particularly interested in sheet aluminum, aluminum plate, and extruded aluminum stock (both channel and angle stock). Most scrap yards will sell this stuff for about 30 to 50 cents a pound. If you attempt to buy this material new, you'll appreciate how much of a savings this is. It is also handy to know that most road signs are made from plate aluminum. When the local highway department replaces old road signs, they will typically sell them to scrap metal companies. Look for road signs while you're at the scrap yard, or you might try contacting your local highway department to ask if they will sell them to you directly.

If your interests happen to lie in robotics, automation, or car sound systems, don't forget your local automobile junk yard. Old automobiles are a good source for powerful electric motors (used in windshield wipers, automatic seat adjusters, etc.), lamps and fixtures, car radios and stereo systems, and miscellaneous hardware.

One of the best ways to be selective, and to obtain all of the salvagable equipment you need, is to advertise in the local "shopper" magazines (often called "advertisers"). For noncommercial individuals, these ads are usually placed free of charge. If your interests happen to be in the digital electronics field (computers and accessories), a typical ad might read as follows:

> Electronics experimenter interested in purchasing obsolete or nonfunctional computers or computer equipment. Call (your telephone number).

Hypothetically, if you placed an ad such as this in a local shopper magazine, and someone called you in response to that ad, try to keep a few points in mind. There are many obsolete and useless computers in homes and businesses today. In many cases, obsolete or defective computers are not even supportable by the companies that manufactured them. The person calling you probably only has two options available; sell the computer stuff to you, or throw it away. You shouldn't pay any more than "scrap" value for equipment of this sort. This is just an example, but the technique should work equally well in any personal electronic interest you might have. But I have one word of caution; do a little research, and know your market before you try this technique.

## What to salvage

Many junked electronic items will contain *subassemblies* that are valuable to the electronic hobbyist and experimenter. Old CD players, computers, and VCRs often contain good power supplies, usable for other projects, or even for lab power supplies. (Two of the power supplies that I use the most often in my lab came from junked equipment.) Junked stereo systems might contain good audio power amplifier subassemblies. These are only a few examples. The point is, always check out the value of equipment subassemblies before tearing everything apart to salvage components.

Any person interested in robotics will find a wealth of electro-mechanical items in old VCRs, including motors, gears, pulleys, belts, limit switches, and optical sensors. Junked CD players often contain functional laser diodes for experimentation or making laser pointers (be very careful to protect your eyes if you experiment with any kind of laser, or laser diode). Junked computers might contain good floppy drives, hard drives, or memory chips. Commonly used electronic hardware (fuse holders, line cords, switches, etc.), is found in almost all electronic equipment.

## Salvaging electronic components

If you are not familiar with the names or appearance of commonly used electronic components, it would be advisable to read the section entitled *Electronic components* in chapter 2 of this book before proceeding.

Now a few words of common sense. An inventory must be organized to be of value to the user. If you have limited inventory space, be practical and selective according to your needs. Every part you salvage will cost you time to remove, and time to enter into your inventory. Although small parts cabinets are not extremely expensive, their cost will add up in time. Try to be more conscious of variety than quantity. Compare the cost of "grab bag" specials from surplus dealers versus your time spent in acquiring the same items by salvaging.

It is usually not a good idea to mix salvaged parts with new "assumed-to-be-good" parts. Salvaged parts are used, and could be defective. Some salvaged parts might be destroyed in trying to remove them. Unless you want to go through exhaustive functional testing procedures, simply keep your salvaged stock separate from your new stock. Then, if the need arises for a salvaged component, you will want to check out that component before using it in a circuit.

The remainder of this section provides some basic guidelines in salvaging electronic components for inventory purposes. These are only suggestions. You might want to do things a little different to meet your specific needs.

**Resistors** Common 1/4-watt and 1/2 watt resistors are probably going to be more trouble than they're worth to salvage. Electronic surplus dealers sell mixed resistors in this size range for as low as a dollar a pound. On the other hand, power resistors (1 watt or higher) are more expensive, and often easier to remove. They are usually a good salvage component.

**Potentiometers** If easily removed, potentiometers are good salvage components. There is a high risk that salvaged potentiometers will function erratically, or contain "dead" spots. They should be thoroughly checked before using in any valuable or critical circuit function.

**Capacitors** Small capacitors typically fall into the same category as small resistors; they're not practical to salvage. Large electrolytic capacitors are practical to salvage, if they're not too old. The functional characteristics of electrolytic capacitors deteriorate with age. Old oil-filled capacitors should not even be brought home; they might contain PCBs! Large nonpolarized capacitors are good salvage items.

**Transformers** Step-down power transformers almost always make good salvage items because of their versatility. Other types of transformers become a matter of choice, depending on your interests.

**Diodes** Small *signal* or *switching* diodes are not practical to salvage because of their low value and ease of availability. Large high-current diodes, high-power zener diodes, and high-current bridge rectifiers are good components to salvage.

**Transistors** Transistors fall into a "gray" area in regards to salvaging. Even small-signal transistors can be practical to salvage if they are marked with generic or easily cross-referenced part numbers. Generally speaking, any transistor marked with a part number that you cannot cross-reference (many manufacturers use "in-house" part numbers), will probably be more trouble than they're worth, considering the time required to analyze them. (In a personal lab, certain types of parameter analysis would not even be possible without damaging the transistor.)

**LEDs** Light-emitting diodes are easily tested and make good salvage items, if their leads are long enough for future use.

**Integrated circuits** There are many "snags" to salvaging ICs. Soldered in ICs are difficult to remove (especially the 40-pin varieties); and, without a professional desoldering station, you stand a good chance of destroying it in the removal process. Once removed, a salvaged IC is difficult to functionally test without very expensive and specialized test equipment. Many ICs are in-house marked. You might run into special cases where it is practical to salvage an IC, such as computer memory chips, but this is not the general rule. If your electronic involvement is very specific, you might find it practical to keep a big box in your lab as a storage place for junk printed circuit boards containing a large quantity of ICs on them. Then, if the need arises, you might find it practical to remove and test the IC.

## Buying from surplus dealers

Before proceeding, lets define the word "surplus" as it applies to electronic parts. *Surplus* means "extra stock." Surplus electronic components and equipment are not substandard or defective, they're just extra stuff that has to be moved out to make room for new. To understand the surplus market, here's how a hypothetical electronic item might get there. Most modern electronic printed circuit boards are manufactured in mass quantities by automated processes. Small "runs" of electronic equipment are not cost-competitive in today's market. If a medium-to-large size electronic manufacturing firm wants to market a new product, they might start out by making 10,000 pieces. If the marketing idea goes sour; or if the product is improved and redesigned; or if the company goes out of business; thousands of these pieces might be left over and sold to a surplus dealership at below manufacturing cost. The surplus dealership can then sell these items far below retail cost and still make a profit.

A typical surplus dealership will sell more than just manufacturers' overruns and excess stock. Because all of the branches of the armed services use considerable electronic equipment (much of it specialized), many surplus dealerships sell *government surplus* equipment as well. Usually, government surplus equipment is used and obsolete, but that doesn't mean its not valuable. Many surplus dealerships sell *factory returns* or *factory refurbished* items. Factory returns are defective items sent back to the manufacturer for repair or replacement. Factory-refurbished items are factory returns that have been repaired by the manufacturer. Surplus dealerships often buy industrial salvage for resale. The surplus dealership might sell industrial subassemblies (control panels, circuit board assemblies, etc.) by listing all of the "goodies" the buyer can get out of it, or they might salvage the subassemblies themselves, and sell the individual components for a greater profit.

Electronic surplus dealerships are great places to buy electronic components. With few exceptions, the components will be new and in "prime" condition. Besides offering low prices on specific components, the grab-bag specials are an excellent way to stock up your general inventory.

If you are lucky, you might have one or more electronic surplus stores in your local area. Because electronic surplus stores do not

*Starting a parts and materials inventory*

cater to the general public, you might have to do a little investigative work to find the ones nearest you, but it will be worth the effort. If you live in a rural area, mail-order surplus is a good alternative. A list of some good mail-order electronic surplus dealerships is included in the appendix of this book.

Electronic surplus is an excellent way to "round-out" your parts and materials inventory, but there are a few cautions and considerations. Don't automatically assume that every item offered for sale by a surplus dealer must be far below retail cost. In some cases, it is not! Before buying or ordering equipment, be aware of its status. It might be new, used, government surplus, factory returned, factory refurbished, or sold "as is" (no guarantee of anything). Grab bag specials will consume a lot of tedious sorting time. Be sure that your eyes and nerves are up to it.

A few additional cautions are relative to mail-order surplus. Most surplus dealerships require a minimum order. Take care to meet this minimum before placing an order. Also, there are *hidden* costs associated with insurance, postage, COD fees, and shipping/handling. Its also wise to verify that the surplus dealership has a lenient return policy, if you are not satisfied for any reason. A good, ethical firm shouldn't have any problem in this area.

# Basic electrical concepts 2

IF YOU ARE A NOVICE IN THE ELECTRICAL/ELECTRONIC fields, you will soon discover that this chapter is not "light reading." I'm not suggesting that it will be difficult to understand, but I am providing an advanced warning that it will be concentrated. This chapter was designed to maximize reading efficiency. Therefore, there are some basic terms used in conjunction with component descriptions that you might be a little confused about. Don't panic! All unexplained terms are covered later in the chapter as it becomes more appropriate to do so.

Much forethought has gone into the structure of this and successive chapters, in order to provide you with the easiest possible way of comprehending the fundamentals of the electrical/electronic fields. It is not always prudent to define every detail of a topic under discussion because the most important aspects might be obscured by issues that could be more clearly explained in a different context. Therefore, I suggest that you read this chapter twice. You will find the confusing terms, of the first reading, to be much easier to understand during the second reading. If, after the second reading, you are still a little "shakey" in a few areas, don't become frustrated. The basics explained in this chapter will be utilized throughout the book. I have tried to be thorough in placing reminders throughout the text as an additional aid to understanding. Also, because you all learn by doing, these same fundamentals will be used in the construction of many practical circuits. In this way, concepts are removed from the realm of theory and put into actual practice. One last suggestion; remember, this is a textbook. Textbooks are not designed or intended to be read in the same way as a good novel; they are designed to be studied. You might have to go back and re-read many chapters as your progress continues.

## Electronic components

The purpose of this section is to provide the reader with a basic concept of the physical structure of individual electronic components, and how to determine their values.

## Resistors

A *resistor*, as the name implies, resists (or opposes) current flow. As you will soon discover, the characteristic of opposing current flow can be used for many purposes. Hence, resistors are the most common *discrete* components found in electronic equipment (the term discrete is used in electronics to mean *nonintegrated*, or standing alone).

Resistors can be divided into two broad categories: fixed and adjustable. *Fixed resistors* are by far the most common. *Adjustable resistors* are called *potentiometers* or *rheostats*.

Most fixed resistors are of carbon composition. They utilize the poor conductivity characteristic of carbon to provide resistance. Other types of commonly manufactured resistors are *carbon film, metal film, molded composition, thick film, and vitreous enamel*. These different types possess various advantages, or disadvantages, relating to such parameters as temperature stability, tolerance, power dissipation, noise characteristics, and cost.

The two most critical resistor parameters are *value* (measured in *ohms*) and *power* (measured in *watts*). Resistors that are larger in physical size, can dissipate (handle) more power than smaller resistors. If a resistor becomes too hot, it can change value or burn up. Consequently, it is very important to use a resistor with adequate power handling capability, as determined by the circuit in which it is placed. Unfortunately, there is not a good, standard way of looking at most resistors, and determining their power rating by some standardized mark or code. Comparative size can be deceiving depending on the resistor construction. For example, a 10-watt wire-wound resistor might be about the same size as a 2-watt carbon composition resistor. Common power ratings of the vast majority of resistors are 1/8, 1/4, 1/2, 1, and 2 watts. Don't worry about being able to determine resistor power ratings at this point. It is largely an experience-oriented talent that you will acquire in time.

The value of most resistors is identified by a series of colored bands which encircle the body of the resistor. Each color represents a number, a multiplier, or a tolerance value. The first digit will always be the colored band closest to one end.

There are two *color-coded systems* in common use today; the four-band and the five-band system. In the *four-band* system, the first band represents the first digit of the resistance value, the second band is the second digit, the third band is the multiplier, and the fourth band is the tolerance (resistors with a tolerance of 20%

will not have a fourth band). The *five-band* system is the same as the four-band, except for the addition of a third band representing a third significant digit. The five-band system is often used for precision resistors requiring the third digit for a higher level of accuracy. There are also a few additional colors used for tolerance identification in the five-band system.

You will need to memorize the following color code table:

**Resistor color codes**

| | |
|---|---|
| Black = 0 | Green = 5 |
| Brown = 1 | Blue = 6 |
| Red = 2 | Violet = 7 |
| Orange = 3 | Gray = 8 |
| Yellow = 4 | White = 9 |
| Brown = 1% tolerance | Yellow = 4% tolerance |
| Red = 2% tolerance | Silver = 10% tolerance |
| Orange = 3% tolerance | Gold = 5% tolerance |

The following are a few examples of how these color code systems work. As stated previously, the color band closest to one end of the resistor body is the first digit. Suppose that you had a resistor marked blue-gray-orange-gold. Because there are only four bands, you know it uses the four-band system. Therefore:

blue = 6　gray = 8　orange = 3　gold = 5% tolerance

The first two digits of the resistor value are defined by the first two bands. Therefore, the two most significant digits will be 68. The third band (multiplier band) indicates how many zeros will be added to the first two digits; in this case it's 3. So, the full value of the resistor is 68,000 ohms. The tolerance band describes how far the actual value can vary. A 5% tolerance means that this resistor can vary as much as 5% more, or 5% less, than the value indicated by the color bands. 5% of 68,000 is 3,400. Therefore, the actual value of this resistor might be as high as 71,400 ohms, or as low as 64,600 ohms.

Another example of the four-band system could be red-violet-brown-silver. Therefore:

red = 2　violet = 7　brown = 1　silver = 10% tolerance

The value of this resistor is 27 with 1 zero added to the end; or 270 ohms. The 10% tolerance indicates it can vary either way by as much as 27 ohms.

Try an example of the five-band system; brown-black-green-brown-red. The first three bands are the first three digits: brown-black-green = 105. The fourth brown band indicates 1 zero should be added to the end of the first three digits resulting in 1050 ohms. The fifth band is tolerance; red = 2%.

Just to make things a little more confusing, some manufacturers add a *temperature coefficient* band to the four-band system resulting in a resistor appearing to be coded in the five-band system. The easy way to recognize this is the fourth band will always be either gold or silver. If this is the case, simply disregard the fifth band. Also, as stated previously, a complete absence of a fourth band indicates a 20% tolerance.

On some precision resistors or larger power resistors, the actual value and tolerance rating might be printed on the resistor body. A variety of typical resistors is illustrated in Fig. 2-1.

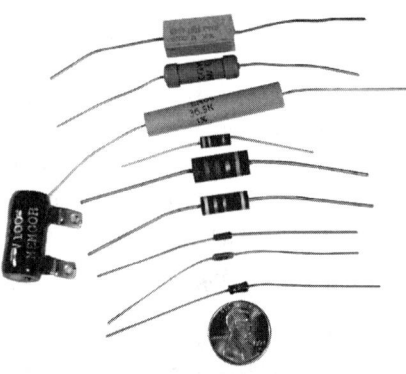

■ **2-1** *A sampling of typical resistor shapes and styles.*

A good way to become very proficient at reading resistor color codes and estimating power ratings (an extremely handy talent to have) is to order one of the large grab-bag resistor specials from one of the mail-order surplus stores listed in the appendix of this book. Use your DVM (set to measure ohms) to verify your color-code interpretations are correct. (Don't hold on to the DVM test lead ends with both hands when measuring; the DVM will include your body resistance in the reading.) You'll wind up with a good assortment of cheap sorted resistors. The experience gained in using your DVM will be a side benefit!

## Potentiometers, rheostats, and resistive devices

Technically speaking, a *potentiometer* is a three-lead continuously variable resistive device. In contrast, a *rheostat* is a two-lead continuously variable resistor. Unfortunately, these two terms are often confused with each other; even by many professionals in the field.

A few good examples of potentiometers can be found in the typical consumer electronic products around the home. Controls labeled as "volume", "level," "balance," "bass," "treble," and "loudness" are almost always potentiometers.

A potentiometer has a round body, about 1/2 to 1 inch in diameter, a rotating shaft extending from the body, and three terminals (or leads) for circuit connection. It consists of a fixed resistor, connected to the two outside terminals, and a *wiper* connected to the center terminal. The wiper is mechanically connected to the rotating shaft, and can be moved to any point along the fixed resistance by rotating the shaft. A rheostat can be made from a potentiometer by connecting either outside terminal to the wiper terminal.

Potentiometers are specified according to their power rating, fixed resistance value, taper, and mechanical design. The fixed resistance value is usually indicated somewhere on the potentiometer body (it can be determined by measuring the resistance value between the two outside terminals with a DVM). If the power rating is not marked on the body, you will probably have to estimate it (if you cannot cross the manufacturer's part number to a parts catalog). Standard size potentiometers (about 1 inch in diameter) will typically be rated at 1 or 2 watts. *Taper* refers to the way the resistance between the three terminals will change in respect to a rotational change of the shaft. A *linear taper* potentiometer will produce proportional resistance changes. For example, rotating the shaft by 50% of its total travel will result in 50% resistance changes between the terminals. *Logarithmic taper* potentiometers are nonlinear, or nonproportional, in their operation. They are often called *audio taper* potentiometers because they are commonly used for audio applications. The sensitivity-to-volume level of the human ear is nonlinear. Logarithmic potentiometers closely approximate this nonlinear sensitivity and are used for most audio volume-level controls, as well as many other nonlinear applications. The mechanical construction of potentiometers will vary according to the intended application. Most are *single-turn*, meaning the shaft will only rotate about 260 degrees. For some precision applications, multiple-turn potentiometers are used.

The term *trim-pot* is used to describe small, single-turn potentiometers intended to be mounted on printed circuit boards and adjusted once, or very infrequently. Figure 2-2 illustrates some common types of potentiometers and trim-pots.

**2-2** *Common potentiometer types.*

If you happen to work with various types of commercial or industrial electrical/electronic equipment, you might run across some *dedicated rheostats*. These are rheostats which were not made from potentiometers. They are usually intended for high-power applications, and are specially designed to dissipate large amounts of heat.

*Adjustable resistors* should not be confused with potentiometers or rheostats. Potentiometers and rheostats are commonly used in applications requiring frequent or precise adjustment. Adjustable resistors are meant to be adjusted once (usually to obtain some hard-to-find resistance value). The body of an adjustable resistor is similar to that of a typical power resistor, with the addition of a metal ring which can be moved back and forth across the resistor body. The position of this metal ring determines the actual resistance value. Once the desired resistance is set, the metal ring is permanently clamped in place.

## Capacitors

Next to resistors, capacitors are probably the most common component in electronics. *Capacitors* are manufactured in a variety of shapes and sizes. In most cases, they are either flat, disc shaped components; or they are tubular in shape. They vary in size from almost microscopic to about the same diameter, and twice the height, of a large coffee cup. Capacitors are two-lead devices, sometimes resembling fixed resistors without the color bands.

Virtually all tubular-shaped capacitors will have a capacitance value and a voltage rating marked on the body. Identification of this type is easy. The small disc or rectangular shaped capacitors will usually be marked with the following code:

### Capacitor markings

| | |
|---|---|
| 01 to 99 - the actual value in picofarads | |
| 101 - 0.0001 microfarad | 331 - 0.00033 microfarad |
| 102 - 0.001 microfarad | 332 - 0.0033 microfarad |
| 103 - 0.01 microfarad | 333 - 0.033 microfarad |
| 104 - 0.1 microfarad | 334 - 0.33 microfarad |
| 221 - 0.00022 microfarad | 471 - 0.00047 microfarad |
| 222 - 0.0022 microfarad | 472 - 0.0047 microfarad |
| 223 - 0.022 microfarad | 473 - 0.047 microfarad |
| 224 - 0.22 microfarad | 474 - 0.47 microfarad |

With the previous coding system, a letter will usually follow the numeric code to define the tolerance rating. *Tolerance*, in the case of capacitors, is the allowable variance in the capacitance value. For example, a 10 microfarad capacitor with a 10% tolerance can vary +/− 1 microfarad. The tolerance coding is as follows:

### Capacitor tolerance markings

| | |
|---|---|
| B = + or − 0.1 picofarad | J = + or − 5% |
| C = + or − 0.25 picofarad | K = + or − 10% |
| D = + or − 0.5 picofarad | M = + or − 20% |
| F = + or − 1% | Z = +80%, −20% |
| G = + or − 2% | |

*Variable capacitors* are, as the name implies, adjustable within a narrow range of capacitance value. The most common type is called the *air dielectric* type and is typically found on the tuning control of an AM radio. Another type of variable capacitor is called a *trimmer* capacitor. These are usually in the range of 5 to 30 picofarads and are used for high-frequency applications.

Capacitor characteristics, functions, and additional parameter information will be discussed in chapter 5. Some common shapes and sizes of capacitors are illustrated in Figure 2-3.

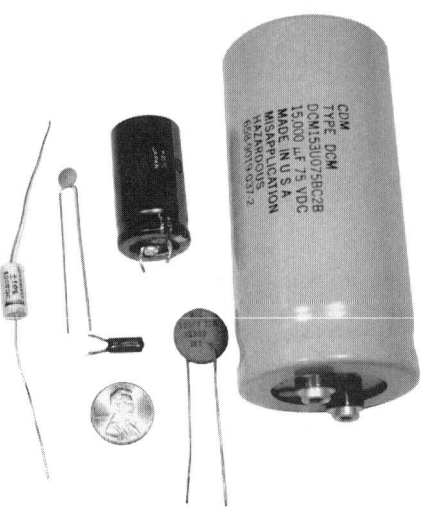

■ **2-3** *Some examples of capacitors. The two larger capacitors are electrolytics.*

### Inductors

*Inductors* are broken down into two main classifications; *transformers* and *chokes* (or *coils*). Transformers designed for power conversion applications, called *power transformers* or *filament transformers*, are usually heavy, block-shaped components consisting of two or more coils of wire wound around rectangular-shaped wafers of iron. Most of these types of transformers are intended to be mounted on a sturdy chassis because of their weight, but some newer types of *split-bobbin* power transformers are designed for

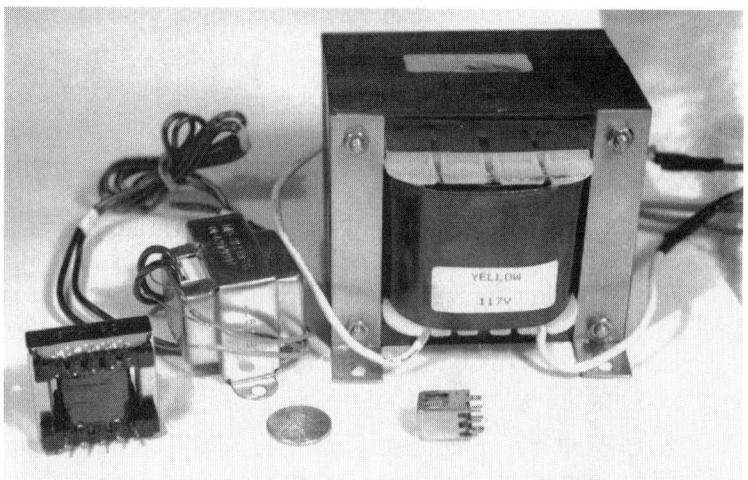

■ **2-4** *A few examples of transformers. The small transformer in front is an adjustable IF transformer.*

*Basic electrical components*

printed circuit board mounting. Transformers designed for signal handling applications are smaller, and are usually mounted on printed circuit boards. Figure 2-4 illustrates several common variations of transformers. Chokes are coils of wire either wound on a metallic core or a non-ferrous form (to hold the shape of the coil).

Some types of small transformers and chokes are manufactured with an adjustable *slug* in the center. The slug is made of a *ferrite* material, and has threads on the outside of it like a screw. Turning the slug will cause it to move further into (or out of) the coil, causing a change in the inductance value of the coil or transformer. (Inductance is discussed in chapter 3.) Depending on their application, these small adjustable inductors are called chokes, traps, IF transformers, and variable core transformers.

## Diodes

*Diodes* are two-lead semiconductor devices with tubular bodies similar to fixed resistors. Their bodies are usually either black or clear, with a single-colored band close to one end.

*Diode bridges* are actually four diodes encased in square or round housings. They have four leads, or connection terminals; two of which will be marked with a horizontal S symbol (sine wave symbol), one will be marked "+" and the other "−". Several types of diodes and a diode bridge are illustrated in Figure 2-5.

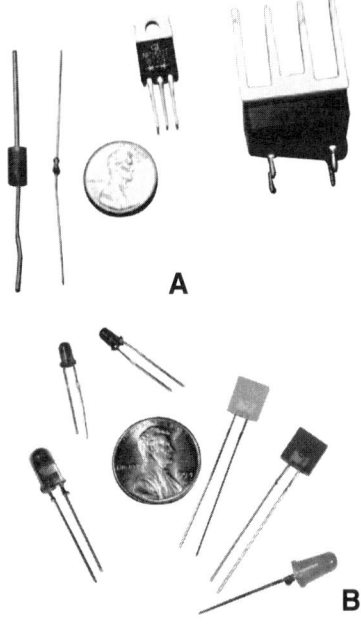

■ **2-5** A. *Two common diode types, a dual diode (3-lead device), and a diode bridge (with the heat sink). B. Common styles of light emitting diodes (LEDs).*

*Stud-mount* diodes are intended for high-power rectification applications. They have one connection terminal at the top, a body shaped in the form of a hex-head bolt (for tightening purposes), and a base with a threaded shank for mounting into a *heatsink*. (Heatsinks are devices intended to conduct heat away from semiconductor power devices. Most are made from aluminum and are covered with extensions, called *fins*, to improve their thermal convection properties.)

Special types of diodes, called *LEDs* (the abbreviation for light emitting diodes), are designed to produce light, and are used for indicators and displays. They are manufactured in a variety of shapes, sizes, and colors, making them attractive in appearance. Figure 2-5 illustrates a sampling of common LEDs. A specific type of LED configuration, called "seven-segment" LEDs, are used for displaying numbers and alphanumeric symbols. The front of their display surface is arranged in a block "8" pattern.

## Transistors

*Transistors* are three-lead semiconductor devices manufactured in a variety of shapes and sizes to accommodate such design parameters as power dissipation, breakdown voltages, and cost. A good sampling of transistor shapes and styles is illustrated in Fig. 2-6.

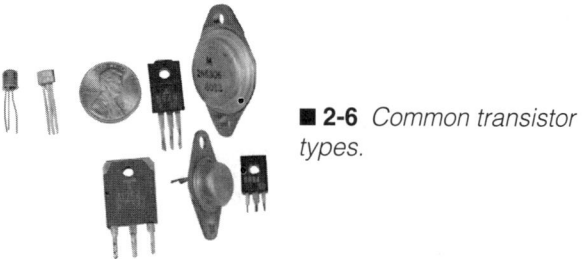

■ **2-6** *Common transistor types.*

Some power transistors have an oval shaped body with two mounting holes on either side (this is called a *TO-3* or *TO-5* case style). At first glance, it appears to only have two leads, but the body itself is the third (collector) lead.

## Integrated circuits

*Integrated circuits*, or ICs, are multicomponent semiconductor devices. A single IC might contain thousands of individual components manufactured through a series of special processes. They have multiple leads, called *pins*, which can vary in number from 8 to 40. ICs can be soldered directly into printed circuit boards, or plugged into specially designed IC sockets. Figure 2-7 illustrates several varieties of integrated circuits.

■ **2-7** *Examples of integrated circuits. These are "dual in-line" (DIP) styles.*

Although the vast majority of integrated circuits have flat, rectangular bodies, a few types of ICs have round "can-type" bodies similar to a common transistor case style. The difference is readily detectable by simply counting the number of lead wires coming from the can; a transistor will have three, but an IC of this style will have eight or more.

### A final note on parts identification

If you think this is all rather complicated, you're right! This section only covers the most common types of components, and the easiest ways of identifying them and ascertaining their values. There are many more styles, types, and configurations than I have time or space to cover. A good way to build on this information is to open up some electronic supply catalogs and skim through them; paying special attention to the pictorial diagrams and dimensions relating to the various components. It will also be a great help to visit your local electronic supply store, and spend some time browsing through the component sales section. And, if its any consolation, I still get stuck on a "what-the-heck-is-it" from time to time!

## Characteristics of electricity

At some point in your life, you have probably felt electricity. If this experience involved a mishap with 120 Vac (common household power from an outlet), you already know that a physical force is associated with electricity. You constantly see the effects of electricity when you use electrical devices such as electric heaters, television sets, electric fans, and other household appliances. You also see the effects of it when you pay the electric bill! But even with the almost constant usage of electricity, it often possesses an aura of mystery because it cannot be seen. In many ways, electricity is similar to other things that you do understand, and to which you can relate. Electricity follows the basic laws of physics, as do all things in our physical universe.

The electrical/electronic fields depend on the manipulation of subatomic particles called *electrons*. Electrons are negatively charged particles that move in orbital patterns, called *shells*, around the nucleus of an atom. Magnetic fields, certain chemical reactions, electrostatic fields, and the conductive properties of various materials affect the movement and behavior of electrons. As you progress through this book, you will learn how to use electron movement and its associated effects to perform a myriad of useful functions in our lives.

As a beginning exercise in understanding electricity, you are going to compare it to something you can easily visualize and understand. Examine the simple water-flow system shown in Fig. 2-8. This hypothetical system consists of a water pump, a valve to adjust the water flow, and a water pipe to connect the whole system together. Assuming that the pump runs continuously and at a constant speed, it will produce a continuous and constant "pressure" to try to force water through the pipe.

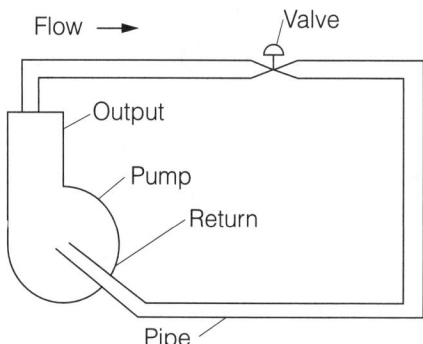

■ **2-8** *Fluid analogy of an electric circuit.*

If the valve is closed, no water will flow through the pipe, but water pressure from the running pump will still be present. The valve is totally "resisting" the flow of water.

If the valve is opened approximately half-way, a "current" of water will begin to flow in the pipe. This will not be the maximum water current possible because the valve is "resisting" about half of the water flow.

If the valve is opened all the way, it will pose no resistance to the flow of water. The maximum water flow for this system will occur, limited by the capacity of the pump and the size (diameter) of the pipe.

The electrical circuit shown in Fig. 2-9 is very much like the water system shown in Fig. 2-8. Note the symbols used to represent a battery, resistor, switch, and wire. The battery is analogous to the pump. It provides an electrical "pressure" that produces an electrical flow through the wire. The electrical pressure is called *voltage*. The electrical flow is called *current*.

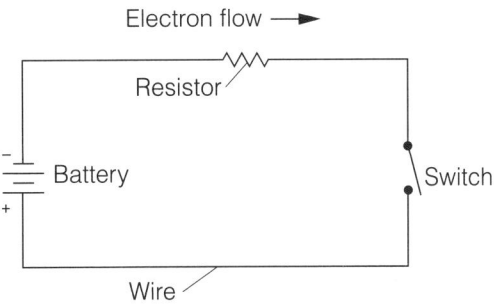

■ **2-9** *Basic electrical circuit.*

If the switch in Fig. 2-9 is open (in the "off" position), it will not allow current to flow because a continuous electrical path will not exist. This condition is analogous to the water valve (Fig. 2-8) being completely closed. Because a continuous electrical path from one side of the battery to the other does not exist with the switch open, this circuit is without *continuity*.

If the switch is closed (in the "on" position), an electrical current will begin to flow through the circuit. This current will start at the negative (−) side of the battery, flow through the resistor (R1), through the closed switch (S1), and return to the positive (+) side of the battery. The wire used to connect the components together is analogous to the pipe (Fig. 2-8). The resistor poses some opposition to current flow, and is analogous to the water valve (Fig. 2-8) being partially closed. The actual amount of current flow that will exist in this circuit cannot be determined because I have not assigned absolute values to the components.

If the resistor was removed, and replaced with a piece of wire (assuming the switch is left in the closed, or "on" position), there would be nothing remaining in the circuit to limit (or resist) the maximum possible current flow. This condition would be analogous to the water valve (Fig. 2-8) being fully opened. A maximum current would flow limited only by the capacity of the battery, and the size (diameter) of the wire.

Notice that in these two comparative illustrations, there is a device producing "pressure" (water pump or battery), a pathway through which a medium could flow (water pipe or wire), a device or devices offering "resistance" to this flow (water valve or resistor/switch combination), and a resultant flow, or "current," limited by certain variables (water pipe or wire diameter; water valve or resistor opposition; and water pump or battery capacity).

Simply stated, in any operational electrical circuit, there will always be three variables to consider: voltage (electrical pressure), current (electrical flow), and resistance (the opposition to current flow). These variables can now be examined in greater detail.

## Voltage

Voltage does not move; it is applied. Going back to the illustration of Fig. 2-8, even with the water valve completely closed blocking all water flow, the water pressure still remained. If the water valve were suddenly opened, a water flow would begin and the water pressure produced by the pump would still exist. In other words, the water valve does not control the water pressure; only the water flow. The water pressure is an applied force promoting movement of the water. The water pressure does not move, only the water moves. The basic concept is the same for the electrical circuit of Fig. 2-9. The voltage (electrical pressure) produced by the battery is essentially independent of the current flow.

To help clarify the previous statements, consider the following comparison. Imagine a one-mile long plastic tube with an inside diameter just large enough to insert a ping-pong ball. If this tube were to be completely filled from end to end with ping-pong balls and you inserted one additional ping-pong ball into one end, a ping-pong ball would immediately fall out of the other end. The force, or pressure, used to insert the one additional ping-pong ball was applied to every ping-pong ball within the tube at the same time. The movement of the ping-pong balls is analogous to electrical current flow. The pressure causing them to move is analogous to voltage. The pressure did not move, it was applied. Only the ping-pong balls moved.

Voltage is often referred to as *electrical potential*, and its proper name is *electromotive force*. It is measured it units called *volts*. Its electrical symbol, as used in formulas and expressions, is $E$.

The level, or *amplitude*, of voltage is usually defined in respect to some common point. For example, the battery in most automobiles provides approximately 12 volts of electrical potential. The negative side of the battery is normally connected to the main body of the automobile, causing all of the metal parts connected to the body and frame to become the common point of reference. When measuring the positive terminal of the battery with respect to the body, a positive 12-volt potential will be seen.

The terms *positive* and *negative* are used to define the polarity of the voltage. *Voltage polarity* determines the direction in which electrical current will flow. Referring to Fig. 2-8, note how the direction of the water flow is dependent upon the direction, or orientation, of the water pump. If the pump had been turned around, so that the output and return were reversed, the direction of water flow would also have been reversed. Another way of looking at this condition is to consider the output side of the pump as having a positive pressure associated with it, and the return has a negative pressure (suction) associated with it. The water will be pushed out of the output and be sucked toward the vacuum.

A similar condition exists with the electrical circuit of Fig. 2-9. The negative terminal of the battery "pushes out" an excess of electrons into an electrical conductor, and the positive terminal "attracts," or sucks in, the same number of electrons as the negative terminal has pushed out. Thus, if the battery terminals are reversed, the direction of current flow will also reverse. (Note the symbol for a battery in Fig. 2-9. A battery symbol will always be drawn with a short line on one end and a longer line on the other. The short line represents the negative terminal; the long line represents the positive terminal.)

## Current

*Electrical current* is the movement, or flow, of electrons through a conductive material. The fluid analogy of Fig. 2-8, showed some common principles of water flow. In the United States, the flow of water (and most other fluids) is measured in units as "gallons per hour" *(gph)*. In the electrical/electronic fields, you need a similar standard to measure the flow of electrons. The "gallon" of electrons is called a *coulomb*. A coulomb is equal to 6,280,000,000,000,000,000 electrons, give or take a few! In scientific notation, that number is written $6.28 \times 10^{18}$ (6.28 with the decimal point moved eighteen places to the right). Whenever one coulomb of electrons flows past a given point in one second of time, the current flow is equal to one *ampere*.

While on the topic of standards, let me provide you with the standard relationship between voltage, current, and resistance. *One volt is the electrical pressure required to push one coulomb of electrons through one ohm of resistance in one second of time.* In other words, 1 volt will cause a 1 ampere current flow in a closed circuit with 1 ohm of resistance. You will understand this relationship more clearly as you begin to work with *ohm's law*.

The direction of current flow will always be from negative to positive. This is referred to as the *electron flow*. As stated previously, electron flow rate is measured in units called amperes, or simply amps. Its electrical symbol, as used in formulas and expressions is $I$.

As illustrated in Fig. 2-10, current can only flow in a *closed circuit*; that is, a circuit that provides a continuous conductive path from the negative potential to the positive potential. If there is a break in this continuous path, such as the open switch (S1) in Fig. 2-11, current cannot flow, and the circuit is said to be *open*. Another way of stating the same principle is to say the circuit of Fig. 2-10 has *continuity* (a continuous path through which current can flow). The circuit in Fig. 2-11 is without continuity.

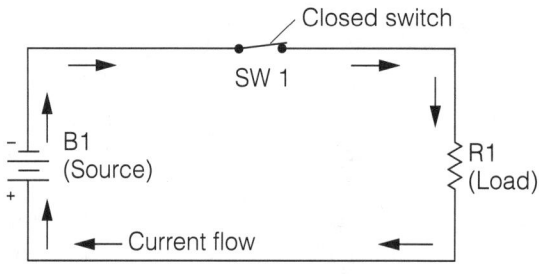

■ **2-10** *Example of a closed circuit.*

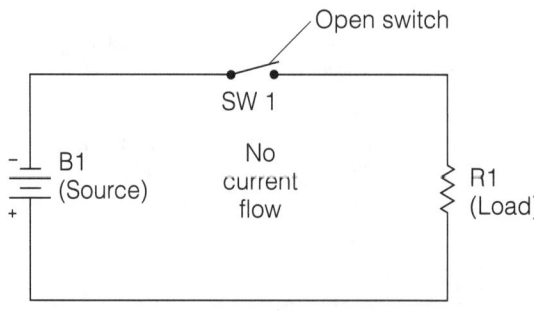

■ **2-11** *Example of an open circuit.*

## Resistance

*Resistance* is the opposition to current flow. The open switch in Fig. 2-11 actually presents an infinite resistance. In other words, its resistance is so high, it doesn't allow any current flow at all. The closed switch shown in Fig. 2-10 is an example of the opposite extreme. For all practical purposes, it doesn't present any resistance to current flow and, therefore, has no effect upon the circuit, as long as it remains closed.

The circuit illustrated in Fig. 2-10 shows the symbol for a resistor (R1) and the current flow ($I$) through the circuit. A resistor will present some resistance to current flow that falls somewhere between the two extremes presented by an open or closed switch. This resistance will normally be much higher than the resistance of the wire used to connect the circuit together. Therefore, under most circumstances, this wire resistance is considered to be negligible. Resistance is measured in units called *ohms*. Its electrical symbol, as used in formulas and expressions, is $R$.

## Alternating current (ac) and direct current (dc)

The periodic reversal of current flow is called *alternating current (ac)*. In ac-powered circuits, the polarity of the voltage changes perpetually at a specified rate, or *frequency*. Because the polarity of the voltage is what determines the direction of the current flow, the current flow changes directions at the same rate. The symbol for an ac voltage is shown in Fig. 2-12.

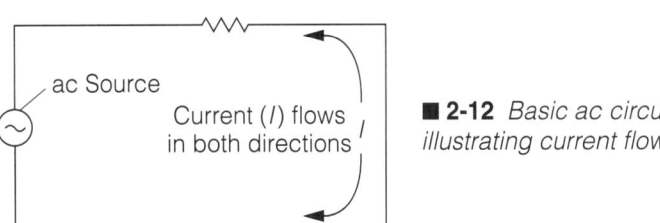

**2-12** *Basic ac circuit illustrating current flow.*

The frequency of current alternations is measured in units called *hertz (Hz)*. A synonym for hertz is *cycles per second (cps)*. Both of these terms define how many times the current reverses direction in a one-second time period. Common power of the average home in the United States has been standardized at 60 Hz. This means the voltage polarity and current flow reverse direction 60 times every second. (See Fig. 2-12.)

In contrast to ac power sources (Fig. 2-12), a *direct current (dc)* power source never changes its voltage polarity. Consequently, the current in a dc powered circuit will never change direction of flow. The battery power sources shown, in Figs. 2-9 through 2-11, are examples of dc power sources, as are all batteries.

## Conductance

Sometimes it is more convenient to consider the amount of current allowed to flow, rather than the amount of current opposed. In these cases, the term conductance is used. *Conductance* is simply the reciprocal of resistance. The unit of conductance is the *mho* (ohm spelled backwards), and its electrical symbol is $G$. The following equations show the relationship between resistance and conductance (for example, 2 ohms of resistance would equal 0.5 mhos of conductance, and vice versa):

$$\frac{1}{G} = R$$

The reciprocal of conductance is resistance.

$$\frac{1}{R} = G$$

The reciprocal of resistance is conductance.

In recent years, the accepted term of conductance, mho, and its associated symbol, $G$, have been replaced with the term *siemen* and its associated symbol $S$. These two terms, with their associated symbols, mean exactly the same thing. However, throughout the remainder of this book, you will continue to use the older term *mho*, which is still the most commonly used method of expressing conductance.

## Power

The amount of energy dissipated (used) in a circuit is called *power*. Power is measured in units called *watts*. Its electrical symbol is $W$.

In circuits such as the one shown in Fig. 2-10, the resistance is often referred to as the *load*. The battery is called the *source*. This is simply a means of explaining the origin and destination of the electrical power used up, or dissipated, by the circuit. For example, in Fig. 2-10, the electrical power comes from the battery. Therefore, the battery is the source of the power. All of this power is being dissipated by the resistor (R1), which is appropriately called the *load*.

# Laws of electricity

As with all other physical forces, physical laws govern electrical energy. Highly complex electrical engineering projects might require the use of several types of high-level mathematics. The electrical design engineer must be well acquainted with physics, geometry, calculus, and algebra.

However, if you do not possess a high degree of proficiency with high-level mathematics, this does not mean you must abandon your goal of becoming proficient in the electrical/electronic fields. Many books are available on specific subjects of interest that simplify the complexities of design into "rule-of-thumb" calculations. Also, if you own a computer system, you can purchase computer programs (software) to perform virtually any kind of design calculations that you will probably ever need. Unfortunately, this doesn't mean that you can ignore electronics math altogether. The electronics math covered in this chapter is necessary for establishing and understanding the definite physical relationships between voltage, current, resistance, and power.

## Ohm's law

The most basic mathematical form for defining electrical relationships is called *Ohm's law*. In the electrical/electronic fields, Ohm's law is a basic tool for comprehending electrical circuits and analyzing problems. Therefore, it is important to memorize, and become familiar, with the proper use of Ohm's law; just as a carpenter must learn how to properly use a saw or hammer. Ohm's law is shown in Equation 2-1.

$$E = IR \qquad \text{EQ. 2-1}$$

This equation tells us that voltage ($E$) is equal to current ($I$) multiplied by resistance ($R$). The simple circuit of Fig. 2-13 illustrates how this equation might be used. The value of R1 is given

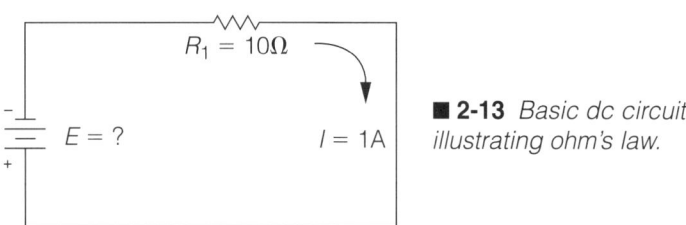

**2-13** Basic dc circuit illustrating ohm's law.

as 10 ohms (note the Greek *omega* symbol is used to abbreviate the word ohm), and the current flow is given as 1 amp, but the battery voltage ($E$) is unknown. By substituting the electrical symbols in Equation 2-1 with the actual values, $E$ can easily be calculated.

$$E = IR$$
$$E = (1 \text{ amp})(10 \text{ ohms})$$
$$E = 10 \text{ volts}$$

Now you can go through a few more examples of using Ohm's law to really get the hang of it. Note the circuit illustrated in Fig. 2-14. I have not assigned any values to this circuit so you can use the same basic circuit format to go through several exercises. If $R$ is 6.8 ohms, and $I$ is 0.5 amp, what is the source voltage?

$$E = IR$$
$$E = (0.5)(6.8)$$
$$E = 3.4 \text{ volts}$$

If $R$ is 22 ohms, and $I$ is 2 amps, what would $E$ be?

$$E = IR$$
$$E = (2)(22)$$
$$E = 44 \text{ volts}$$

If $R$ is 10,000 ohms, and $I$ is 0.001 amp, what would $E$ be?

$$E = IR$$
$$E = (0.001)(10,000)$$
$$E = 10 \text{ volts}$$

Now consider some different ways you can use this same equation. An important rule in algebra is that you might do whatever you want to one side of an equation, as long as you do the same to the other side of the equation. The principle is the same as with a balanced set of scales. As long as you add or subtract equal amounts of weight on both sides of the scales, the scales will remain balanced. Likewise, you can add, subtract, multiply, or divide by

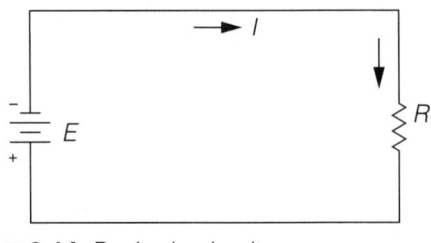

■ 2-14 *Basic dc circuit.*

equal amounts to both sides of Equation 2-1 without destroying the validity of the equation. For example, divide both sides of Equation 2-1 by $R$:

$$\frac{E}{R} = \frac{IR}{R}$$

The two $R$s on the right side of the equation cancel each other out. Therefore:

$$\frac{E}{R} = I \text{ or } I = \frac{E}{R} \qquad \text{EQ. 2-2}$$

Equation 2-2 shows that if you know the voltage ($E$) and resistance ($R$) values in a circuit, you can calculate the current flow ($I$). Referring back to Fig. 2-13, you were given the resistance value of 10 ohms and the current flow of 1 amp. From these values you calculated the source voltage to be 10 volts. Assume, for the moment, you don't know the current flow in this circuit. Equation 2-2 will allow you to calculate it:

$$I = \frac{E}{R} = \frac{10 \text{ volts}}{10 \text{ ohms}} = 1 \text{ amp}$$

Now go back to the previous three exercises where you calculated voltage using Fig. 2-14 as the basic circuit. Assume the current flow value to be unknown, and using the known voltage and resistance, solve for current. In each exercise, the calculated current flow value should be the same as the given value.

Similarly, you could rearrange Equation 2-1 to solve for resistance, if you knew the values for current flow and voltage. Note that each side of the equation must be divided by $I$:

$$\frac{E}{I} = \frac{IR}{I}$$

The two $I$s on the right side of the equation cancel out each other leaving:

$$\frac{E}{I} = R \text{ or } R = \frac{E}{I} \qquad \text{EQ. 2-3}$$

Referring back to Fig. 2-13, assume the resistance value of R1 is not given. Using Equation 2-3, you can solve for $R$:

$$R = \frac{E}{I} = \frac{10 \text{ volts}}{1 \text{ amp}} = 10 \text{ ohms}$$

Once again, refer back to the exercises associated with Fig. 2-14, and assume the value of resistance is not given in each case. Using

Equation 2-3 and the known values for current flow and voltage, solve for $R$. In each exercise, the calculated value of $R$ should equal the previously given value.

At this point, you should be starting to understand what is meant by a proportional relationship between voltage, current, and resistance. If one of the values is changed, one of the other values must also change. For example, if the source voltage is increased, the current flow must also increase (assuming that the resistance remains constant). If the resistance is increased, the current flow must decrease (assuming that the source voltage remains constant), and so forth. For practice, you might try substituting other values for current, voltage, and resistance, using Fig. 2-14 as the basic circuit. It is vitally important to become very familiar with how this relationship works.

## Parallel circuit analysis

If Fig. 2-14 was an example of the most complex circuit you would ever have to analyze, you wouldn't need to go any further in circuit analysis. Unfortunately, electrical circuits become much more complicated in real world applications. But don't become discouraged; even extremely complicated circuits can usually be broken down into simpler equivalent circuits for analysis purposes.

Figure 2-15 is an example of a simple *parallel circuit*. A parallel circuit is one in which two or more electrical components are electrically connected across each other. Note that resistors R1 and R2 are wired across each other, or in parallel. In a parallel circuit, the total current supplied by the source will divide between the parallel components. Note that there are two parallel paths through which the current might flow.

Before covering the calculations for determining current flow in parallel circuits, consider these functional aspects of Fig. 2-15.

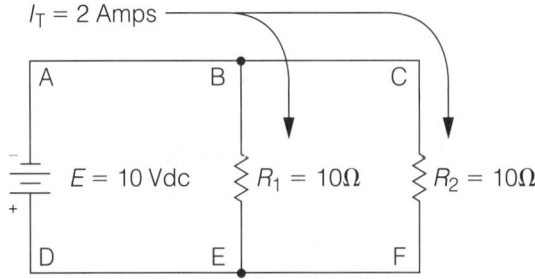

■ **2-15** *Simple parallel dc circuit.*

Electrically speaking, points A, B, and C are all the same point. Likewise, points D, E, and F are also at the same *electrical* point in the circuit. Although this might seem confusing at first, remember that the wire shown in an electrical circuit diagram (electrical circuit diagrams are called *schematics*) is considered to have negligible resistance as compared to the rest of the components in the circuit. In other words, for convenience sake, consider the wire to be a perfect conductor. This means that point A is the exact same electrical point as point B because there is only wire between them. Point B is also the same electrical point as point C. From a circuit analysis perspective, the top of R1 is connected directly to the negative terminal of the battery. Likewise, the top of R2 is also connected directly to the negative terminal of the battery. This means that Fig. 2-15 can be redrawn into the form shown in Fig. 2-16. It is important to recognize that Fig. 2-15 and Fig. 2-16 are exactly the same circuit. (Because this concept is a little abstract, you might need to re-read this paragraph and study Figs. 2-15 and 2-16 several times.)

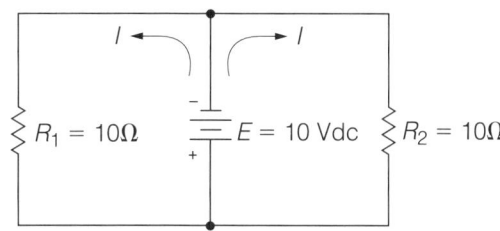

■ **2-16** *The equivalent circuit of Fig. 2-15.*

Figure 2-16 makes it easy to see that both resistors are actually connected directly across the battery. Because the battery voltage is 10 Vdc, the voltage across both resistors will also be 10 Vdc. Now that two circuit variables relating to each resistor are known (the voltage across each resistor and its resistance value), Ohm's law can be used to calculate the current flow through R1:

$$I = \frac{E}{R} = \frac{10 \text{ volts}}{10 \text{ ohms}} = 1 \text{ amp}$$

The previous calculation was for determining the current flow through R1 only. R2 is also drawing a current flow from the battery. It should be obvious that, because R2's resistance is the same as R1, and they both have 10 Vdc across them, the calculation for determining R2's current flow would be the same as R1. Therefore, you can conclude that 1 amp of current is flowing through R1, and

1 amp of current is flowing through R2. The battery in this circuit is a single source that must provide the electrical pressure to promote the total electron flow for both resistors. Therefore, 2 amps of total current must leave the negative terminal of the battery, divide into two separate flows of 1 amp through each resistor, and finally re-combine into the 2-amp total current flow before entering the positive terminal of the battery. This 2-amp total current flow is illustrated in Fig. 2-15.

Try another example. Referring back to Fig. 2-16, assume R1 is 5 ohms instead of 10 ohms. Note that the voltage across R1 will not change (regardless of what value R1 happens to be), because it is still connected directly across the battery. The current flow through R1 would be:

$$I = \frac{E}{R} = \frac{10 \text{ volts}}{5 \text{ ohms}} = 2 \text{ amps}$$

Now you have 2 amps flowing through R1, plus the 1 amp flowing through R2 (assuming the value of R2 remained at 10 ohms). The total circuit current being drawn from the source must now be 3 amps; because 2 amps are flowing through R1, and 1 amp is still flowing through R2.

As the previous examples illustrate, there are a couple of general rules that apply to parallel circuits:

In a closed parallel circuit, the applied voltage will be equal across all parallel legs (or *loops*, as you think of them)

In a closed parallel circuit, the current will divide *inversely proportionally* to the resistance of the individual loops. (The term inversely proportional simply means the current will increase as the resistance decreases, and vice versa.)

In a closed parallel circuit, the total current flow will equal the sum of the individual loop currents.

When analyzing parallel circuits, it is often necessary to know the combined, or *equivalent*, effect of the circuit instead of the individual variables in each loop. Referring back to Fig. 2-15, it is possible to consider the combined resistance of R1 and R2 by calculating the value of an imaginary resistor, $R_{equivalent}$. The following equation can be used to calculate the equivalent resistance of any two parallel resistances:

$$R_{equiv} = \frac{(R_1)(R_2)}{(R_1) + (R_2)} \qquad \text{EQ. 2-4}$$

By inserting the values given in Fig. 2-15:

$$R_{equiv} = \frac{(R_1)(R_2)}{(R_1) + (R_2)} = \frac{(10)(10)}{(10) + (10)} = \frac{100}{20} = 5 \text{ ohms}$$

The 5 ohm ($R_{equiv}$) represents the total combined resistive effect of both R1 and R2. Because you already know the total circuit current is 2 amps, you can use Ohm's law to prove the calculation for $R_{equiv}$ is correct. Using Equation 2-1:

$$E = IR = (2 \text{ amps})(5 \text{ ohms}) = 10 \text{ volts}$$

Because it is true that the source voltage is 10 Vdc, you know the calculation for $R_{equiv}$ is correct.

In circuits where there are three or more parallel resistances to consider, the calculation for $R_{equiv}$ becomes a little more involved. In these circumstances, a calculator is a handy little tool to have! You must calculate the individual conductance values of each resistor in the parallel network, add the conductance values together, and then calculate the reciprocal of the total conductance. This sounds more complicated than it really is. Referring to the complex parallel circuit illustrated in Fig. 2-17, note there are four

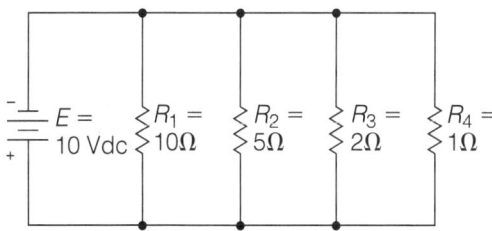

■ **2-17** *Complex parallel circuit.*

resistors in parallel. To calculate $R_{equiv}$, the first step is to find the individual conductance values for each resistor. For example, to find the conductance of R1:

$$G = \frac{1}{R_1} = \frac{1}{10} = 0.1 \text{ mho}$$

The conductance of the remaining resistors should be calculated in the same way. If you perform these calculations correctly, your conductance values should be:

$G(R1) = 0.1$ (previous calculation)

$G(R2) = 0.2$

$$G(R3) = 0.5$$

$$G(R4) = 1$$

By adding all of the conductance values together, the total conductance value comes out to 1.8 mhos. The final step is to calculate the reciprocal of the total conductance, which is:

$$\frac{1}{1.8} = 0.5555 \text{ ohm}$$

$R_{equiv}$ for Fig. 2-17 is approximately 0.5555 ohms. Note that in the previous examples, $R_{equiv}$ is less than the lowest value of any single resistance loop in the parallel network. This will always be true for any parallel network.

The procedure for calculating $R_{equiv}$ in parallel circuits having three or more resistances, can be put in an equation form:

$$R_{equiv} = \frac{1}{\frac{1}{R_1} + \frac{1}{R_2} + \frac{1}{R_3} \cdots \frac{1}{R(n)}} \qquad \text{EQ. 2-5}$$

When using Equation 2-5, remember to perform the steps in the proper order as previously shown in the example with Fig. 2-17.

## Series circuit analysis

The circuit in Fig. 2-18 is an example of a simple *series* circuit. Notice the difference between this circuit and the circuit shown in Fig. 2-15. In Fig. 2-18, the current must first flow through R1, and then through R2, in order to return to the positive side of the battery. There can only be one current flow, and this flow must pass through both resistors.

In a series circuit, the combined effect of the components in series is simply the sum of the components. For example, in Fig. 2-18,

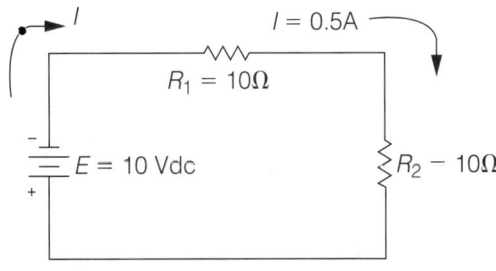

■ **2-18** *Simple series dc circuit.*

the combined resistive effect of R1 and R2 is the sum of their individual resistive values: 10 ohms + 10 ohms = 20 ohms. In series circuits, this combined resistive effect is usually called $R_{total}$, although it might be correctly referred to as $R_{equiv}$.

Knowing the total circuit resistance to be 20 ohms, and the source voltage to be 10 Vdc, Ohm's law can be used to calculate the circuit current ($I_{total}$):

$$I = \frac{E}{R} = \frac{10 \text{ volts}}{20 \text{ ohms}} = 0.5 \text{ amp}$$

In a series circuit, the current is the same through all components, but the division of the voltage is *proportional to the resistance*. This rule can be proven through the application of Ohm's law. The resistance of R1 (10 ohms) and the current flowing through it ($I_{total}$ = 0.5 amp) give us two known variables relating to R1. Therefore, R1's voltage would be:

$$E = IR = (0.5 \text{ amp})(10 \text{ ohms}) = 5 \text{ volts}$$

The 5 volts that appears across R1 in this circuit is commonly called its voltage *drop*. R2's voltage drop can be calculated in the same manner as R1's:

$$E = IR = (0.5 \text{ amp})(10 \text{ ohms}) = 5 \text{ volts}$$

The fact that the two voltage drops are equal should come as no surprise, considering that the same current must flow through both resistors, and they both have the same resistive value.

If the voltage dropped across R1 is added to the voltage dropped across R2, *the sum is equal to the source voltage*. This condition is applicable to all series circuits.

A few general rules can now be stated regarding series circuits:

> In a closed series circuit, the sum of the individual voltage drops must equal the source voltage.
>
> In a closed series circuit, the current will be the same through all of the series components, but the voltage will divide proportionally to the resistance.

As has been noted previously, the resistors in Fig. 2-18 are of equal resistive value. However, in most cases, components in a series circuit will not present equal resistances. To calculate the voltage drops across the individual resistors, you might use Ohm's law as in the previous example. Another way of calculating the same drops is by using the *ratio method*. To use the ratio method, calculate $R_{total}$ by adding all of the individual resistive values. Make a division problem with $R_{total}$ the divisor, and the dividend the resistive value of the resistor for which the voltage drop is being calculated. Perform this division, and then multiply the quotient by the value of the source voltage. The answer is the value of the unknown voltage drop.

Referring to Fig. 2-19, assume you desire to calculate the voltage drop across R1 using the ratio method. Start by finding $R_{total}$:

$$R_{total} = R_1 + R_2 + R_3 = 2 \text{ ohms} + 3 \text{ ohms} + 5 \text{ ohms} = 10 \text{ ohms}$$

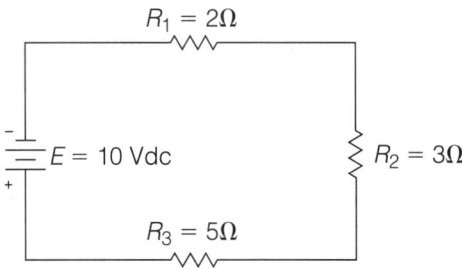

■ **2-19** *Series circuit with unequal resistances.*

Next, make a division problem with $R_{total}$ as the divisor and $R_1$ as the dividend and perform the division:

$$\frac{2 \text{ ohms}}{10 \text{ ohms}} = 0.2$$

Finally, multiply this quotient by the value of the source voltage:

$$(10 \text{ volts})(0.2) = 2 \text{ volts}$$

The voltage dropped across R1 in Fig. 2-19 will be 2 volts. As an exercise, calculate the voltage drop across R1 using Ohm's law. You should come up with the same answer.

### Series-parallel circuits

In many situations, series and parallel circuits are combined to form *series-parallel* circuits (Fig. 2-20). In this circuit, R2 and R3 are in parallel, but R1 is in series with the parallel network of R2 and R3. If this is confusing, notice that the total circuit current must flow through R1, indicating that R1 is in series with the remaining circuit components. However, in the case of R2 and R3, the total circuit current can branch; with part of it flowing through

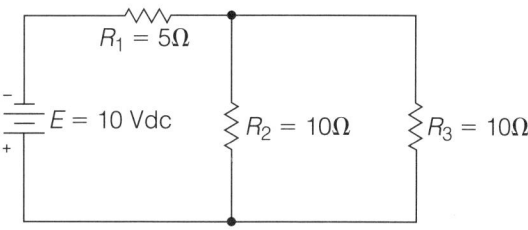

**2-20** *Example of a series-parallel circuit.*

R2, and part of it flowing through R3. This is the indication that these two components are in parallel.

To analyze the circuit of Fig. 2-20, you would start by simplifying the circuit into a form that's easier to work with. As previously explained regarding parallel circuits, resistors in parallel can be converted to an $R_{equiv}$ value. In this case, R2 and R3 are in parallel, so Equation 2-4 might be used to calculate $R_{equiv}$:

$$R_{equiv} = \frac{(R_1)(R_2)}{(R_1) + (R_2)} = \frac{(10)(10)}{(10) + (10)} = \frac{100}{20} = 5 \text{ ohms}$$

Once the value of $R_{equiv}$ is known, the circuit of Fig. 2-20 can be redrawn into the form illustrated in Fig. 2-21. You should recognize this as a simple series circuit. $R_{total}$ can be calculated by adding the values of $R_1$ and $R_{equiv}$:

$$R_{total} = R_1 + R_{equiv} = 5 \text{ ohms} + 5 \text{ ohms} = 10 \text{ ohms}$$

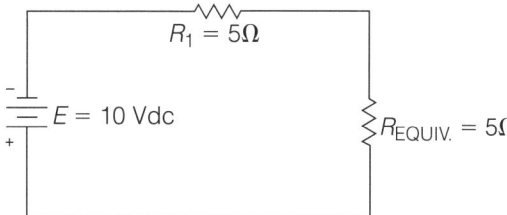

**2-21** *Simplified equivalent circuit of Fig. 2-20.*

Using the values of $R_{total}$, and the source voltage, Ohm's law can be used to calculate the total circuit current ($I_{total}$):

$$I = \frac{E}{R} = \frac{10 \text{ volts}}{10 \text{ ohms}} = 1 \text{ amp}$$

The voltage drop across R1 can be calculated using the variables associated with R1 ($I_{total}$ and $R_1$):

$$E = IR = (1 \text{ amp})(5 \text{ ohms}) = 5 \text{ volts}$$

In this particular circuit, you can take a short-cut in determining the voltage drop across $R_{equiv}$. Because you know the individual voltage drops must add up to equal the source voltage in a series circuit, you can simply subtract the voltage drop across R1 from the source voltage, and the remainder must be the voltage drop across $R_{equiv}$:

$$10 \text{ volts (source)} - 5 \text{ volts (R1)} = 5 \text{ volts } (R_{equiv})$$

As stated previously, the applied voltage (in a parallel circuit) is the same for each loop of the circuit. Because you know 5 volts is being dropped across $R_{equiv}$, this actually means 5 volts is being applied to both R2 and R3 in Fig. 2-20. Now that two variables are known for R2 and R3 (the voltage across them, and their resistive value), the current flow through them can be calculated. Calculating the current flow through R2 first:

$$I = \frac{E}{R} = \frac{5 \text{ volts}}{10 \text{ ohms}} = 0.5 \text{ amp}$$

Obviously, with the same resistive value and the same applied voltage, the current flow through R3 would be the same as R2. Knowing that the individual current flows in a parallel circuit must add up to equal the total current flow, you can add the current flow through R2 and R3 to double-check our previous calculation of $I_{total}$:

$$0.5 \text{ amp } (R_2) + 0.5 \text{ amp } (R_3) = 1 \text{ amp}$$

## Power

An important variable to consider in most electrical/electronic circuits is power. *Power* is the variable which defines a circuit's ability to perform work. Electrical power is dissipated (used up) in the forms of heat and work. For example, the majority of electrical power used by a home stereo is dissipated as heat, but a good percentage is converted to *acoustic energy* (varying pressure waves in the air which is called sound).

It is often necessary to calculate the amount of power that must be dissipated in a circuit or component to keep from destroying something. Also, if you are supplying the power to a circuit, you need to know the amount of power to supply. The following three equations are used for power calculations:

$$P = IE \qquad \text{EQ. 2-6}$$

Power is equal to the current multiplied by the voltage.

$$P = I^2R \qquad \text{EQ. 2-7}$$

Power is equal to the current squared, multiplied by the resistance.

$$P = \frac{E^2}{R} \qquad \text{EQ. 2-8}$$

Power is equal to the voltage squared, divided by the resistance.

Examine how these equations could be used to calculate the power dissipated by R1 in Fig. 2-18. (Remember, you calculated the voltage drop across R1 to be 5 volts in an earlier problem.) Using Equation 2-6 and the known variables for R1:

$$P = IE = (0.5 \text{ amp})(5 \text{ volts}) = 2.5 \text{ watts}$$

Using Equation 2-7 and the known variables for R1:

$$P = I^2R = (0.5 \text{ amp})(0.5 \text{ amp})(10 \text{ ohm}) = 2.5 \text{ watts}$$

Using Equation 2-8 and the known variables for R1:

$$P = \frac{E^2}{R} = \frac{(5 \text{ volts})(5 \text{ volts})}{10 \text{ ohms}} = 2.5 \text{ watts}$$

As you can see from the previous calculations, you only need to know two of three common circuit variables (voltage, current, and resistance) to solve for power.

There is a very important point to keep in mind when performing any electrical/electronic calculation; *don't mix up circuit variables with component variables.* For example, in the previous power calculations, the power dissipated by R1 was solved for. To do this, you used the voltage across R1, the current flowing through R1, and the resistance value of R1. In other words, every variable you used was "associated with R1". If you wanted to calculate the power dissipated by the entire circuit of Fig. 2-18, you would have to use the variables associated with the entire circuit. For example, using Equation 2-8 and the known circuit variables:

$$P = \frac{E^2}{R} = \frac{(10 \text{ volts})(10 \text{ volts})}{20 \text{ ohms}} = \frac{100}{20} = 5 \text{ watts}$$

Notice that the voltage used in the previous equation is the source voltage (the total voltage applied to the circuit), and the resistance value used is $R_{total}$ ($R_1 + R_2$; the total circuit resistance). Our answer, therefore, is the power dissipated by the *total circuit.*

## Common electronic prefixes

Throughout this chapter, the example problems and illustrations have used low value, easy-to-calculate values for the circuit variables. In the real world, however, you will need to work with extremely large and extremely small quantities of these circuit variables. Because it is cumbersome, and both space and time consuming, to try to work with numbers that might have 12 or more zeros in them, a system of prefixes has been standardized. These prefixes, together with their associated symbols and values are:

### Electronic prefixes

pico (symbol "$p$") = 1/1,000,000,000,000
nano (symbol "$n$") = 1/1,000,000,000
micro (symbol "$\mu$") = 1/1,000,000
milli (symbol "$m$") = 1/1,000
kilo (symbol "$k$") = 1,000
mega (symbol "$M$") = 1,000,000

Notice that prefixes for numbers greater than one are symbolized by capital letters, but symbols for prefixes of less than one are lower case letters. Also, note that the Greek symbol $\mu$ (pronounced "mu") is used as the symbol for micro, instead of "$m$," to avoid confusion with the symbol used for the prefix "milli."

The following list provides some examples of how prefixes would be used in conjunction with common circuit variables. The abbreviation provided with each example is the way that value would actually be written on schematics or other technical information.

**Examples:**

1 picovolt      = 0.000000000001 volt  (abbrev. 1 pV)
15 nanoamps     = 0.000000015 amp      (abbrev. 15 nA)
200 microvolts  = 0.0002 volt          (abbrev. 200 $\mu$V)
78 milliamps    = 0.078 amp            (abbrev. 78 mA)
400 kilowatts   = 400,000 watts        (abbrev. 400 kW)
3 megawatts     = 3,000,000 watts      (abbrev. 3 MW)

# The transformer and ac power

# 3

DESCRIBED BRIEFLY IN CHAPTER 2, THE TERM *ALTERNATing current* (abbreviated ac) refers to current flow that periodically changes direction. This condition is related to the periodic changes in the polarity of the applied voltage. Alternating voltage and current are very important because of a phenomenon known as "transformer action." Transformer action enables the efficient transmission of large quantities of power over long distances. This is why all common household current is ac.

## Ac waveshapes

Ac voltages and currents can be thought of as having waveshapes. *Waveshapes* are the visual, or graphical, representations of ac voltage or current amplitudes (or levels) relating to time. Common household ac has a type of waveshape called *sinusoidal* (*sine wave* for short). An illustration of a sine wave is shown in Fig. 3-1. Note how it is similar to a circle that is split in half, and joined together at opposite ends on the zero reference line. The sine wave is the most common type of ac waveshape you will be working with in the electrical/electronic fields.

Any repetitive, cyclic condition can be represented in the same manner as the illustration of the sine wave in Fig. 3-1. On the horizontal plane (that is, reading from left to right), a certain period of time is represented. On the vertical plane (from top to bottom), levels or amplitudes are indicated. The amplitudes above the zero reference usually denote positive variations, and the amplitudes below are usually negative. A *cycle* is defined as one complete periodic variation. In other words, if the illustration in Fig. 3-1 were to be carried out any further, it would begin to repeat itself. A "half-cycle" is either the positive half, or the negative half of a full cycle. Notice how the positive half-cycle would be identical to the negative half-cycle if it were to be turned upside-down.

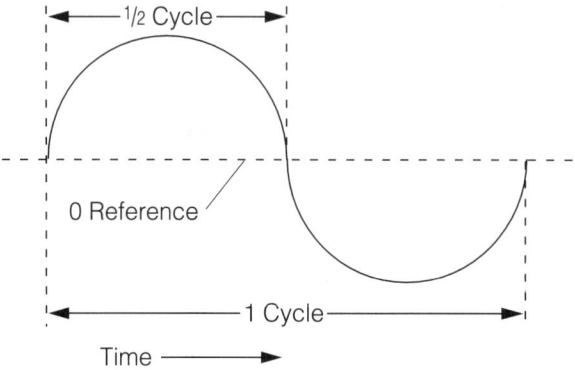

**3-1** *One cycle of a sinusoidal (sine) ac waveshape.*

## Ac frequency

The speed or rate at which an ac voltage or current waveform repeats itself is called its *frequency*. Frequency is measured in units called *hertz*; which is a method of relating frequency to time. One hertz is one cycle of ac occurring in a one-second time period.

If you have never been exposed to the concept of ac waveforms, this all might be rather abstract to you. To aid in understanding Fig. 3-1, assume it to be one cycle of common household ac power. Common household ac is usually labeled *120 Vac, 60 Hz*. This means the usable amplitude of the power is 120 volts, and it cycles at a rate of 60 times a second. If Fig. 3-1 represented this type of power, the time period from the beginning to the ending of one complete cycle would be about 16.6 milliseconds. The time period of a half-cycle would be about one-half of the time period for a full cycle, or about 8.3 milliseconds.

There is an *inversely proportional* (opposite) relationship between frequency and the time period of one cycle. In other words, as frequency increases, the time period per cycle decreases, and vice versa. Equations 3-1 and 3-2 show how time is calculated from frequency, and frequency from time.

$$Time\ period = \frac{1}{frequency} \qquad \text{EQ. 3-1}$$

$$Frequency = \frac{1}{time\ period} \qquad \text{EQ. 3-2}$$

For example, to convert the common 60-Hz power frequency to its associated time period, you would use Equation 3-1:

$$Time\ period = \frac{1}{frequency} = \frac{1}{60\ Hz} = 0.0166\ second$$

0.0166 second is commonly referred to as 16.6 milliseconds. If you knew the time period of one cycle, and wanted to calculate the frequency, you would use Equation 3-2:

$$Frequency = \frac{1}{time\ period} = \frac{1}{0.0166} = 60\ Hz$$

Note: Anytime a number is divided into 1, the result is called the *reciprocal* of that number. Therefore, it is proper to state the time period as being the reciprocal of frequency. Likewise, frequency is the reciprocal of the time period.

## Ac amplitude

As previously stated, one-half of a complete cycle is appropriately called a "half-cycle". As shown in Fig. 3-1, for each half-cycle, the voltage or current actually passes through zero. In fact, it must pass through zero to change polarity or direction, respectively. If Fig. 3-1 is a voltage waveform, the voltage is shown to be positive during the first half-cycle, and negative during the second half-cycle.

The horizontal dotted line representing zero is called the *zero reference line*. As shown in Fig. 3-2, the maximum amplitude of either the positive, or negative, half-cycle (as measured from the zero reference line) is called the *peak voltage*. The total deviation from the negative peak to the positive peak is called the *peak-to-peak voltage*.

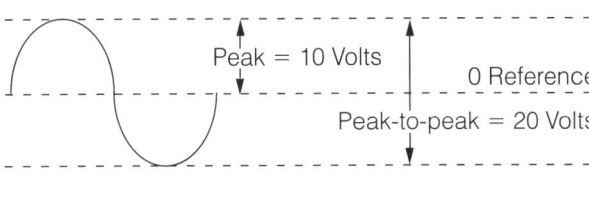

■ **3-2** *Relationship of peak, peak-to-peak, average, and RMS voltages relative to a sinusoidal waveshape.*

Defining the amplitude of pure dc voltages and currents is easy, because the level is constant and continuous. Defining ac voltage or current amplitudes is a little more complicated. As shown in Fig. 3-2, the peak voltage of this waveform is 10 volts. This represents an *instantaneous* voltage level. At other points in the waveform, the voltage level is less than 10 volts (sometimes even zero), and it can be of positive or negative polarity. For this reason, there are numerous ways to accurately define alternating voltage and current amplitudes, depending on the application.

One, somewhat controversial, method of defining an ac amplitude is by algebraically adding the negative half-cycle to the positive half-cycle. Because the positive half-cycle of a true sine wave is the "exact opposite and equal" of the negative half-cycle, the two cancel each other out resulting in an average of zero. If this is difficult to understand, consider this analogy. A dc voltage or current is like driving a car in one direction. The passengers in this car will go somewhere, and end up in a different location than where they started. In contrast, an ac voltage or current is like driving a car back and forth in a driveway. The passengers in this car can ride all day long, and still end up where they started. This is why a dc voltmeter (a test instrument for measuring dc voltage levels) will measure 0 volts if a pure 60-Hz ac voltage is applied to the test probes. The dc voltmeter cannot respond to the rapid ac polarity reversals. Therefore, it indicates the average level, or zero. In the beginning of this paragraph, I stated that this is a controversial definition.

In many electronics textbooks, the *average* ac amplitude for a sine wave is defined as *63.7% of the peak amplitude*. Technically speaking, this is incorrect; but it was generally adopted because many older type ac voltmeters would indicate this value when measuring a sine-wave voltage.

Regardless of definition, don't concern yourself with this situation. Average values of ac voltage are seldom used for any practical purpose. I only included this explanation to aid in understanding the operational physics of an ac sine wave.

If the ac voltage shown in Fig. 3-2 was applied to a load, as shown in Fig. 3-3, current would flow in a back-and-forth motion, proportionally following the voltage variations. Some quantity of power would be *dissipated* because current is flowing through the load. The power dissipated, while the current flows in one direction, does not negate the power dissipated when the current

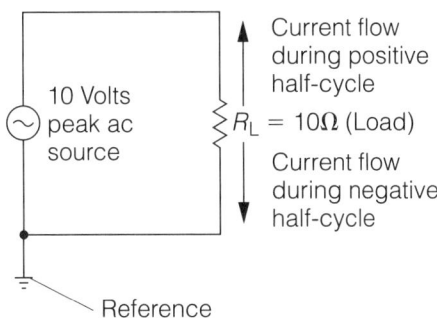

**3-3** *Simple ac circuit illustrating current flow reversal.*

reverses (there is no such thing as negative power). Therefore, power is being dissipated continuously. Consider the previous analogy of the car driving back and forth in the driveway. Although it ended up where it began, energy (gasoline) was consumed; and work was performed (back and forth movement of the car).

A practical method of comparing ac voltage and current to a dc equivalent is the result of a complicated mathematical analysis called *root mean square* (abbreviated *rms*). In other words, ac rms sources can be directly compared to dc sources in their ability to provide power, or to perform work. For example, the common 120 Vac from household outlets is an rms rating. It can perform just as much work as 120 volts of dc. Values of rms are often called *effective* values.

To calculate the rms value of a sine wave, simply *multiply the peak value by 0.707*. Referring to Fig. 3-2, the peak amplitude of the sine wave is 10 volts. To find the rms voltage, the 10 volts is multiplied by 0.707 providing the result of 7.07 volts. Figure 3-3 illustrates this same ac voltage being applied to a load. The power this load dissipates is the same power it would dissipate if a 7.07-volt dc source was applied to it.

## Ac calculations

Note: All of the following equations and calculation methods are only applicable to sinusoidal wave voltages and currents. Virtually all ac used for power sources will fall into this category, so you will be using this information frequently.

If you know the peak, or peak-to-peak, value of a sine wave ac voltage or current, you can calculate the rms value using Equations 3-3 or 3-4:

$$rms = (peak\ value)0.707 \qquad \text{EQ. 3-3}$$

$$rms = (1/2\ peak\text{-}to\text{-}peak\ value)0.707 \qquad \text{EQ. 3-4}$$

For example, using Equation 3-4 to calculate the rms value of the peak-to-peak voltage shown in Fig. 3-2:

$$rms = (1/2\ peak\text{-}to\text{-}peak)0.707$$

$$rms = (1/2\ of\ 20\ volts)0.707$$

$$rms = (10\ volts)0.707$$

$$rms = 7.07\ volts$$

The ac voltage shown in Fig. 3-2 can be accurately defined as being 7.07 volts$_{rms}$. This means it is equivalent to 7.07 Vdc. Rms current calculations are performed in the same manner.

If an rms voltage or current is known, and you wish to calculate the peak or peak-to-peak value, Equations 3-5 and 3-6 might be used:

$$peak = (rms\ value)1.414 \qquad \text{EQ. 3-5}$$

$$peak\text{-}to\text{-}peak = (peak\ value)2 \qquad \text{EQ. 3-6}$$

For example, common household outlet voltage is 120 Vac rms. The peak and peak-to-peak values might be calculated as follows:

$peak = (rms\ value)1.414$

$peak = (120\ volts)1.414 = 169.68\ volts$

$peak\text{-}to\text{-}peak = (peak\ value)2$

$peak\text{-}to\text{-}peak = (169.68\ volts)2 = 339.36\ volts$

You can use any of the ac voltage or current terms in the familiar Ohm's law equations. The only stipulation is "they must be in common". In other words, you cannot mix up rms, peak, and peak-to-peak values, and still arrive at the correct answer. For example, if you wanted to use Ohm's law ($E = IR$) to calculate a peak voltage, you must use the associated peak current value. Resistance is a constant value; therefore, it cannot be expressed in terms such as peak, peak-to-peak, or rms. For the following examples, refer to Fig. 3-3.

**Example 1:** Calculate the rms voltage applied to the load.

$$rms = (peak\ value)0.707 = (10)0.707 = 7.07\ V_{rms}$$

**Example 2:** Calculate the rms current flowing through the load.

$$I_{rms} = \frac{E_{rms}}{R} = \frac{7.07 \text{ volts}}{10 \text{ ohms}} = 0.707 \text{ amps}$$

Note that the rms voltage had to be used, in the previous equation, to calculate the rms current.

**Example 3:** Calculate the peak current flowing through the load.

$$I_{peak} = \frac{E_{peak}}{R} = \frac{10 \text{ volts}}{10 \text{ ohms}} = 1 \text{ amp peak}$$

Note that the peak voltage had to be used, in the previous equation, to calculate the peak current.

**Example 4:** Calculate the peak power being dissipated by the load.

$$P_{peak} = (I_{peak})(E_{peak}) = (1 \text{ amp})(10 \text{ volts}) = 10 \text{ watts peak}$$

The previous equation is simply the familiar $P = IE$ power formula, with the "peak" labels inserted.

**Example 5:** Calculate the rms (effective) power being dissipated by the load.

$$P_{rms} = (I_{rms})(E_{rms}) = (0.707 \text{ amp})(7.07 \text{ volts}) = 5 \text{ watts rms}$$

## Inductance

At some point in your life (possibly in a school science class), you might have built a small electromagnet by twisting some insulated electrical wire around a nail, and then connecting the ends of the wire to a flashlight battery, as shown in Fig. 3-4. An electromagnet is an example of an inductor. An *inductor* is simply a coil of wire.

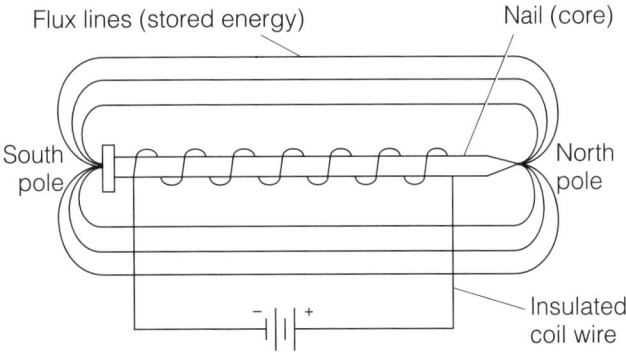

■ **3-4** *Simple electromagnet.*

A typical inductor will consist of a coil (or multiple coils) wound on a metallic, ferrite, or phenolic core. The *metallic* and *ferromagnetic cores* concentrate the magnetic flux lines (lines of force), thus increasing the inductance value. Some coils used in high-frequency applications contain tunable ferrite slugs which allow adjustment of the inductance value. Other types of high-frequency, or special purpose, inductors do not contain metallic cores. These are referred to as *air core* coils.

The electromagnet shown in Fig. 3-4 illustrates how a coil will develop an electromagnetic field around it. An *electromagnetic field* is made up of many lines of force called *flux lines*. The flux lines flow through the nail causing it to become a temporary magnet. Electromagnetic flux lines are stationary only as long as the current flow through the coil is constant. If the current flow through the coil is increased, the electromagnetic field will expand, causing the flux lines to move outward. If the current flow is decreased, the electromagnetic field will collapse, causing the flux lines to move inward. The most important thing to understand is that the field will move (expand and contract) if the current flow changes. Remember this foundational rule of electricity:

> "Whenever an electrical conductor (wire) cuts magnetic flux lines, an electrical potential (voltage) will be induced in the conductor."

It also stands to reason that if the conductor in the previous statement is part of a closed circuit, an induced current will flow whenever an induced voltage is generated. This is the basic principle behind the operation of any electrical generator. To generate electricity, magnetic flux lines must be cut by a conductor. This can be accomplished by moving either the conductor or the electromagnetic field.

As stated previously, if the current flow through a coil (inductor) is varied, the electromagnetic field will move in proportion to the variance. The movement of the electromagnetic field will cause the coil conductors to be cut by the changing flux lines. Consider these effects under the following three conditions:

1. A coil has a steady and continuous current flow through it; the applied voltage to the coil is held constant. Thus, the electromagnetic field surrounding the coil is also constant and stationary.
2. The applied voltage to the coil is reduced. The current flow through the coil tries to decrease causing the electromagnetic field surrounding the coil to collapse by some undetermined

amount. As the field collapses, the flux lines cut the coil wire generating a voltage which opposes the decrease in the applied voltage. The end result is that the coil tries to maintain the same current flow (for a period of time) by using the stored energy in the electromagnetic field to supplement the decreased current flow.

3. The applied voltage to the coil is increased. The current flow through the coil tries to increase causing the electromagnetic field surrounding the coil to expand. As the field expansion occurs, the flux lines cutting the coil wire generate an opposing voltage. The end result is that the coil tries to maintain the same current flow (for a period of time) by storing energy into the electromagnetic field.

In essence, an inductor tries to maintain a constant current flow through itself, by either expanding or contracting its associated electromagnetic field. This expansion or contraction of the electromagnetic field generates a voltage that opposes any changes in the applied voltage. This opposing voltage is referred to as *counter-electromotive force* (abbreviated *cemf*). Electromotive force is another name for voltage.

Inductors are storage devices; electrical energy is stored in the electromagnetic field surrounding an inductor. The quantity of energy that an inductor is capable of storing is called its *inductance* value. Inductance is measured in units called henrys. One *henry* is the inductance value if a current change of 1 ampere per second produces a *cemf* of 1 volt. The electrical symbol for inductance is $L$.

Although this might seem somewhat complicated at first, consider the circuit shown in Fig. 3-5. (Note the symbol for an inductor. The vertical lines to the right of the wire turns denote an iron core.) With the switch (S1) in the open position, obviously there can not be any current flow through the inductor. The graph illustrated in Fig. 3-6 shows the response of the current flow immediately after the switch in Fig. 3-5 is closed. The inductor tries to maintain the same current flow that existed prior to the closing of the switch (which was zero in this case), by storing energy in its associated electromagnetic field. While the electromagnetic field is expanding, a reverse voltage (*cemf*) is being generated by the coil opposing the applied voltage. The end result, as shown in Fig. 3-6, is that the current takes time to change even though the voltage is applied instantaneously.

The time period required for the current to reach its maximum value is measured in seconds, and is defined by a unit called the

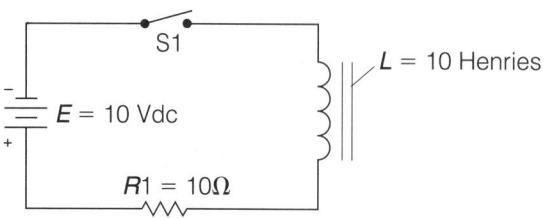

**3-5** Basic LR (inductive-resistive) circuit.

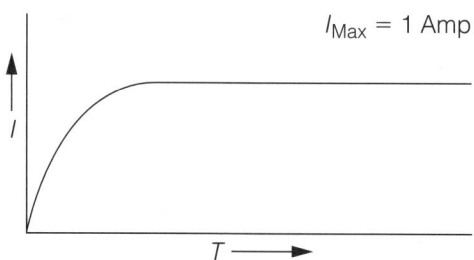

**3-6** Circuit current response of Fig. 3-5.

time constant. The *time constant* is the time required (in seconds) for the current flowing through the inductor to reach approximately 63% of its maximum value. The time constant is found through dividing the inductance value by the circuit resistance. For example, the time constant for the circuit shown in Fig. 3-5 would be:

$$Tc = \frac{L}{R} = \frac{10 \text{ henries}}{10 \text{ ohms}} = 1 \text{ second}$$

Assume that the maximum obtainable current of the circuit shown in Fig. 3-5 is 1 amp. According to the previous time constant calculation, the circuit current will climb to a value of 0.63 amp (63% of 1 amp) after the switch has been closed for 1 second. During the next 1 second time interval, the current will increase by another 63% of the *difference between its present level and the maximum obtainable level*. In other words, if the current rose to 0.63 amp after the first time constant, that leaves a 0.37 amp difference between its present level and the maximum level of 1 amp (1 amp − 0.63 amp = 0.37 amp). Therefore, during the second time constant, there would be a current increase of 63% of 0.37 amp; or approximately 0.23 amp. This results in a circuit current of approximately 0.86 amp after two time constants. Likewise, during the third time constant, there would be another current increase of 63% of the difference between the present level and the maximum obtainable level, and so on.

In theory, the maximum steady-state current of an LR circuit can never be obtained; it would always be increasing by 63% of some negligible current value. In a practical sense, the maximum circuit current value is usually assumed to have been reached after five time constants. The current flow of the circuit shown in Fig. 3-5 would reach its maximum 1-amp level in approximately 5 seconds after S1 is closed.

The effect of the current taking time to catch up with the applied voltage is called the *current lag*. When the applied voltage is ac (changing constantly), the current will always lag behind the voltage.

Periodic ac voltages are divided up into *degrees*, like a circle. Just as a complete circle will contain 360 degrees, one complete cycle of ac is considered as having 360 degrees. Figure 3-7 illustrates how one complete cycle of sine-wave ac voltage is divided into 360 degrees. Note how the positive peak is at 90 degrees; the end of the first half-cycle occurs at 180 degrees; the negative peak is at 270 degrees; and the end of the cycle occurs at the 360-degree point. The waveform below the voltage waveform is the current waveform. Note how the current lags behind the voltage; its first peak occurs at 180 degrees; the end of the first half-cycle is at 270 degrees; and so forth. In other words, the current is *out of phase* with the voltage by 90 degrees.

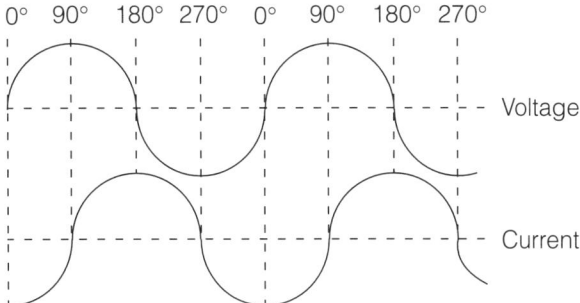

■ **3-7** *Voltage and current phase relationship in an inductive circuit.*

The waveforms shown in Fig. 3-7 are characteristic of any "purely inductive circuit" (a circuit consisting only of an ac source and an inductor, or multiple inductors). In a purely inductive circuit, the current will always lag the voltage by 90 degrees, regardless of the applied ac frequency. This voltage-current phase differential is often referred to as the *phase angle*, and it is also stated in terms of degrees.

An interesting consideration in purely inductive circuits is the *power consumption*—there is none! To understand why, refer back to Fig. 3-7. For discussion sake, assume the peak voltage waveform levels to be 10 volts, and the peak current waveform levels to be 10 amps. At the 90-degree point, the voltage is at its peak level (10 volts) and the current is at zero. Using the power equation $P = IE$, the *instantaneous* power dissipation can be calculated:

$$P = IE = (0)(10 \text{ volts}) = 0$$

At 180 degrees, the voltage is at zero, and the current is at its peak of 10 amps. Using the same power equation:

$$P = IE = (10 \text{ amps})(0) = 0$$

At 270 degrees, the voltage is at its negative peak of $-10$ volts, and the current is zero. Again, using the same power equation:

$$P = IE = (0)(10 \text{ volts}) = 0$$

Because of the 90-degree phase differential between voltage and current, the power dissipation in any purely inductive circuit is considered to be virtually zero. The inductor continually stores energy and then regenerates this energy back into the source.

Assume that you were examining a purely inductive circuit possessing the voltage and current waveforms shown in Fig. 3-7. Again, assume the peak voltage amplitude to be 10 volts, and the peak current amplitude to be 10 amps. If you used a voltmeter to measure the applied voltage, it would read 7.07 volts; the rms value of 10 volts peak. Likewise, if you used an ammeter to measure the current flow, the reading would be 7.07 amps; the rms value of 10 amps peak. If you multiplied the measured voltage value by the measured current value, the answer should be the *rms (or effective) power* value $(P = IE)$. The answer would be 50 watts rms (7.07 volts times 7.07 amps = 50 watts). This contradicts the previous statement regarding zero power dissipation in a purely inductive circuit. The reason for this discrepancy is that you did not take the voltage and current phase differential into consideration. In a purely inductive circuit, the power calculation based on the measured voltage and current values is called the *apparent power*. The actual power dissipated in an inductive circuit when the voltage and current phase differential are taken into consideration is called the *true power*.

If you wish to pursue higher mathematics, true power is calculated by finding the apparent power, and then multiplying it by the cosine of the differential phase angle. For any purely inductive circuit, the differential phase angle (as stated previously) is 90 degrees. The cosine of 90 degrees is zero. Therefore, zero times any apparent power calculation will always equal zero. (If you don't understand this, don't worry about it. Depending on your interests, you might never need to perform these calculations; but if you do, a good electronics math book will explain it in "easy" detail.)

Another term relating to inductive circuits is called the *power factor*. The power factor of any circuit is simply the true power divided by the apparent power.

$$Power\ factor = \frac{true\ power}{apparent\ power}$$

## Dc resistance

Earlier in this chapter, the time constant of the circuit in Fig. 3-5 was examined and the maximum obtainable current flow was found to be approximately 1 amp. The analysis and calculations of this circuit were performed by assuming the circuit to be "ideal;" that is, all components were considered to be perfect. With any circuit, and with all components, there are shortcomings which cause them to be something less than perfect in their operation. To become very technically accurate regarding the circuit in Fig. 3-5, you would have to consider such factors as the internal resistance of the battery (source resistance), the wiring resistance, the resistance of the switch contacts of S1, and the dc resistance of the inductor. In the majority of design situations, all of these factors are so low that they are considered negligible. But to fully understand the operation of inductors, the *dc resistance factor* should be discussed.

Referring back to Fig. 3-5, it was stated that after 5 time constants, the circuit current would reach its maximum amplitude. From this point on (assuming S1 remains closed), both the voltage and current will remain stable. Therefore, the electromagnetic field around the coil will also remain stable. If the electromagnetic field does not move, the wire making up the coil cannot cut flux lines. This means there cannot be any generation of *cemf*. In effect, the only opposition to current flow posed by the inductor is the wire resistance of the coil. The coil's wire resistance (usually called the dc resistance) is normally very low and might be disregarded in most applications.

A term for defining the dc resistance of any coil relative to its inductance value is $Q$ (abbreviation for quality). The $Q$ of an inductor is the inductance value (in henries) divided by the dc resistance (in ohms):

$$Q = \frac{L}{R}$$

For most inductors, this value will be 10 or higher.

## Transformers

An inductor with two or more coils wound in close proximity to each other is called a *transformer*. (A single coil inductor is often called a "choke.") If an ac voltage is applied to one coil of a transformer, its associated moving magnetic field will cause the magnetic flux lines to be cut by itself (causing *cemf*) and any other coil near it. As the other coils are cut by flux lines, an ac voltage is also induced in them. This is the basic principle behind *transformer action*.

If the multiple coils of a transformer are wound on a common iron core, the transfer of electrical energy through the moving electromagnetic field becomes very efficient. Iron-core transformers are designed for efficient transfer of power, and are consequently called *power transformers* or *filament transformers* (an older term from the vacuum tube era). Properly designed, power transformers can obtain efficiencies as high as 99 percent.

Figure 3-8 illustrates some examples of transformer symbols. The coil on which the ac voltage is applied is called the *primary*. The transfer of power is *induced* into the *secondary*. Transformers might have multiple primaries and secondaries. The primaries and secondaries might also contain "taps" to provide multiple voltage outputs or to adapt the transformer to various input voltage am-

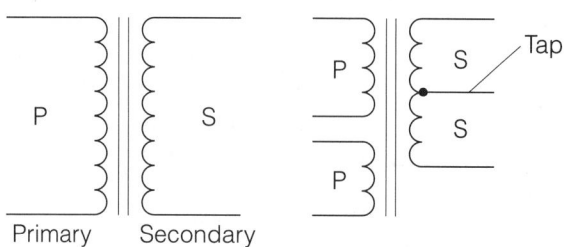

■ **3-8** *Transformer symbols.*

plitudes. The example shown on the left side of Fig. 3-8 is a transformer with one primary and one secondary. The illustration on the right side shows a transformer that has two primaries, one secondary, and a secondary tap. If a tap is placed in the exact center of a transformer coil, that coil is said to be *center-tapped* (abbreviated *ct*).

The most important attribute of a power transformer is its ability to increase or decrease ac voltage amplitudes without a significant loss of energy. This is accomplished by means of the *turns ratio* designed into the transformer. The turns ratio is simply the ratio of the number of turns on the primary to the number of turns on the secondary. For example, if a transformer has a 1:1 turns ratio, it has the same number of turns on the primary, as it has on the secondary. A transformer with a 2:1 turns ratio has twice the number of turns on the primary, as on the secondary. A transformer with a 1:12 turns ratio has twelve times the number of turns on the secondary, as on the primary. In the ratio expression, the primary is always represented by the first number of the ratio, and the secondary is the second.

The turns ratio has a *directly proportional relationship* to the transformer's voltage amplitudes. If the secondary of a transformer has more turns than the primary, the secondary voltage will be proportionally higher in amplitude, or "stepped up". If the secondary contains less turns than the primary, the secondary voltage will be proportionally lower, or "stepped down". For example, if a transformer has a 2:1 turns ratio, the secondary voltage will be one-half the amplitude of the applied voltage to the primary. This is because there is only one-half the number of turns on the secondary as compared to the primary. The reverse is also true. If the transformer has a 1:2 turns ratio, the secondary voltage will be twice that of the primary.

A basic law of physics tells us that: it is impossible to obtain more energy from anything than is originally put into it. For this reason, if the secondary voltage of a transformer is doubled, the secondary current rating will be one-half the value of the primary. Similarly, if the secondary voltage is only one-half the primary voltage, the secondary current rating will be twice the value of the primary. In this way, a transformer will always maintain an equilibrium of *power transfer ability* on both sides. The maximum power transfer ability of a transformer is called its *volt-amp rating* (abbreviated *VA*).

This principle is demonstrated in Fig. 3-9. Notice the transformer has a 10:1 turns ratio. This means whatever voltage is applied to

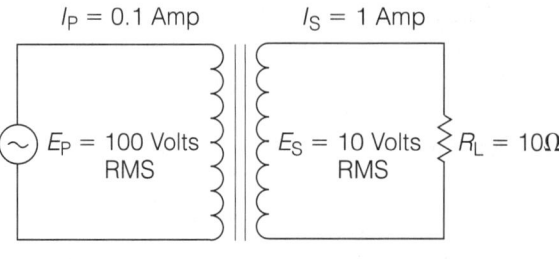

**3-9** *Transformer with secondary load demonstrating the relative primary and secondary currents.*

the primary ($E_p$) will be reduced by a factor of 10 on the secondary ($E_s$). The current flow is just the opposite. The current flowing through the load resistor ($I_s$) will be ten times greater than the primary current flow ($I_p$). In this illustration, $E_p$ is 100 volts rms. Because there is only one-tenth the number of turns on the secondary, $E_s$ is 10 volts rms. According to Ohm's law, the current flow through the load resistor will be 10 volts rms divided by 10 ohms, or 1 amp rms. Therefore, the power transferred to the secondary load will be:

$$P = IE = (1 \text{ amp rms})(10 \text{ volts rms}) = 10 \text{ watts rms}$$

Now consider the power being delivered to the primary. The current in the primary will be one-tenth that of the secondary, or 0.1 amp rms. because $E_p$ is 100 volts rms:

$$P = IE = (0.1 \text{ amp rms})(100 \text{ volts rms}) = 10 \text{ watts rms}$$

It is important to understand that the transformer is not dissipating 10 watts rms of power. It is transferring 10 watts rms of power from the primary to the secondary. The transformer itself is dissipating a negligible amount of power. If the secondary circuit were opened, the current flow in the secondary would consequently drop to zero. Similarly, the current flow in the primary would also drop. (A smaller current flow would still remain in the primary. The explanation and calculations for this condition will be covered in chapter 15.)

## Soldering

Enough of the theory stuff—if you please!! If you're thinking that about now, it just means you're human. If you have comprehended "most" of the theory covered in chapter 2 and in this chapter, you are doing just fine. The "fuzzy" areas of understanding will clear up as you begin to physically work with components and circuits. The remainder of this chapter will deal with the beginning steps for constructing a laboratory quality power supply. This power supply will become the power source for many of the future projects in this book and, hopefully, many more of your own.

Before beginning any kind of electrical/electronic construction, it is essential to become proficient at soldering. By now, you should have accumulated the basic tools needed to solder (as covered in chapter 1). They are:

- [ ] A soldering iron with several size tips
- [ ] A soldering iron holder
- [ ] A damp sponge for tip cleaning
- [ ] A roll of good grade, 60/40 rosin core solder (if you have any "acid core" solder, throw it away or give it to a plumber!)
- [ ] A desoldering tool

In addition to the specific items just listed, you will need a few small hand tools, a comfortable, well-lighted place to work, a little patience, and a worktable (please don't try this on an expensive piece of furniture!). A little steel wool and some isopropyl alcohol might come in handy for cleaning purposes.

## Soldering overview

*Soldering* is a process by which conductive materials are electrically and mechanically bonded together with a tin-lead alloy by the application of heat. For the bonding process to occur properly, the solder and the materials being soldered must be of sufficient temperature. The materials being soldered must also be clean and free of corrosion, oil, and dirt.

During the actual soldering process, the soldering iron is used to heat the material to be soldered (such as a component lead, bare wire, or the copper "artwork" on a printed circuit board) until it reaches a temperature above the melting point of the solder. Rosin-core solder is then placed on the material to be soldered and allowed to melt. As it melts, it will flow outward (called *wetting*) toward the soldering iron tip. When sufficient solder has flowed for a good electrical and mechanical bond, the soldering iron tip and the unused solder are removed simultaneously. The newly formed solder connection is given time to cool and solidify. After cooling, it should appear bright, shiny, and smooth. A rough or grainy looking joint (a *cold joint*) is questionable, and should be re-done.

## Soldering for the first time

If you have never tried soldering, I'm sure you probably think it is more difficult than it is. My experience in working with beginning students has taught me that if I just give them a few "pointers" and leave them alone, they will quickly get the hang of it.

Safety note: wear safety glasses or goggles to protect your eyes when soldering. Also, work in a well-ventilated area, and don't inhale the smoke.

To begin, you will need something to practice on. I recommend a scrap of printed circuit board. *Printed circuit boards* (commonly called *PC boards*) are thin sheets of phenolic or fiberglass material with electronic components mounted on them. The electronic components interconnect on the board by means of copper foil "artwork," called *traces*, glued to the PC board. These are easy to come by. Virtually any kind of junk electronic equipment will contain at least one PC board. If it is difficult to obtain a scrap PC board, you can purchase a general-purpose "grid board" from your local electronics supply store, together with a bag of miscellaneous electronic components. Buy the cheapest you can find; the board and the components will probably be scrap by the time you're comfortable with soldering.

When you've collected the tools, materials, and something to practice on, its time to try your first soldering exercises. Plug in the soldering iron and allow it time to warm up. If you have a soldering iron with an adjustable tip temperature, set it to about 670 degrees. If your soldering iron is adjustable, but doesn't have a readout (in degrees), initially set it to a low temperature, and adjust it by trial and error. The tip should be hot enough to readily melt the solder, but not hot enough to "instantly" melt the solder while producing a lot of smoke.

If the tip of the soldering iron is not silver and shiny, it will need to be "tinned". Clean the tip with some steel wool and apply a little rosin core solder all around the tip and about a half-inch up on the tip. Make sure the solder flows evenly around the tip and then wipe the tip on the damp sponge to remove any excess solder. As the soldering iron is used, it will be necessary to periodically re-tin the tip whenever flux or other contaminates build up on the tip causing it to turn dark. If you remember to clean the tip frequently with the damp sponge, the need for re-tinning will be reduced. If the tip becomes dark quickly, even when not in use, the tip temperature is probably too high.

## Soldering procedure

If you have a scrap PC board to practice on, you will probably have to remove a few components before soldering them back. Skip this section and go on to the next section entitled "Desoldering Procedure." Return to this section after you have finished desoldering.

Make up a few practice solder joints by placing some components randomly in the PC board. Before trying to solder, examine the joint to be soldered. It should be clean, free of corrosion, and mechanically stable (held in place). Place the soldering iron tip on "all" of the material to be soldered. If you are soldering on a PC board, this means the soldering iron tip should be placed on the PC board foil material, and on the component lead at the same time. All of the material to be soldered must be heated or the solder will not adhere properly. With the joint properly heated, apply the solder "to the joint" (not the soldering iron tip). Allow the solder to flow evenly around the joint, and eventually flow to the tip. Remove the unused solder and the soldering iron from the joint, and allow it to cool.

Examine the newly soldered joint. It should be shiny and smooth, and the connection should be mechanically strong. If the solder "balled-up" on the joint, it probably wasn't hot enough. A rough, gray-looking joint is also indicative of improper heating. If the soldering iron tip is not properly tinned, it will not conduct heat adequately to the solderable material. A solder joint with poor electrical integrity is referred to as being "cold." A *cold solder joint* will not be mechanically solid, either. Wiggle the newly soldered component to verify it is tightly bonded together.

The real skill in soldering relates to the speed at which it can be performed. You should be able to solder a typical connection in well under 5 seconds. Heating a connection involving solid-state components for too long can destroy the components. Most electronic supply stores sell aluminum "clamp-on" *heatsinks* to conduct most of the heat away from the component when soldering. But the usefulness of these devices is limited. Most integrated circuits, and many miniature components, simply do not have an available lead length onto which to clamp.

Another point to consider, when soldering, is *heat build-up*. Suppose that you are soldering a component with eight leads into a PC board. If you solder all eight connections, one immediately after another, the component might become very hot because of heat build-up. It becomes progressively warmer with each soldering because it did not thoroughly cool from the previous one. The solution is to allow sufficient time for the component to cool between solderings.

### Desoldering procedure

*Desoldering*, of course, is the opposite of soldering. Desoldering is required to change defective components, to correct construction errors, to redo wiring jobs, and to salvage parts.

One commonly used method of desoldering utilizes pre-fluxed copper braid to absorb molten solder by capillary action. The *desoldering braid*, or wicking, goes by a variety of brand names and is easy to use. The braid is simply placed in the molten solder, and the capillary action draws the excess solder up into the braid. The used piece of braid is then cut off and discarded. For smaller solder joints, the action is often improved by sandwiching the braid between the soldering iron tip and the joint to be desoldered.

Most people prefer to use a vacuum desoldering tool as described in chapter 1. Desoldering is accomplished by "cocking" the tool: holding the tip close to the joint to be desoldered, melting the solder with a soldering iron, and pressing the "trigger" on the tool. The rapid suction action sucks the excess solder up into a holding chamber, from where it is removed later. This tool is often called a *solder sucker*.

If you plan on doing a lot of desoldering (for salvage purposes, for example), you might want to consider investing in a dedicated desoldering station. These units contain a dedicated vacuum pump and a specially designed soldering iron with a hollow tip. The molten solder is sucked up through the hollow tip of the iron and held in a filtered container for removal. The suction action is triggered by a foot or knee pedal. Unfortunately, dedicated desoldering stations are very expensive. Be sure you can justify the cost before investing in one.

## Assembling and testing the first section of a lab power supply

Now comes the fun part! In this section you will begin to build a lab-quality power supply that you can use as a power source for many other projects throughout this book. The building of this supply is also a learning process. You will perform checks and tests to "drive home" the theories and principles you have learned thus far. The experiments performed will also verify the accuracy of your understanding. The same procedure will be followed for virtually every project in this book, so upon completion, you not only have some "neat" projects to show off, but you also have the practical and theoretical knowledge to go with them (which is more important by far!).

## Safety is emphasized throughout this book!

Please don't think I sound like your mother, but if that's what it takes to save even one reader from having an unfortunate accident, the price is well worth it. You learn by doing, but you live to tell about it if you do it right! **Never compromise on safety.**

Before beginning construction, let me briefly explain some of the features that this power supply will have upon completion. This supply will have two independently adjustable, voltage-regulated outputs; one positive, and one negative. The adjustment range for each will be from about 4 to 15 volts. It will be short-circuit protected (so you can make a "goof" without destroying part of the supply), and it will go into *current limit mode* at about 1.5 amps. The current limit feature allows the power supply to double as a battery charger (with certain precautions). A handy feature will be two additional fuse protected "raw" dc outputs (+ and − 34 volts) for testing audio amplifiers, or powering higher voltage projects. In short, you will discover it to be a versatile and valuable piece of test equipment.

### Materials needed for the completion of this section

Note: Throughout the remainder of this book, I will always assume that you have a DVM (or equivalent), soldering tools, alligator clip leads, hook-up wire, mounting hardware, and the necessary hand tools. The materials list for each project will only list the materials and components actually used in the project itself.

| Quantity | Item description |
| --- | --- |
| 1 | 9″ wide × 9″ deep × 3″ high (or larger) metal project box |
| 1 | grounded (3-conductor) power cord |
| 1 | power cord strain relief |
| 1 | SPST 10-amp on/off switch |
| 2 | 24-volt at 2-amp transformers (120-volt primary) |
| 1 | 3AG size (1/4″ × 1 1/4″) fuse block or fuse holder |
| 1 | 2-amp 250-volt slow-blow fuse |
| 1 | locking terminal solder lug (see text) |

The project box doesn't have to be fancy and expensive unless you're the type that wants to go first class all the way. The enclosure from a junked CD player or VCR should do nicely, providing you re-work the front panel. (Front panels can be modified to accommodate almost any need by placing a sheet of chassis aluminum over the face of the original front panel. This hides the original front panel artwork, covers the holes, and provides a solid mounting for new hardware. Because the front panels on most consumer electronic equipment is made from plastic, the aluminum sheet can be easily mounted with self-tapping screws.) The project box can be much larger than specified; it is strictly a matter of personal preference. The bottom part of it must be metal, however, because it will be used for "heatsinking" purposes in a later construction section.

The power cord must have a ground wire; that is, it must have 3 prongs (ground, neutral, and hot). Also, be sure to use the proper size *strain relief* for the power cord. Do not use a rubber grommet in place of a strain relief; this could be dangerous in the long run.

Try to find an on/off switch that comes complete with a backplate to indicate its position, or some other means of indicating its status. It's easy to forget which way is ON and which way is OFF with a bench full of parts and equipment.

This power supply design is a little unusual because it incorporates the use of two 24-volt transformers to form a single 48-volt center-tapped transformer. I decided to design it this way because it might be a little difficult to find a 48-volt ct transformer; however, 24-volt, 2-amp transformers are common.

The specified fuse is the common glass type, as used in a variety of electrical/electronic equipment, and in many automobiles (1/4 inch in diameter and 1 1/4 inch long). Be sure the fuse holder or fuse block is made for that type of fuse. I would personally recommend fuse blocks because they are easier to mount, but that will depend on your preference, and what you might already have in the junk box.

A *locking terminal solder lug* looks like a small internal tooth lock washer with an arm, or extension, sticking out of it. The extension will have one or two holes in it for wire connections. They can be bolted down to a chassis or PC board in the same manner as any other lock washer, but the extension provides a convenient solderable tab for connection purposes. One of these will be used in this project for connecting the ground lead of the power cord to the chassis.

For this particular project, the component placement, enclosure size, and various component construction parameters (such as physical dimension, mounting holes, color, etc.) are not critical to the functional operation of the finished project. This allows you the freedom to be creative and save some money, depending on what you might already have, or what can be obtained through salvaging.

I must assume, at this point in the book, you are inexperienced at buying electronic parts (if I'm wrong, please forgive me). Chances are, you will purchase some (or all) of this material list from a local electronic supply store. This might be slightly traumatic for some people, so let me give you some encouragement and advice. Don't feel intimidated because you happen to be a novice! You are to be commended because you are learning and accomplishing something that many people are afraid to try. The personnel who work in an electronic supply store should be in total agreement with me on that point. Never be reserved about asking any questions you might have about purchasing the right materials and components for your projects. If any sales person ever makes you feel "low" because of your inexperience, I hope you bring this book into the store, and let that person read this paragraph. Then walk out, and take your hard-earned money to a different store that employs sales personnel with the right attitude.

## Mounting the hardware

Read this entire paragraph before physically mounting any components. Lay out the components in the enclosure for mounting. Figure 3-10 illustrates a top view of the approximate way the components should be mounted to the bottom of the enclosure; it is assumed the top cover has been removed. The transformers should be placed side-by-side in either rear corner. Most transformers of this size are constructed with the primary connections on one side, and the secondary connections on the other. Be sure that the primary connection sides are facing the rear panel. The transformers should be placed far enough away from the rear panel to allow sufficient room for the fuse block, or the fuse holder, to be mounted. If you are using a fuse holder instead of a block, pay careful attention to how far it extends "into" the enclosure after mounting; this distance tends to be more than you would guess at first glance. If you are using a fuse block, it should be mounted close to the primary of each transformer. The power cord and strain relief assembly should come into a convenient location close to the fuse block or fuse holder.

*Safety is emphasized throughout this book!*

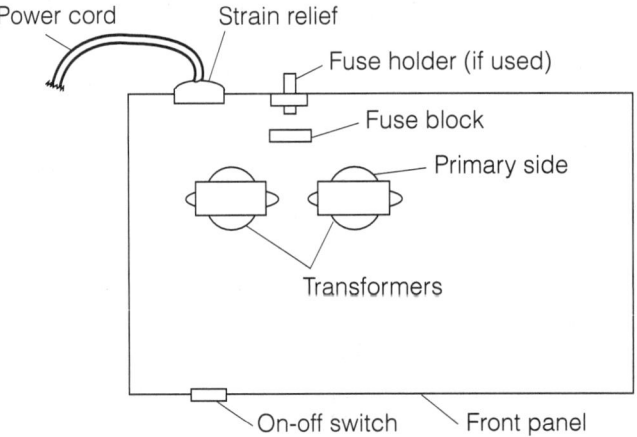

■ **3-10** *Approximate component layout for first section of power supply project.*

The on/off switch is mounted to the front panel. Note how it should be on the same side (in the front) as are the transformers in the back. In regards to height, the power cord and strain relief assembly, the fuse holder (if used), and the on/off switch should be placed about half-way up the height of the enclosure. Doublecheck all of the mounting considerations in this paragraph; and if everything looks good, go ahead and mount the components in the enclosure, using whatever hardware is applicable (nuts, bolts, screws, washers, etc.)

### Wiring and testing procedure

Do not attempt to apply power to the power supply until all of the tests are performed and understood. *Do not attempt to perform any tests using your DVM until you have read and understood the owner's manual for your particular DVM.* Even though a general procedure for using a DVM will be provided throughout this section, this is not a substitute for thoroughly understanding the particularities of your personal instrument.

Refer to Fig. 3-11 during the course of this wiring procedure. Look at the male outlet plug on the power cord. It should have three prongs; two will be flat "blade" type prongs, and the third will be round. The round prong is the *earth ground* connection. The green wire on the other end of the power cord attaches to this round prong. Verify this using the DVM in the "ohms" position. Touch one DVM test probe to the round prong and the other to the green wire; you should read very close to zero ohms. Using a DVM

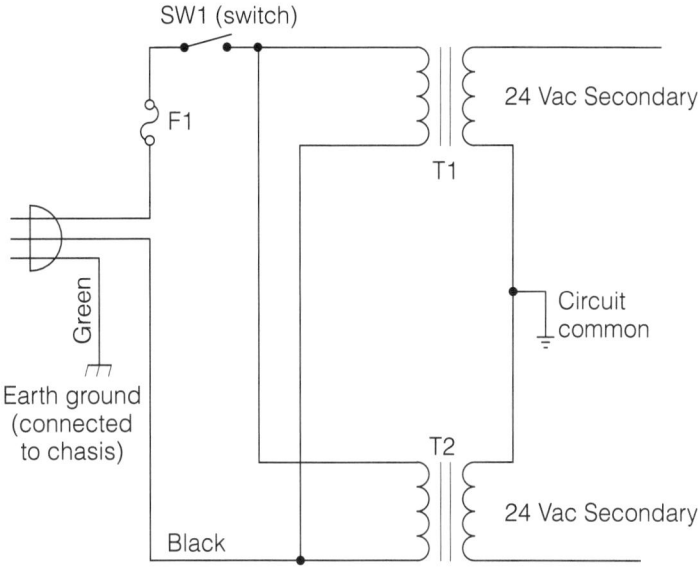

■ 3-11 *Schematic diagram of first section of lab power supply.*

in this way is called testing for *continuity*. In other words, you have proven that there is a continuous electrical path from the round prong, on the plug end of the power cord, to the green wire on the other end. If, by chance, you have a power cord without color-coded wires, test all three wires on the chassis end of the power cord to determine the one connected to the round prong. Solder the wire (connected to the round prong) to a locking terminal solder lug, run the cable through the strain relief hole, and bolt the solder lug to the metal chassis bottom, underneath the power cord strain relief.

Obtain some 18- to 20-gauge hook-up wire with an insulation rating of at least 200 volts. Cut two lengths of this wire long enough to reach from the on/off switch to about three inches beyond the rear of the chassis, so they can be routed along the chassis bottom. Solder these two wires to the on/off switch connections. Connect one of the switch wires to one of the fuse block lugs. In a later step, the other fuse lug will be used to connect the power cord's hot lead.

Connect the other switch wire to one side of the T1 primary lugs, or wires. Some units will have metal lugs; others will have insulated and color-coded wire leads. Connect another piece of hook-up wire from this same point (T1) to the same side of the T2 primary. Solder both of these connections. Now take a few mo-

*Safety is emphasized throughout this book!*

ments to compare your wiring progress so far, with the schematic diagram illustrated in Fig. 3-11.

If you must connect two or more wires together to make these connections, be sure that the connection is well insulated with either insulated mechanical connectors, or soldered and covered with heat shrink tubing. You might use an insulated "butt splice" connector, or an insulated "wire nut" type connector. When soldering two wires together and using a length of heat shrink tubing to insulate the connection, remember to install the tubing onto one of the wires before you twist and solder the ends. The tubing must be just large enough to fit over the twisted and soldered splice. Heat the tubing with a hair dryer, or a heat gun, until it shrinks tightly around the splice; tight enough to remain in position permanently. I do not recommend using electrical tape to insulate this type of connection.

Refer to Fig. 3-11 and note how one side of the F1 fuse block (or holder) is connected to SW1. Connect the other side of the F1 fuse block (or holder) to the black-wired hot side of the ac power cable. This black wire should show continuity to the smaller one of the two flat blades at the other end of the power cable.

You should have one white (neutral) power cord wire left. Connect it to the unused primary lug (or wire) of T2. Also, connect a piece of hook-up wire from this connection to the unused side of T1. Doublecheck all wiring with the schematic diagram in Fig. 3-11, and solder all connections.

Note that the primaries of T1 and T2 are wired in parallel (Fig. 3-11). The bottom of T1's primary is connected to the bottom of T2's primary. And, the top of T1's primary is connected to the top of T2's primary. The secondaries of T1 and T2 are to be wired in series. The bottom of T1's secondary is connected to the top of T2's secondary. You must relate this to the way your transformers are constructed.

Let me attempt to "walk" you through this. Look at your transformer's secondary connections. If there are three connections for each secondary, it means they are *center-tapped* secondaries. Disregard the center-tap connections; they will not be used for this project. That leaves you with two connections for each secondary. Most transformers are constructed so that one connection will be on the left side of the transformer, and one on the right side. Simply connect the left side of one secondary to the right side of the other, and they should be in series, as shown in Fig. 3-11. Make this wiring connection to the best of your ability. If it's wrong, it won't hurt anything; and you can correct it in the following section.

If your transformers have leads, these wires will be coded. The two (usually) black wires are for the primaries. The secondary leads will normally exit the unit on the opposite side from the primary leads. These leads might be any solid color (i.e., red, green, yellow, etc.). If three leads are there, the multi-colored (red/yellow; red wire with yellow stripes) lead is the center tap. Connect the "left-hand" red lead of one unit, to the "right-hand" red lead of the other unit. This will *stack* them in series, and with the correct *phase*. Insulate each of the center tap leads from each other, and from other circuit components, with wire nuts, or with shrink tubing.

### Testing the first section of the lab power supply

There are two purposes for testing your work thus far. It is important, in regard to function and safety, to be sure the construction you have completed to this point is correct. But, it is just as important, in regard to theory and practicality, to understand the testing procedure. If you become confused, go back and review the area of your confusion. It will be time well spent.

Do not apply power to the power supply yet. If you have installed fuse F1 into the fuse block (or holder), remove it. Place the on/off switch in the OFF position. Adjust your DVM to read resistance (ohms position), and set it for the lowest resistance range available. Measure the resistance across the two blade prongs of the power cord. Take the same type of measurement from each blade to the round earth ground prong. All three of these readings should be "infinite" (meaning an open circuit or no continuity). Take the same measurement from the round earth ground prong to any point on the metal chassis. This reading should be very close to zero. This is because you connected the earth ground wire directly to the chassis through the locking terminal solder lug.

Now place SW1 in the ON position. Again, take a resistance reading between the two blade prongs on the power cord. Again, this reading should be infinite because the fuse is not installed. Install F1 into its fuse block (or holder). Again, measure the resistance between the two blade prongs. This time, you should get a very low resistance reading (about 4 to 7 ohms). Referring back to Fig. 3-11, with SW1 closed and F1 installed, you are actually reading the parallel dc resistance of the T1/T2 primaries.

The resistance of each of these primaries is actually twice what this reading has indicated. To illustrate this fact, take two equal value resistors; 100 ohms each, for instance. Measure each resistor; they should each register approximately 100 ohms. Now, twist the ends of the resistors together, so that they are connected in parallel. The measured resistance should now be about 50 ohms. The paralleled resistors represent the paralleled primaries of the power transformers. Your blade-to-blade measurement across the two primaries, just as across the two resistors, shows a value of one-half of the actual primary resistance, and it demonstrates continuity. Your reading is the equivalent resistance ($R_{equiv}$) of the primaries.

Now, turning your attention to the transformer secondaries, measure the resistance between T1's secondary connections. This should be a very low reading; usually less than 1 ohm. Record this value. Similarly, measure T2's secondary resistance. Record this value. Now, measure the resistance from the unconnected side of the T1 secondary to the unconnected side of the T2 secondary. Because the two secondaries are in series, the last measurement should be the sum of the first two secondary measurements. As discussed in chapter 2, to calculate the total series resistance, you simply add the individual resistance values.

As a final safety measure, take a resistance measurement between either secondary connection and the primary connection of T1/T2 at the power switch. This reading should be infinite. This proves there is isolation (no direct wiring connection) between the primaries and the secondaries of T1 and T2. Do the same test between the secondary leads and the chassis ground, and verify that you achieve the same results. This verifies that the secondaries are not shorted to ground. Consider the following safety tips before proceeding.

**CAUTION: Serious injuries, resulting from electrical shock, can occur in several ways.**

One form of injury is through the "indirect" effect of electrocution. Upon receipt of a high-voltage shock, the human body's automatic response is to involuntarily contract the muscles in and around the affected tissue. In other words, you jerk away from it really quick! If, for example, you were to receive a nasty shock to your arm, and your arm happened to be close to some sharp metal edges, you are likely to cut a major size gash in your arm as it jerks back against the sharp metal. This is an *indirect injury*. The arm wasn't actually cut by the electricity, but the effect of the electrical shock caused the injury. I learned a long time ago that you don't bend over and curiously watch someone working on a TV from behind. You could get your jaw broken!

Serious injuries resulting directly from electrical shock can be in the form of *burns*. This occurs more often in very high-voltage environments, such as those encountered by power company employees and industrial electricians.

But the most dangerous form of injury occurs when an accident or circumstance arises causing electrical current to pass through a person's vital organs; this is *electrocution*. This kills! (Contrary to what a lot of people think, voltage does not kill; current is what you need to fear.) I recommend the practice of the "*one hand*" *technique*. The principle is very simple. Electrical current needs a closed circuit to flow through. As long as a person is working close to a high-voltage source "with only one hand" (assuming no other body extremity is grounded), it is impossible for a current to flow through his, or her, vital organs. A small capacitive current might "bite" your hand, but you'll live to talk about it. Therefore, when working with high-voltage sources (120 Vac is definitely considered high voltage), get into a habit of putting one hand in your pocket, or laying it on your lap. Here are a few more safety tips to remember:

- [ ] Don't "snake" your hands and arms into tight fitting places where electrocution could occur; there are tools available to perform those types of jobs.
- [ ] When working on equipment with power applied, always move slowly and deliberately; pay attention to what you are doing. Don't try to catch any falling objects.
- [ ] Never do any kind of electrical work while sitting on a metal seat of any kind, standing on metal grating or on a wet floor, or leaning or holding to any conductive or wet object.
- [ ] Don't stand too close to anyone working on high-voltage equipment.
- [ ] Read and follow the safety guidelines in the owner's manuals for all electronic test equipment you use.
- [ ] Perform "under voltage" tests only when accompanied by an assistant who knows the location of the power shut-off switch, and who knows how to use it!

There are, of course, more safety considerations to follow than the previously listed ones, depending on the work environment. Check them out according to your personal situation. These considerations involve tool-related injuries, wire punctures, accidents that damage equipment, etc. Just use common sense and forethought. Your brain and experience are two of the most important safety tools that you own. Use them well and often.

*Safety is emphasized throughout this book!*

If you have understood everything thus far, all of the resistance measurements have been correct, and you feel lucky (only joking), its time to make the final tests with the power applied. For safety sake, please *don't jump ahead* in the following steps. And, if a fault is detected, *don't proceed until it is corrected.* Before applying power, be sure F1 is installed, and that SW1 is in the OFF position. If your transformers have leads, rather than lugs, place a piece of cardboard on the floor of the chassis, in such a manner that: it will *keep the free ends of the secondary leads from shorting to the chassis.* Tape the leads to the cardboard so that they cannot short to each other. As discussed in chapter 1, you should have an outlet strip with a 15 amp circuit breaker and on/off switch attached to the work bench. Place the outlet strip switch in the OFF position. Plug the power supply circuit into the outlet strip.

Using only one hand, turn on the outlet strip. Using the same hand, place SW1 in the on position. Place SW1 back in the OFF position and observe the circuit. The fuse should not have blown, and there shouldn't have been any visual or audible indication of a fault. Place the outlet strip switch back in the OFF position.

Unplug the circuit from the outlet strip. Even though the outlet strip is turned off, unplugging a piece of electrical equipment, before working on it, is a good safety habit. Temporarily attach one end of an insulated alligator clip lead to the connection you made between the two secondaries of the transformers. This point is referred to as the circuit common in Fig. 3-11. Attach the other end of the clip lead to one test lead of your DVM. Adjust the DVM to read "ac Volts", and set it to the 200 volt range, or higher. (If you have an auto-ranging DVM, it will automatically set its range upon taking the voltage measurement.) Be sure that there is no possibility of an accidental short occurring by verifying the temporary connections are secure, and not extremely close to any other conductive points. Plug the circuit back into the outlet strip, and place the outlet strip switch to the ON position. Place SW1 in the ON position.

Using the DVM test lead not attached to the circuit common, touch it to the "free" T1 secondary lead. The DVM should read about 24 Vac (it is typical for this reading to be a little higher; possibly 26 to 29 Vac). This is the T1 secondary voltage. Now, touch the DVM test lead to the T2 secondary lead. This is the T2 secondary voltage. T2's secondary voltage should be very close in value to T1's secondary voltage. Turn off the circuit and the outlet strip, and unplug the circuit.

Leave one end of the alligator clip lead attached to the DVM test probe, and attach the other end of the clip lead to the T1 secondary lead not connected to the circuit common. Plug the circuit back into the outlet strip; turn the outlet strip on; and place SW1 back in the ON position. Using the DVM test lead that is not attached to the clip lead, touch it to the "free" T2 secondary lead that is not connected to circuit common. This voltage reading should be the sum of the two secondary voltages, or about 50 volts.

If this last voltage measurement was close to zero, it means you have connected the two transformers *out of phase* with each other. In other words, either the primaries, or the secondaries, are connected in such a way that the ac voltage output of one secondary is "up-side down" relative to the other. Another way of putting it is to say that the two secondaries are *bucking* each other, and that they are 180 degrees out of phase with each other. The easiest way to understand this phenomenon is to look at an instantaneous point of time in the transformer's operation.

Referring to Fig. 3-12, note the transformer secondaries shown in the left side of the illustration. Although you do not normally associate any *polarity* with ac voltages, assume this drawing represents an "instant" of time when both secondaries are at a peak voltage output. The polarity of the secondary voltages shown would add to each other—in the same way flashlight batteries add to each other in a flashlight. Positive to negative, positive to negative, etc.

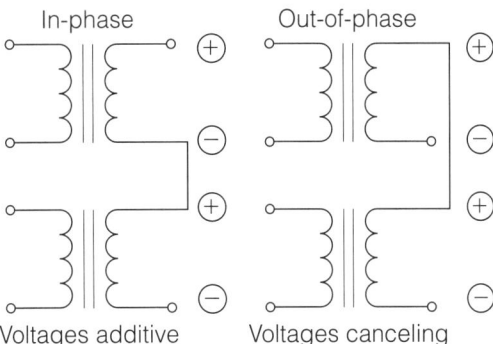

■ **3-12** *Possible phase relationship of two transformer secondaries.*

However, if you accidently wired the transformer secondaries as shown on the right side of Fig. 3-12, the two currents would cancel each other out, and no voltage would be regenerated. Note how the positive is connected to positive. The ability of a transformer to turn an ac voltage up-side down (called *inversion*) is beneficial in many applications; but in our present application, it would render the power supply inoperative. If you measured approximately zero volts across both secondaries in your last reading, the correction is easy. Be certain that all applied power is off, and that the circuit is unplugged. Then, simply reverse the wiring connection on *only one* transformer secondary. Finally, repeat the last voltage check to verify that the 50 Vac is present.

Turn off all applied power, and unplug the circuit. Going back to the resistance measurements performed earlier, the actual dc resistance of The transformer primaries. was a very low value. For ease of calculation, assume it to have been about 12 ohms. When power was applied to the circuit, about 120 Vac was applied to both primaries. If you used Ohm's law to calculate the primary current flow, the result would be:

$$I_{primary} = \frac{E_{primary}}{R_{primary}} = \frac{120 \text{ Vac}}{12 \text{ ohms}} = 10 \text{ amps??}$$

Obviously, this is not correct. F1 is only a 1-amp fuse, and it would have blown instantly, if the primary current flow had been that high.

Transformers are inductors. As discussed previously, inductors try to maintain a constant current flow, by storing and releasing energy from their associated electromagnetic field. Because the 120 volts applied to the primaries is an ac voltage, the storing and releasing of energy from their electromagnetic field is a constant and continuous process. This results in the generation of a *cemf* (counter-electromotive force) that opposes the applied voltage and reduces the primary current flow. The opposition an inductor poses to an ac current flow is called *inductive reactance*. More about inductive reactance will be covered in chapter 15. The sum of this reactance and the dc resistance combine to limit the primary current to a much lower level.

Just as a point of interest, I measured the ac current flow through the primary of T1 in this circuit. It was about 37 milliamps. The ac current flow through the primary of T2 should be the same. The total current flowing through the fuse (F1) would then be about 74 mA, or 0.074 A.

# Rectification

EARLIER IN THIS BOOK YOU LEARNED THE BASICS OF electron flow. You discovered that electrical current is actually a flow of negative charge carriers called *electrons*. Electrons orbit around the nucleus of an atom, just as the earth orbits around the sun. Electrons are held in their orbital paths by their attraction to the positive nucleus, which contains the positive charge carriers called *protons*. Electrons are attracted to the positive nucleus due to a basic law of physics which states "unlike charges attract, and like charges repel." This same principle can be demonstrated by observing the attraction between two permanent magnets. The two north poles of the magnets will repel each other, but a north and south pole will attract. In the same way, the unlike charges of the negative electron and the positive proton attract one another.

## Introduction to solid-state devices

Because one electron (negative charge carrier) will equalize the effect of one proton (positive charge carrier), a normal atom will be balanced in reference to the number of electrons and protons it contains. For example, if an atom has 11 electrons, it will also contain eleven protons. The end result is the negative charge of the electrons will be canceled out by the positive charge of the protons, and the atom will not present any external charge.

The orbital paths of the electrons around the nucleus follow a definite pattern of circular *shells* or *rings*. The maximum number of electrons in each shell is defined by the formula:

$$2(n^2) \text{ where: } n = \text{shell number}$$

For example, if you wanted to calculate the maximum number of electrons for the first shell, $n$ would be 1. Therefore:

$$2(1 \times 1) = 2$$

The maximum number of electrons in the first shell of any atom is 2, regardless of the total number of electrons in that particular atom. Similarly, the maximum number of electrons in the second shell is:

$$2(2 \times 2) = 8$$

The maximum number of electrons in the second shell of any atom is 8. This goes on and on, with the maximum number of electrons in the third, fourth, and fifth shells being 18, 32, and 50 respectively.

In electronics, only the outermost shell of an atom is important because all electron flow occurs with electrons from this shell. The outermost shell of any atom is called the *valence shell*. If the atomic structure of any substance is made up of atoms containing the maximum number of electrons in their valence shells, the substance is said to be an *insulator* because the electrons are rigidly bonded together. If a substance is made up of atoms with valence shells far from being full, the electrons are easily loosened from their orbital bonds, and the substance is said to be a *conductor*. The principle can be compared to buying a new bottle of aspirin. In the beginning, when the aspirin bottle is full, you can shake the bottle, but produce very little movement of the pills because they are tightly packed together. But when you only have a few pills left in the bottle—even the slightest movement will produce a chorus of rattling and rolling pills. Think of this illustration the next time that the electrical/electronic field gives you a headache!

The atoms in some types of crystalline substances (such as silicon) fill their valence shells by overlapping the orbital paths of neighboring atoms. An isolated silicon atom contains four electrons in its valence shell. When silicon atoms combine to form a solid crystal, each atom positions itself between four other silicon atoms in such a way that the valence shells overlap from one atom to another. This causes each individual valence electron to be shared by two atoms. By sharing the electrons from four other atoms, each individual silicon atom appears to have eight electrons in its valence shell. This condition of sharing valence electrons is called *covalent bonding*.

In its pure state, silicon is an insulator because the covalent bonding rigidly holds all of the electrons leaving no (easily loosened) *free electrons* to conduct electrical current. If an impurity is injected into the pure silicon, having a valence shell containing five electrons, it cannot fit into the covalent bonding pattern of the silicon. The result is one free electron per impurity atom that can readily move and conduct current.

Similarly, if an impurity is injected containing only three electrons in its valence shell, the absence of the fourth electron (needed for proper covalent bonding) causes a free positive charge. (A free positive charge is another way of describing a *hole*, or the absence of an electron.)

In both cases, a *semiconductor* has been formed. The term semiconductor simply indicates the substance is neither a good insulator, nor a good conductor; it is somewhere in between. (The term "semi-insulator" would be just as accurate as the term semiconductor.) The process of injecting an impurity into a substance to form a semiconductor is called *doping*.

In addition to becoming a semiconductor, the impure silicon will also possess a unique property; depending upon whether it has been doped with a *pentavalent* impurity (an impurity with five electrons in its valence shell), or with a *trivalent* impurity (an impurity with three electrons in its valence shell). Silicon doped with a pentavalent impurity will become *N-type* material, and it will contain an excess of negative charge carriers (one free electron per impurity atom). Silicon doped with a trivalent impurity will become *P-type* material, and it will contain an excess of positive charge carriers (one "hole" per impurity atom).

## Diode principles

When N-type semiconductor material is sandwiched with P-type material, the resulting component is called a *diode*. A diode is a two-layer device that has an extremely low resistance to current flow in one direction, and an extremely high resistance to current flow in the other. Because it is a two-layer device, it can also be considered a *single-junction device* because there is only one junction between the P and N material, as shown in Fig. 4-1. A diode is often called a *rectifier*.

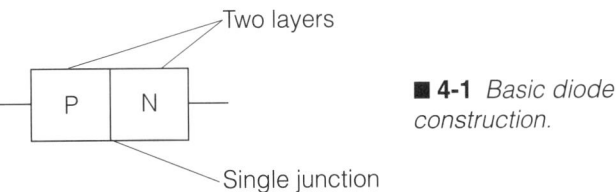

■ 4-1 *Basic diode construction.*

Ideally, you can consider a diode as being capable of passing current in only one direction. If the P side voltage is positive, relative to the N side, by an amount greater than its *forward threshold voltage* (about 0.7 volt if silicon, and 0.3 volt if germanium), the diode will freely pass current almost like a closed switch. This diode is said to be *forward-biased*. If the P side is negative, relative to the N side, virtually no current will be allowed to flow, unless and until the device's breakdown voltage is reached. This condition is referred to as being *reverse-biased*. If the *reverse-breakdown voltage* is exceeded (the point at which reverse-biased current starts to flow) in most normal diodes, the diode may be destroyed.

The P side of a diode is called the *anode*. The N side is called the *cathode* (Fig. 4-2).

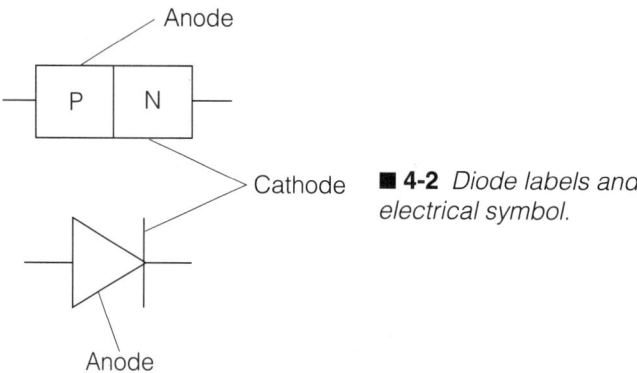

**4-2** *Diode labels and electrical symbol.*

The principle behind diode operation is shown in Fig. 4-3. The diagram of a *forward-biased diode* demonstrates the operation of a diode in the *forward conduction mode* (freely passing current). With the polarity, or bias, of the voltages shown, the forward-biased diode will conduct current as if it were a closed switch. As stated previously, like charges repel each other. If a positive voltage is applied to the P material, the free positive charge carriers will be repelled and move away from the positive potential toward the junction.

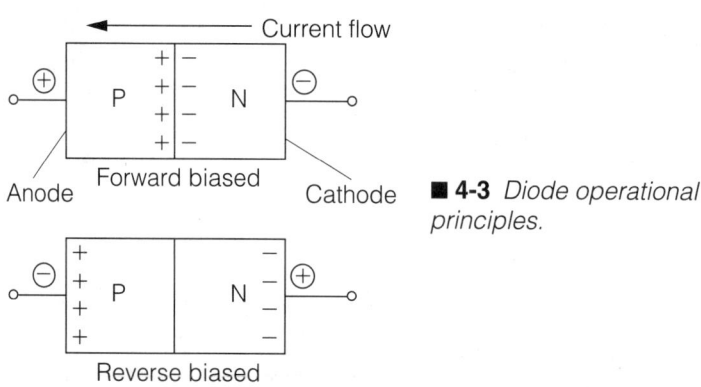

**4-3** *Diode operational principles.*

Similarly, the negative potential applied to the N material will cause the free negative charge carriers to move away from the negative potential toward the junction. When the positive and negative charge carriers arrive at the junction, they will attract (unlike charges attract) and combine. As each positive and negative charge carrier combine at the junction, a new positive and negative charge carrier will be introduced to the semiconductor material from the source voltage providing the bias. As these new charge carriers enter the semiconductor material, they will move toward the junction and combine. Thus, current flow is established, and will continue for as long as the bias voltage remains above the forward bias threshold.

The *forward-threshold voltage* must be exceeded before a forward-biased diode will conduct. The forward-threshold voltage must be high enough to loosen the charge carriers from their atomic orbit and push them through the junction barrier. With *silicon diodes*, this forward-threshold voltage is approximately 0.7 volt. With *germanium diodes*, the forward-threshold voltage is approximately 0.3 to 0.4 volt.

The *maximum forward current* rating of a diode is based on its physical size and construction. Diode manufacturers will typically specify this rating in two ways; the "maximum continuous (or average) forward current" and the "peak forward surge current." The *maximum continuous forward current* is precisely what the name implies; the maximum forward current the diode can conduct on a constant basis. The *peak forward surge current* is the maximum forward current a diode can conduct for 8.3 milliseconds. This last specification can be between 5 and 50 times higher than the continuous rating. 8.3 milliseconds is used as a standard reference for this rating, because it is the time period for one half-cycle of 60-Hz ac. Diodes used for power rectification experience a *surge current* upon the initial application of power (covered further in chapter 5).

Figure 4-3 also illustrates a diagram of a *reverse-biased diode*. As might be expected, the opposite effect occurs if the P material is negative-biased, relative to the N material. In this case, the negative potential applied to the P material attracts the positive charge carriers, drawing them away from the junction. Similarly, the positive potential applied to the N material draws the negative charge carriers toward it, and away from the junction. This leaves the junction area depleted; virtually no charge carriers exist there. Therefore, the junction area becomes an insulator, and current flow is inhibited.

The reverse-bias potential might be increased to the reverse-breakdown voltage for which the particular diode is rated. As in the case of the maximum forward current rating, the *reverse-breakdown voltage* is specified by the manufacturer. The reverse-breakdown voltage is much higher than the forward threshold voltage. A typical general-purpose diode might be specified as having a forward-threshold voltage of 0.7 volt, and a reverse-breakdown voltage of 400 volts. Exceeding the reverse-breakdown voltage is destructive to a general-purpose diode. (Some manufacturers refer to the reverse-breakdown voltage as the *peak-inverse voltage*, PIV, or as the *peak reverse voltage*, PRV.)

Diodes are commonly used to convert alternating current (ac) to direct current (dc). This process is called *rectification*. A single diode used for rectification is called a *half-wave rectifier*. When four diodes are connected together, and are used to redirect both the positive and negative alternations of ac to dc, the four diode configuration is called a *diode bridge*, or a *bridge rectifier*. These configurations can be demonstrated in a few common types of circuits.

Figure 4-4 shows a simple *half-wave rectifier circuit*. Common household power (120 Vac rms) is applied to the primary of a step-down transformer (T1). The secondary of T1 steps down the 120 Vac rms to 12 Vac rms. The diode (D1) will only allow current to flow in the direction shown (from cathode to anode). Diode D1 will be forward-biased during each positive half-cycle (relative to common). When the circuit current tries to flow in the opposite direction, the diode will be reverse-biased (positive on the cathode, negative on the anode), causing the diode to act like an open switch. As shown in Fig. 4-5, this results in a pulsating dc voltage applied across the load resistor ($R_{load}$). Because common household power cycles at a 60-Hz frequency, the pulses seen across $R_{load}$ will also be at 60 Hz. Figure 4-5 also shows the voltage waveform across D1.

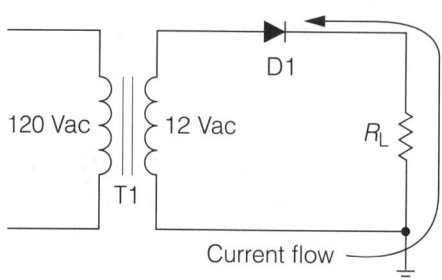

■ **4-4** *Half-wave rectifier circuit.*

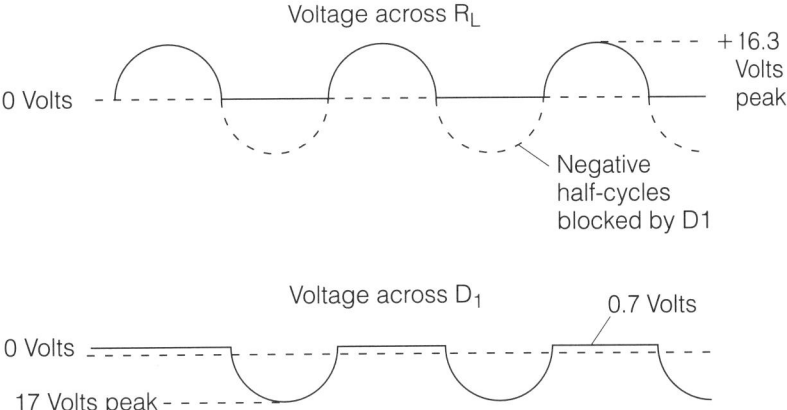

**4-5** *Waveshapes across D1 and RL from circuit illustrated in Fig. 4-4.*

During the positive half-cycle, D1 will drop the 0.7 volt forward-threshold voltage. $R_{load}$ will drop the majority of the voltage; about 11.3 volts in this case. The term *drop* refers to the fact that a certain voltage difference appears across an electrical device, or component, as a current flows through it. In this example, 0.7 + 11.3 = 12 Vdc total. The entire negative half-cycle will be dropped across D1 while it is reverse-biased. The negative half-cycle is dropped across D1 because it looks like an open switch when reverse-biased. An open switch is an infinitely high resistance. Note that D1 is in series with $R_{load}$. As previously discussed, in a series circuit, the higher the resistance value of a component, the more of the source voltage it will drop. Because D1 looks like an infinitely high resistance when reverse-biased, it will drop the total source voltage (output of T1's secondary) during the negative half-cycle.

Consider the amplitude of the voltage developed across $R_{load}$. As shown in Fig. 4-4, the secondary of T1 is 12 Vac rms (ac voltages are always assumed to be V rms values unless otherwise stated). Therefore, the peak voltage output from the T1 secondary is:

$$Peak = (rms\ value)1.414 = (12\ Vac)1.414 = 16.968\ \text{volts peak}$$

For discussion sake, the 16.968 volts can be rounded off to 17 volts. The previous calculation tells us that for each full cycle, the T1 secondary will output one positive 17-volt peak half-cycle, and one negative 17 volt peak half-cycle. The negative half-cycles are blocked by D1, allowing $R_{load}$ to receive only the positive half-cycles. The actual peak voltage across $R_{load}$ will be the 17-volt positive peak, minus the 0.7-volt forward-threshold voltage being dropped by D1. In other words, 16.3-volt positive peaks will be applied to $R_{load}$, as shown in Fig. 4-5.

The diode circuit illustrated in Fig. 4-4 is called a *half-wave rectifier*, because only one-half of the full ac cycle is actually applied to the load. However, it would be much more desirable to utilize the full ac cycle. The circuits shown in Fig. 4-6 are designed to accomplish this, and they are called full-wave rectifiers.

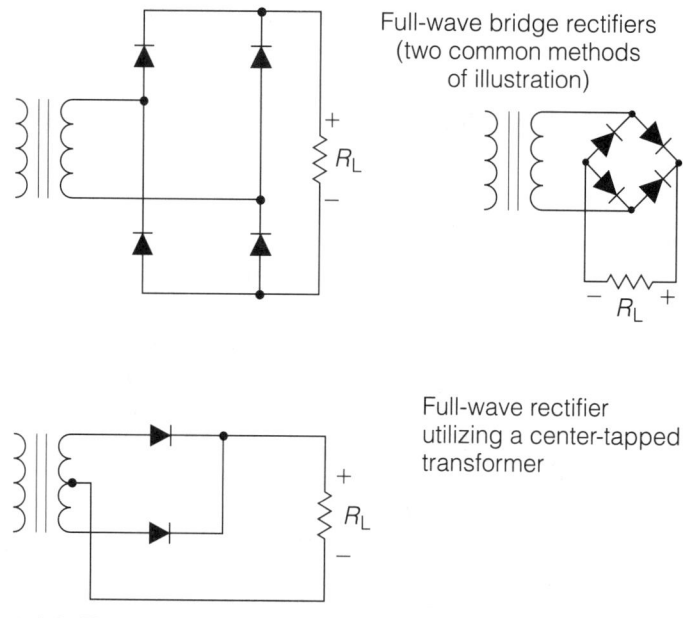

■ **4-6** *Two types of full-wave rectifiers.*

The *full-wave bridge rectifier* illustrated in Fig. 4-6 is the more common type of ac rectifier. It consists of four diodes. The operation of a full wave bridge rectifier can be examined by referring to Fig. 4-7. When the transformer secondary outputs a half-cycle with the polarity shown in the top illustration, the current will follow the path indicated by the arrows. (Remember, forward current always flows through a diode from cathode to anode.) With this polarity applied to the bridge, diodes D1 and D4 are forward-biased, while D2 and D3 are reverse-biased.

The lower illustration of Fig. 4-7 shows the current path when the polarity on the transformer secondary reverses. With this polarity applied to the bridge, diodes D2 and D3 are forward-biased, while D1 and D4 are reverse-biased. In either case, the current always

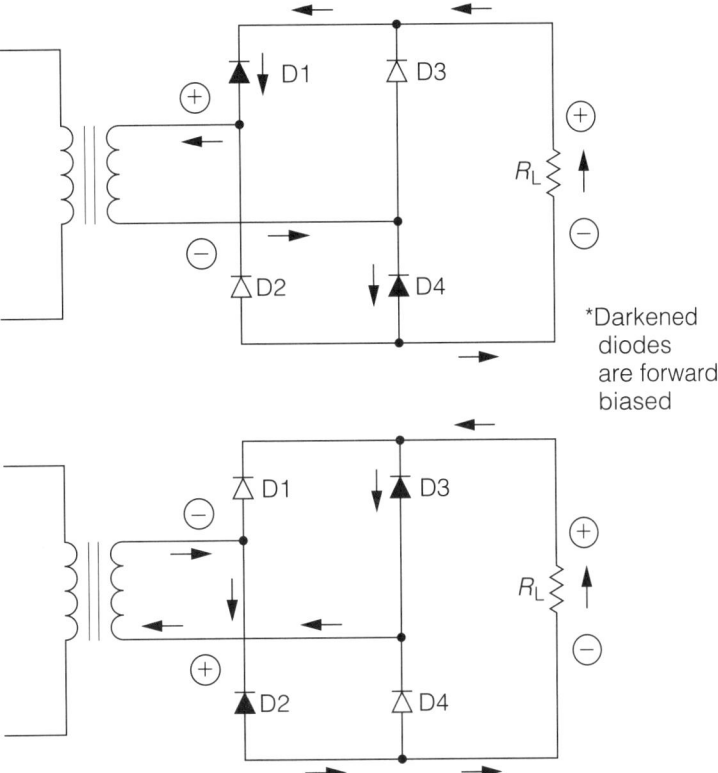

**4-7** *Current flow through a full-wave bridge rectifier.*

flows through the load resistor, $R_{load}$, in the same direction; meaning that the voltage polarity across $R_{load}$ does not change. Essentially, the switching action of the diode bridge actually turns the negative half-cycle upside-down.

Referring back to the circuit of Fig. 4-4, and to its associated waveforms shown in Fig. 4-5, note that $R_{load}$ only receives the positive half-cycles of the ac being output by T1's secondary. These positive half-cycles occur at a 60-Hz rate because the full ac cycle is at a 60-Hz rate. Compare this to the circuit of Fig. 4-7, and to its associated waveshapes shown in Fig. 4-9. Note how $R_{load}$ will now receive the negative half-cycle turned upside-down (converted to another positive half-cycle) as well as the regular positive pulse. In a sinusoidal waveform, the negative half-cycle is the exact inversion of the positive half-cycle; thus, the waveform across $R_{load}$ will begin to repeat itself at the end the first half-cycle. In simple terms, the frequency will double. The frequency of the trans-

*Introduction to solid-state devices*

former secondary is 60-Hz ac, but the frequency at the output of the diode bridge is 120-Hz *pulsating* dc.

You might be confused about the difference between ac and pulsating dc. As stated in the previous chapter, ac is characterized by a voltage polarity and current flow reversal. With pulsating dc, there is a large ac component, usually called *ripple*, but the current flow never changes direction through the load. Obviously, this means that the voltage polarity never changes either. In Fig. 4-9, note that the pulsating dc never crosses the *zero reference line* into the negative region.

The functional diagram of a *full-wave rectifier*, utilizing a transformer secondary with a center-tap, is illustrated in Fig. 4-8. This functions just as well as the bridge rectifier discussed previously, and all of the waveshapes shown in Fig. 4-9 are applicable to this circuit as well. The secondary center tap becomes the circuit common (or circuit reference). As shown in Fig. 4-10, the two outputs from each side of the secondary are exactly opposite to each other, in reference to the center tap (180 degrees out of phase). If one output is in the positive half-cycle, the other output must be in the negative half-cycle and vice versa. Thus, during each half-cycle (of the applied ac to the primary), one of the two secondary outputs will output a positive half-cycle.

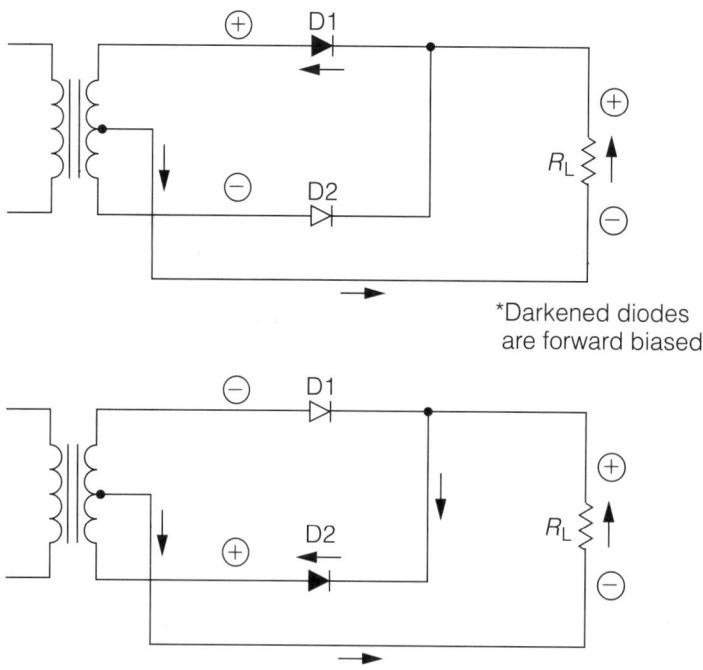

*Darkened diodes are forward biased

■ **4-8** *Current flow through a center-tapped transformer full-wave rectifier.*

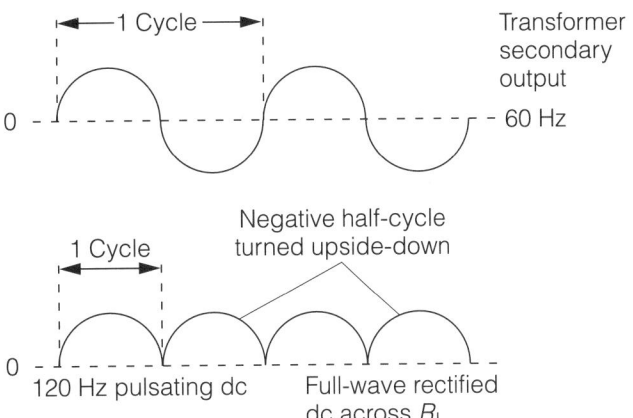

■ **4-9** *Applied 60-Hz ac waveshape compared to the 120-Hz pulsating dc output of a full-wave rectifier.*

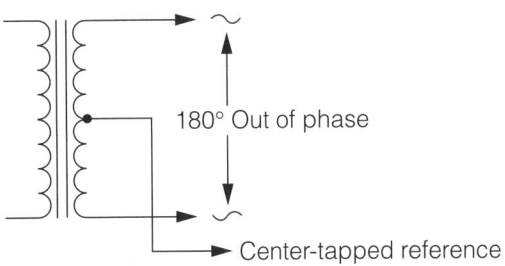

■ **4-10** *Two leads of a center-tapped transformer 180 degrees out of phase when ct is used as the reference.*

The circuit of Fig. 4-8 is actually two half-wave rectifiers connected to the same load. The two outputs from the transformer secondary are *inverted* (opposite) from each other. Therefore, if D1 is forward-biased, D2 will be reverse-biased. When D1 becomes reverse-biased, D2 becomes forward-biased. The end result is that the inverted action of the two half-wave rectifiers are combined to form one full-wave rectifier. The effect is the same as that achieved with the circuit of Fig. 4-7. However, to obtain the same dc voltage amplitude to the load, the transformer's rated secondary voltage, in Fig. 4-8, must be twice the value of the transformer shown in Fig. 4-7. This is because the center-tap will divide the voltage in half when it is used as the common, or reference. For example, a 24-volt, center-tapped transformer secondary will measure 12 volts rms from the center-tap to either side of the secondary.

*Introduction to solid-state devices*

A third common full-wave rectification circuit is illustrated in Fig. 4-11. This circuit is actually the same center-tapped full-wave rectifier circuit as shown in Fig. 4-8, but with two additional diodes incorporated. Referring to Fig. 4-8, the top diagram illustrates how the upper half of the secondary output is applied to $R_{load}$, while the lower half of the secondary output is blocked by D2. The reverse occurs in the lower diagram. In both cases, half of the secondary output is not used.

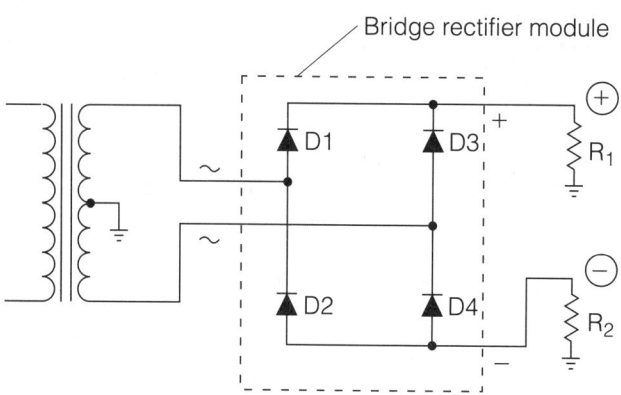

■ **4-11** *Dual-voltage rectification circuit.*

The circuit in Fig. 4-12 utilizes this unused portion of the secondary output to form another dc output of the opposite polarity. This can be more easily understood by referring back to Figs. 4-11 and 4-8. Figure 4-12 is exactly the same circuit (with the same component labeling) as illustrated in Fig. 4-11. Compare Fig. 4-12 with Fig. 4-8. Notice how the positive full-wave rectification section of Fig. 4-12 is identical to the circuit illustrated in Fig. 4-8. The circuit of Fig. 4-12 incorporates two additional diodes (D4 and D2) to form another full-

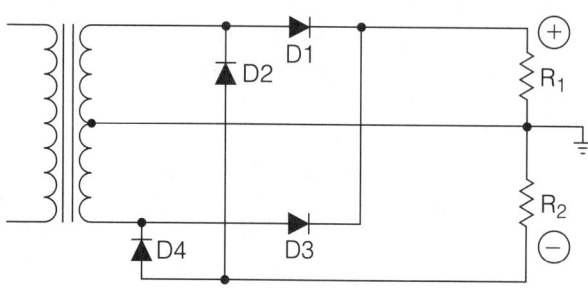

■ **4-12** *Re-drawn schematic of Fig. 4-11.*

wave output from the unused half-cycles. Because of the orientation of the diodes, it will be negative in respect to the circuit common. *Dual polarity power supplies* are very common because of their extensive use in operational amplifier and audio circuits; both of which will be discussed later in this book. As in the case of the center-tapped full-wave rectifier circuit of Fig. 4-8, the voltages across R1 and R2 will only be half of the amplitude of the secondary rating.

## Referencing

By now, you might be beginning to appreciate the importance of waveshapes in the electrical/electronic fields. All of the waveshapes shown in the previous chapters are shown just as they would appear if they were viewed with an oscilloscope. When discussing voltage waveshapes, they are always said to exist with respect to some point of "reference." As pointed out in chapter 2, voltage readings are always taken in respect to some *common point of reference*.

To help in understanding this, consider the following analogy. If an airplane is said to be flying at an altitude of 10,000 feet, it is always assumed that the point of reference is sea level. If the point of reference was changed to a 5,000-foot mountain top, the plane would then be said to be flying at a 5,000-foot altitude. If the plane were to be flying close to the ground, it could be said the plane was flying at nearly a "negative" 5,000-foot altitude.

Referring back to Fig. 4-10, the center-tap essentially splits the transformer secondary into two halves when it is used as the common reference. In the same way that you could make the airplane fly at a negative altitude by moving the reference point up, you can provide an inverted output by moving the circuit reference up to the half-way point of the secondary output.

Unless otherwise stipulated, voltage amplitudes and waveshapes will always be in reference to the circuit common (usually the chassis "ground") on all schematics and electrical drawings.

## Assembly and testing of the second section of a lab power supply

Materials Needed:

| Quantity | Description |
|---|---|
| 1 | 6-amp, 200-volt PIV bridge rectifier module |
| 2 | 10-Kohm, 1/2-watt resistors |

The type of bridge rectifier module specified is commonly available. Although the case styles vary somewhat with different manufacturers, it should be a square or rectangular block, about an inch on each side, with a mounting hole in the middle. (There are some rectifiers with these ratings, meant for PC board installation; they do not have a hole in the middle, and they will be harder to mount). The (2) 10-kohm resistors will be used for testing purposes only. Their tolerance rating, and other parameters, are not important.

## Testing bridge rectifier modules

Before mounting the bridge rectifier module, it will be good practice to test its functional operation. This will help you to understand it and provide you with more experience at using your DVM.

A DVM measures resistance by applying a low voltage to the unknown resistance value, measuring the current through the unknown resistance, and converting the current reading to a resistance reading. The older forms of DVMs, usually called VTVMs (*vacuum tube voltmeters*), would apply about 1.5 volts to the unknown resistance value for measurement purposes. When checking resistance values on solid-state equipment, it is often undesirable to use 1.5 volts as a measurement standard, because this is above the forward threshold voltage of junction devices (diodes, transistors, etc.); causing them to "turn on," and to interact with the component being measured. Interaction of this sort will cause errors in resistance measurements. For this reason, most modern DVMs use only a few tenths of a volt as the measurement standard. When the need arises to check a semiconductor junction, the DVM must be set to the "diode test" position. In this position, the DVM will apply about 1.5 volts to the semiconductor junction under test. The black DVM test lead is of negative polarity, and the red lead is positive (Be sure you have the DVM test leads plugged into the correct holes on the instrument; black to common).

When testing a diode, touch the black DVM test lead to the cathode (the "banded" side of the diode body), and the red lead to the anode. This forward biases the diode, and the resistance reading should be low (the actual resistance reading is irrelevant; when testing semiconductor junctions, you are only concerned with "high [or infinite]" verses "low" readings). By reversing the DVM leads (red on cathode; black on anode), the diode is reverse biased, and the resistance reading should be infinitely high.

Be aware that some recent VOMs (*volt-ohm-milliammeters*) are reverse-polarized for their resistance checks; the positive dc potential being on the black probe. In this instance, your readings will always be reversed. Test about a dozen common diodes (1N4001 or 1N4148/1N914). If they all test "good" only when the probes are reversed, then you have + volts on your black probe.

The bridge rectifier module you obtained for this project actually contains four diodes connected in a bridge configuration, as shown in Fig. 4-11. Four individual diodes would function just as well. The four internal-module diodes can be tested with a DVM just as though they were four standard diodes.

The bridge rectifier module should have four connection terminals, or leads, extending from it. Each terminal, or lead, should have an identification symbol associated with it, on the case of the module. Two of the terminals will be marked with an *"ac"* or the symbol for a sine wave; these are the ac input terminals from the transformer. The other two terminals will be specified with polarity symbols (*"−" or "+"*); these are the dc output terminals. Figure 4-11 illustrates the actual external and internal connections associated with the bridge.

Referring to Fig. 4-11, if you wanted to test the internal diodes labeled D1 and D3, you would place the black (negative) DVM lead on the + terminal of the diode module, and the red (positive) DVM lead to either one of the ac terminals of the module. A low resistance reading would indicate either D1 or D3, respectively, was functioning correctly. You actually wouldn't know which one of the diodes you were testing because the module will not specify any difference in the ac terminals. The issue is totally unimportant; if either one of the diodes is defective, the whole bridge module must be replaced.

Rather than detailing each individual diode measurement, the following chart will make it easy for you to check any bridge rectifier module. Use this table, in conjunction with Fig. 4-11, to understand how the measurements will check each internal diode.

■ Table 4-1 Testing Bridge Diodes

| Bridge terminals (or leads) | | | | |
|---|---|---|---|---|
| (+) | (-) | (ac) | (ac) | Results |
| Black | | Red | | Low resistance |
| Black | | | Red | Low resistance |
| Red | | Black | | Infinite resistance |
| Red | | | Black | Infinite resistance |
| | Red | Black | | Low resistance |
| | Red | | Black | Low resistance |
| | Black | Red | | Infinite resistance |
| | Black | | Red | Infinite resistance |

*Assembly and testing of the second section of a lab power supply*

Assuming there were no problems with the bridge rectifier module, mount it, close to the T1 and T2 secondaries, with the appropriate hardware. Wire and solder the connections from the T1 and T2 secondaries to the ac input terminals on the bridge module as illustrated in Fig. 4-13. Using an alligator clip lead, temporarily connect one side of test resistor $R_{T1}$ to the positive terminal of the bridge rectifier. Connect the other side of $R_{T1}$ to the circuit common connection between the two transformer secondaries with another clip lead. Using two more clip leads, connect one side of test resistor $R_{T2}$ to the negative terminal of the bridge, and the other side to the same circuit common point as $R_{T1}$. Verify your wiring is the same as is shown in Fig. 4-13. *Be sure all four clip leads are secure, and not touching anything except the desired connection point.*

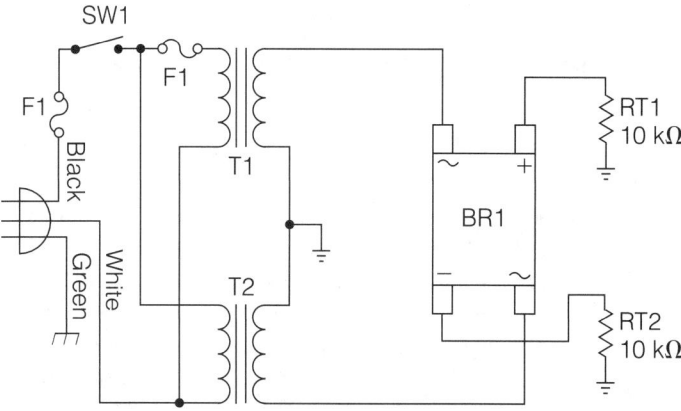

■ **4-13** *Schematic diagram of the first and second sections of the lab power supply.*

Turn OFF your lab outlet strip, and set SW1 to the OFF position. Plug the circuit into the outlet strip, and (using only one hand) turn ON the outlet strip. Again, using only one hand, set SW1 to the ON position. If F1 blows, turn OFF all power immediately; unplug the circuit; and double-check all wiring. If the wiring is good, the fault must be in the bridge rectifier (assuming that you followed the test procedures for the first section of the power supply in the previous chapter).

Hopefully, F1 didn't blow, and you can leave the power applied to the circuit and continue with the test. Set your DVM to indicate "dc volts" on at least the 100-volt range. Use one more clip lead (and only one hand!) to connect the black DVM test lead to circuit common. Touch the red DVM test lead to the side of $R_{T1}$ that is connected to the positive terminal of the bridge rectifier. You should read about +24 Vdc. Touch the red lead to the side of $R_{T2}$ that is connected to the negative terminal of the bridge rectifier. You should read about $-24$ Vdc. Leave the black DVM test lead connected to the circuit common, and set the DVM to indicate "ac volts" on the same range. Again, measure the voltage across $R_{T1}$, and then $R_{T2}$. Both readings should be about 12 Vac. Turn SW1 OFF, the outlet strip OFF, and unplug the circuit.

Referring back to Fig. 4-9, the dc voltage across $R_{T1}$ and $R_{T2}$ was pulsating dc as shown in the bottom waveform illustration. The DVM did not measure the peaks of the dc pulses; it read the effective dc level. You might think of it as "cutting off the peaks, and using that energy to fill in the valleys". The ac voltage across the resistors is called the *ac component*, or the *ripple*. The DVM measured the rms value of this ac component.

If you own an oscilloscope, read the owner's manual, and be sure you understand how to operate it correctly and safely. Then, use it, in conjunction with this power supply circuit, to observe these waveforms. They should appear just as shown in the various illustrations. *Do not try to view the waveforms of the transformer primaries without using an isolation transformer.*

$R_{T1}$ and $R_{T2}$ were used for testing purposes only. They can now be removed from the circuit, together with the alligator clip leads used to temporarily connect them.

# Capacitance

A CAPACITOR IS ONE OF THE MOST COMMON COMPONENTS used in electronics, but is probably one of the least understood. As in the case of inductors, a capacitor is a storage device. An inductor stores electrical energy in the form of an electromagnetic field, which collapses or expands to try to maintain a constant current flow through the coil. In comparison, a *capacitor* stores an electrostatic charge, which increases or decreases to try to maintain a constant voltage across the capacitor.

## Capacitor types and construction

A capacitor consists of two conductive plates with an insulator placed between them called the *dielectric*. The size of the plates, the thickness of the dielectric, and its dielectric constant all combine to determine the *capacity*, or energy storage capability, of the capacitor. The capacity can be increased through the use of larger conductive plates, thinner dielectric material, or a dielectric material with a higher dielectric constant. Thinner dielectric yields higher capacity; but it also lowers the maximum voltage rating. Because it is impractical to manufacture (or to try to use) extremely large metal plates, capacitors are usually manufactured by rolling the foil plates (with the plastic dielectric sheets interleaved between them) into a round tubular form. Alternatively, layers of foil plates can be stacked and sandwiched, as when shuffling a deck of cards, with the dielectric sheets (usually mica, ceramic, or plastic) interleaved between them.

Basically, all capacitors can be divided into two categories; polarized and non-polarized. If a capacitor is polarized (a step in the manufacturing process), the correct voltage polarity must be maintained when using it. A *polarized capacitor* will indicate which lead is to be connected to a positive polarity, and/or which lead is to be connected to a negative polarity, by means of labels or symbols imprinted on its body. Accidental reversal of the indicated polarity will destroy a polarized capacitor, and **it could cause it**

**to explode**. *Non-polarized capacitors* do not require any observance of voltage polarity.

The vast majority of capacitors use conductive metal foil as the plate material. The only exceptions are adjustable capacitors and trimmers. The big difference in capacitor types is based on the dielectric material used. The two important properties of dielectric materials are called dielectric strength and dielectric constant. The *dielectric strength* defines the insulating quality of the material, and is a key factor in determining the capacitor's voltage rating. Dielectric constant will be explained a little later in this chapter.

Paper and mica were the standard dielectric materials for many years. Mica is used in special applications, and paper is still used quite often for general-purpose use. The paper is impregnated with a wax, or a special oil, to reduce air pockets and moisture absorption.

Plastic films of polycarbonate, polystyrene, polypropylene, and polysulfone are used in many of the newer *large-capacity, small-size capacitors*. Each film has its own special characteristic, and is chosen to be used for various applications according to its unique property.

Ceramic is the most versatile of all the dielectric materials because many variations of capacity can be created by altering it. Special capacitors (that increase their value, stay the same, or decrease value with temperature changes) can be made using ceramics. If a ceramic disc capacitor is marked with a letter **P** (positive change), such as **P100**, then the value of the capacitor will increase 100 parts per million per degree centigrade increase in temperature. If the capacitor is marked **NPO** (neg/pos/zero) or **COG** (change zero), then the value of capacity will remain relatively constant with an increase or decrease in temperature. If it is marked with an **N** (negative), such as **N1500**, it will decrease in capacity as the temperature increases.

The term defining the manner in which a component is affected by changes in temperature is called the *temperature coefficient*. If a component has a *negative temperature coefficient*, its value decreases as the temperature increases, and vice versa. If it has a *positive temperature coefficient*, its value increases as the temperature increases, and vice versa. A capacitor's temperature coefficient is critical for circuits in which minor changes of capacitance can adversely affect the circuit operation. One of the reasons for ceramic capacitors being the most commonly used type is the versatility of their different temperature coefficients. The other main reason for their widespread use is cost; they are very inexpensive to manufacture.

A ceramic capacitor marked **GMV** means the marked value on the capacitor is the "guaranteed minimum value" of capacitance at room temperature. The actual value of the capacitor can be much higher. This type of capacitor is used for applications in which the actual value of capacitance is not critical.

*Aluminum electrolytic capacitors* are very popular because they provide a large value of capacitance in a small space. Electrolytic capacitors are polarized and the correct polarity must always be observed when using or replacing these devices. For special-purpose applications (such as crossover networks in audio speaker systems and electric motors), *non-polarized electrolytics* are available, which will operate in an ac environment.

The aluminum electrolytic capacitor is constructed with pure aluminum foil wound with a paper soaked in a liquid electrolyte. When a voltage is applied during the manufacturing stage, a thin layer of aluminum-oxide film forms on the pure aluminum. This oxide film becomes the dielectric because it is a good insulator. As long as the electrolyte remains liquid, the capacitor is good; or it can be "reformed" by applying a dc voltage to it for a period of time (while observing the correct polarity). If the electrolyte dries out, the leakage increases and the capacitor loses capacity. This undesirable condition is called *dielectric absorption.*

Dielectric absorption can happen to aluminum electrolytics even in storage. Sometimes, if an electrolytic capacitor has been sitting on the shelf (in storage) for a long period of time, it can need to be reformed to build up the oxidation layer. This can be easily accomplished by connecting it to a dc power supply for approximately an hour. Remember to observe the correct polarity, and do not exceed its voltage rating, or the capacitor can explode. (Reversing the polarity on an electrolytic capacitor causes it to look "resistive" and build up internal heat. The heat causes the electrolyte to boil, and the steam builds up pressure. The pressure can cause the capacitor to explode if it is not equipped with a "safety plug".) Most electrolytic capacitor manufacturers do not guarantee a shelf-life of more than five years.

Another type of polarized electrolytic capacitor is called the *tantalum capacitor*. Because they have the physical shape of a water drop, they are commonly referred to as "tear-drop" capacitors. Tantalum capacitors have several advantages over aluminum electrolytics; lower leakage, tighter tolerances, smaller size for an equivalent capacity, and a much higher immunity to electrical "noise" (undesirable "stray" electrical interference from a variety of sources). Unfortunately, their maximum capacity values are limited and they are more expensive than comparable aluminum electrolytics. For applications requiring large capacitance values, aluminum electrolytics are still preferred.

## Basic capacitor principles

When a dc voltage is placed across the two plates of a capacitor, a certain number of electrons are drawn from one plate, and flow into the positive terminal of the source (the positive potential attracts the negative charge carriers). At the same time, the same number of electrons flow out of the negative terminal of the source and are pushed into the other plate of the capacitor (the negative potential repels the negative charge carriers). This process continues until the capacitor is charged to the same potential as the source. When the full charge is completed, all current flow ceases, because the dielectric (an insulator) will not allow current to flow from one capacitor plate to the other. In this state, the capacitor has produced an *electrostatic field*; that is, an excess of electrons on one plate and an absence of electrons on the other.

Figure 5-1 illustrates a capacitor in series with a resistor and an applied source voltage of 10 Vdc. Assuming that C1 is not initially charged, when the switch (SW1) is first closed, a circuit current will begin to flow. The rate of the current flow (which determines how rapidly the capacitor will charge) will be limited by the series resistance in the circuit, and the difference in potential between the capacitor and the source. As the capacitor begins to charge (build up electrical pressure), the rate of current flow from the source to the capacitor begins to decrease. The graph of Fig. 5-2 illustrates the voltage across the capacitor (C1) relative to time. The capacitor eventually charges to the full source potential of 10 volts. When this happens, all circuit current will cease because there can be no cur-

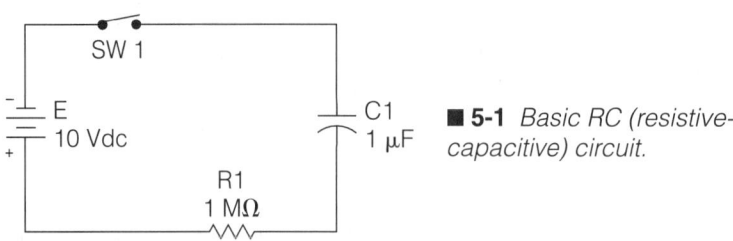

■ **5-1** *Basic RC (resistive-capacitive) circuit.*

rent flow through the dielectric. If SW1 is opened at this time, the capacitor will hold this static charge until given a discharge path of lesser potential. (The word *static* means "stationary"; hence, the term *electrostatic field* means an electrically stationary field.)

As stated previously, with SW1 closed, the capacitor will eventually charge to the source potential. The *capacity*, or the quantity of charge it can hold at this potential, is determined by the physical characteristics of the capacitor.

Here is an analogy to help clarify the previous functional aspects of capacitor theory; electrical references will be to the circuit shown in Fig. 5-1. Imagine you have a very large tank of water filled to a 10-foot level (this is analogous to the battery at a 10-volt potential). From this large tank, you wish to fill a small tank to the same 10-foot level (the small tank is analogous to the capacitor). You connect a water pipe and valve from the bottom of the large tank to the bottom of the small "empty" tank. When you first open the water valve to allow water to flow from tank to tank, the flow rate will be at its highest because the level differential will be at its greatest (the water flow is analogous to electrical current flow). As the water begins to fill up the small tank, it also begins to exert a downward pressure opposing the flow of incoming water. Consequently, the flow rate begins to decrease. As the water level in the small tank gets higher, the flow rate continues to decrease until the small tank is at the same 10-foot water level as the large tank. As soon as the two levels are equal, all water flow from tank to tank will cease. As the old proverb states, "water seeks its own level".

If you were to monitor the level increase in the small tank with respect to time, you would notice that the level does not rise in a linear fashion; that is, it rises at a slower rate as it approaches the 10-foot top. If you charted the rise in "level versus time" in the form of a graph, the curve would be identical to the curve shown in Fig. 5-2.

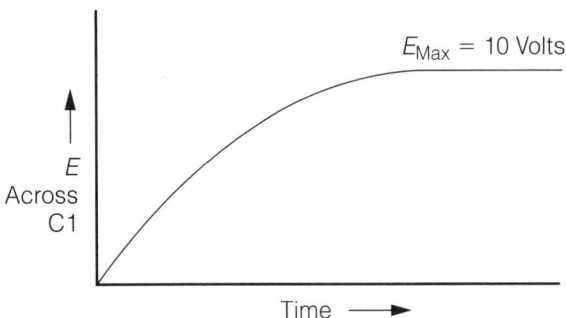

■ **5-2** *Capacitor voltage response of Fig. 5-1.*

The small water tank will have a holding capacity associated with it. For instance, it can be specified as capable of holding 40 gallons. Capacitors are also rated in respect to the quantity of charge they can hold.

The *capacity* (quantity of charge) of a capacitor is measured in units called farads. A *farad* is the amount of capacitance required to store one coulomb of electrical energy at a 1 volt potential. A *coulomb* is a volume measurement unit of electrical energy (charge). It is analogous to other volume measurement units such as quart, pint, or gallon. A gallon represents 4 quarts. A coulomb represents $6.28 \times 10^{18}$ electrons (or 6,280,000,000,000,000,000 electrons). The basic unit of current flow, the *ampere*, can be defined in terms of coulombs. If one coulomb passes through a conductor in one second, this is defined as one ampere of current flow. Thus, 1 amp = 1 coulomb/sec.

A farad is generally too large of a quantity of energy to be stored by only one capacitor. In the 1930s, a Gernsback magazine calculated the size requirements for building a 1-farad paper-foil capacitor. Completely fill the Empire State Building, from bottom to the top, with a stack of paper and foil layers!

Therefore, the capacity of most capacitors is defined in terms of *microfarads* (1 $\mu$F = 0.000,001 farad) or *picofarads* (1 pF = 0.000,000,000,001 farad). As stated previously, capacitors also have an associated voltage rating. If this *voltage rating* is exceeded, the capacitor could develop an internal short (the term *short* means an undesired path of current flow. In reference to capacitors, the short would occur through the dielectric, destroying the capacitor in the process).

As the graph in Fig. 5-2 illustrates, the capacitor in the circuit of Fig. 5-1 charges "exponentially". This means it charges in a nonlinear fashion. An *exponential curve* (like the charge curve in Fig. 5-2) is one that can be expressed mathematically as a number repeatedly multiplied by itself. The exponential curve of the voltage across a charging capacitor is identical to the exponential curve of the current increase in an LR circuit as shown in chapter 3 (Figs. 3-5 and 3-6).

As discussed in chapter 3, the current change in an LR (inductive-resistive) circuit, relative to time, is defined by the *time constant* and expressed in seconds. Similarly, in an RC (resistive-capacitive) circuit, the voltage change across the capacitor is defined by the time constant, and it is also expressed in seconds. An *RC time constant* is defined as the amount of time required for the voltage across the capacitor to reach a value of approximately 63% of the applied source voltage. The RC time constant is calculated by multiplying the capacitance value (in farads) times the resistance value (in ohms). For example, the time constant of the circuit shown in Fig. 5-1 would be:

$$Tc = RC = (1{,}000{,}000 \text{ ohms})(0.000{,}001 \text{ farad}) = 1 \text{ second}$$

The principle of the time constant, that applies to inductance, also applies to capacitance. During the first time constant, the capacitor charges to approximately 63% of the applied voltage. The capacitor in Fig. 5-1 would charge to approximately 6.3 volts in one second after SW1 is closed. This would leave a remaining voltage differential between the battery and capacitor of 3.7 volts (10 volt − 6.3 volts = 3.7 volts). During the next time constant, the capacitor voltage would increase by an additional 63% of the 3.7-volt differential. 63% of 3.7 volts is approximately 2.3 volts. Therefore, at the end of two time constants, the voltage across the capacitor would be 8.6 volts (6.3 volts + 2.3 volts = 8.6 volts). Five time constants are required for the voltage across the capacitor to reach the value usually considered to be the same as the source voltage.

In the circuit shown in Fig. 5-1, the approximate source voltage (10 volts) would be reached across the capacitor in 5 seconds. If SW1 is opened, after C1 is fully charged, it would hold the stored energy (1 microfarad) at a 10-volt potential for a long period of time. A "perfect" capacitor would hold the charge indefinitely; but in the real world, perfection is hard to come by. All capacitors have internal and external leakage characteristics, which are undesirable. It would be reasonable, however, to expect a well-made capacitor to hold a charge for several weeks, or even months.

Consider the current and voltage relationship in Fig. 5-1 when SW1 is first closed (assuming C1 is discharged). Immediately after SW1 is closed, the capacitor offers virtually no opposition to current flow (just like the small empty water tank in the earlier analogy). This maximizes the current flow, and minimizes the voltage across the capacitor. As the capacitor begins to charge, the current flow begins to decrease. At the same time, the voltage across the capacitor begins to increase. As stated earlier, this process continues until the voltage across the capacitor is equal to the source voltage, and all current flow stops.

The important point to understand is that the peak current occurs before the peak voltage is developed across the capacitor. Simply stated, the *voltage lags behind the current in a capacitive circuit*. As a matter of convenience, professionals in the electrical/

electronic fields tend to describe this phenomenon as "the current leading the voltage." This is very much akin to deciding how to describe a glass containing water at the 50% level; is it half-full or half-empty?

As discussed in chapter 3, the current lags the voltage by 90 degrees in a purely inductive circuit. In comparison, the current leads the voltage by 90 degrees in a purely capacitive circuit. As in the case of purely inductive circuits, purely capacitive circuits do not dissipate any "true" power because of the 90-degree phase differential between voltage and current.

### Filter capacitors

Capacitors used in filter applications remove an ac component from a dc voltage. Some filter capacitors are implemented to remove only a frequency-dependent part of an ac signal, but these applications will be discussed in a later chapter. In this section, you will examine how filter capacitors are used in dc power supply applications.

Figure 5-3 illustrates a simple half-wave rectifier circuit with R1 and C1 acting as a filter network. For a moment, review what you already know about this circuit. T1's secondary is rated at 12 Vac. This is an *rms* voltage, because no other specification is given. The peak voltage output from this secondary will be about 17 volts (12 volts × 1.414 = 16.968 volts). The negative half-cycles (in reference to circuit common) will be blocked by D1, but it will act like a closed switch to the positive half-cycles, and apply them to the filter network (R1 and C1). The amplitude of the applied positive half-cycles will not be the full 17 volts, because the 0.7-volt forward threshold voltage must be dropped across D1. This means that the positive half-cycles applied to the filter network will be about 16.3 volts in peak amplitude.

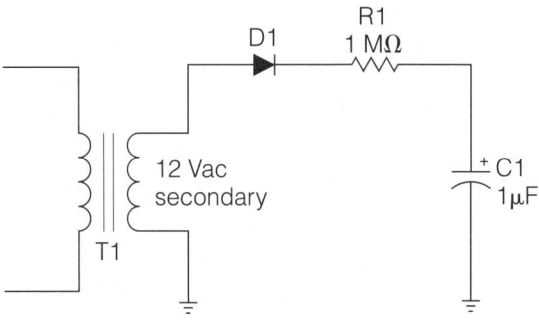

■ **5-3** *A dc filter circuit.*

These positive half-cycles will occur every 16.6 milliseconds (the reciprocal of the 60-Hz power line frequency). The positive *half-cycle duration* will be about 8.3 milliseconds, because it only represents one-half of the full ac cycle. (Refer back to chapter 4 (Fig. 4-5) and the related text if this is confusing.) Going back to Fig. 5-1, the calculated time constant for the circuit was 1 second. Because the same component values are used in the filter network of Fig. 5-3, the time constant of this filter is also 1 second. In power supply circuits, this time constant is called the *source time constant*.

Assuming C1 is fully discharged, examine the circuit operation of Fig. 5-3 from the moment that power is first applied to the primary of T1. When D1 is forward-biased by the first positive half-cycle from the T1 secondary, the +16.3-V peak half-cycle is applied to the filter network of R1 and C1. C1 will begin to charge to the full +16.3-V peak amplitude, but it will not have time to do so. Because the filter's RC (resistor-capacitor) time constant is 1 second, and the positive half-cycle only lasts for 8.3 milliseconds, C1 will only be able to charge to a very small percentage of the full peak level.

During the negative half-cycle, while D1 is reverse-biased, C1 cannot discharge back through D1; because C1's charged polarity reverse-biases D1. Therefore, C1 holds its small charge until the next positive half-cycle is applied to the filter network. During the second positive half-cycle, C1 charges a little more. This process will continue, with C1 charging to a little higher amplitude during the application of each positive half-cycle, until C1 charges to the full +16.3-V peak potential. When C1 is fully charged, all current flow within the circuit will cease; C1 cannot charge any higher than the positive peak, and D1 blocks all current flow during the negative half-cycles. At this point, the voltage across C1 is pure dc. The ac component (ripple) has been removed, because C1 remains charged to the peak amplitude during the time periods when the applied voltage is less. Even if all circuit power is removed from the T1 primary, C1 will still remain charged because it doesn't have a discharge path. Notice that it could not discharge back through D1, because its charged polarity is holding D1 in a reverse-biased state.

The operation of the circuit illustrated in Fig. 5-3 can be compared to pumping up an automobile tire with a hand pump. With each downward stroke of the hand pump, the pressure in the tire increases a little bit. Think of the tire as being C1. Air is inhibited from flowing back out of the tire by a small "one-way" air valve in the base of the pump. The one-way air valve only allows air to flow in one direction; into the tire. D1 only allows current to flow in one direction; causing C1 to charge. When the tire is pumped up to the desired pressure, it will hold this pressure even if the pump is removed. Once C1 is charged up to the peak electrical pressure (voltage), it will retain this charge, even if the circuit power is removed.

Although the circuit of Fig. 5-3 is good for demonstration purposes, it is not very practical as a power supply. To understand why, examine the circuit illustrated in Fig. 5-4. You should recognize T1 and the four diode network as being a *full-wave bridge rectifier*. As discussed in chapter 4, for each half-cycle output of T1's secondary, two diodes within the bridge will be forward-biased, and thus conduct "charging" current in only one direction through C1. With SW1 opened, C1 will charge to the peak voltage, minus 1.4 volts (since two bridge diodes must drop their forward threshold voltages at the same time; 0.7 volt + 0.7 volt = 1.4 volts). The peak voltage of a 12-Vac secondary is about 17 volts, so C1 will charge to 17 volts − 1.4 volts, or about 15.6 volts. When C1 is fully charged, the voltage across it will be pure dc.

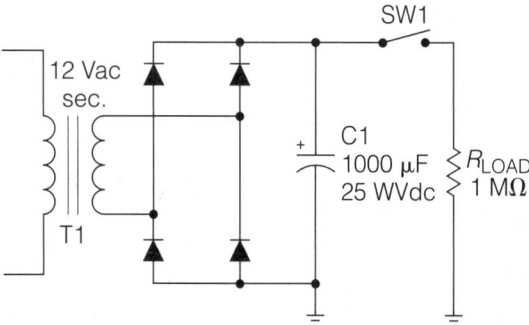

■ **5-4** *Full-wave filtered dc power supply.*

Note the symbol used for C1 in Fig. 5-4. The positive symbol close to one plate indicates that it is an electrolytic capacitor, meaning that it is polarized. The value also indicates this capacitor type because a capacity of 1000 μF is too large for a conventional, non-polarized capacitor. In this circuit, the negative side is connected to circuit common, which is the most negative point in the circuit. C1 also has an associated voltage rating of 25 WVdc. WVdc stands for *working Vdc*; and that means that this is the highest direct voltage that can be safely applied to the capacitor. Because the peak voltage applied to it will only be +15.6 volts, you are well within the safe operating parameters in this circuit.

Referring back to Fig. 5-3, you calculated the source time constant by multiplying the resistance value of R1 by the capacitance value of C1. This RC time constant was 1 second. Going back to Fig. 5-4, it might appear that C1 would charge "instantly" because there doesn't seem to be any resistance in series with it. In an "ideal" (perfect) circuit, this would be true. In reality, the dc resistance of the T1 secondary, the wiring resistance, and a small, nonlinear "current-dependent" resistance presented by the diode bridge will be in series with C1. Of these three "real-world" factors, the only one that is really practical to consider is the dc resistance of the T1 secondary.

For illustration, assume the dc resistance of T1's secondary winding is 1 ohm. To calculate the source time constant, the 1 ohm resistance would be multiplied by the capacity of C1:

$$Tc_{source} = (1 \text{ ohm})(0.001 \text{ farad}) = 0.001 \text{ second (or) 1 millisecond}$$

Because the time duration of each half-cycle is approximately 8.3 milliseconds, for all practical purposes, you can say C1 will be fully charged by the end of the first positive half-cycle. During the rapid charging of C1, a very high *surge current* will flow through the diode pair that happens to be in the forward conduction mode at the time. This is because C1 will initially "look" like a direct short until it is charged to a sufficient level to begin opposing a substantial portion of the current flow.

As you can recall, one of the forward current ratings for diodes (discussed briefly in chapter 4) was called the *peak forward surge current* and was based on an 8.3 millisecond time period (for 60-Hz service). The purpose for such a rating should now become apparent. Diodes, used as power supply rectifiers, will be subjected to high surge currents every time the circuit power is initially applied. 8.3 milliseconds is used as a basis for the peak forward surge current rating, because that is the time period of one half-cycle of 60-Hz ac. A "rule-of-thumb" method for calculating the peak forward surge current is to estimate the maximum *short-circuit secondary current* of T1 based on its peak output voltage and dc resistance. The peak output voltage of a 12-volt secondary is about 17 volts, and you have assumed the dc resistance to be 1 ohm. Therefore, using Ohm's law:

$$I_{peak\ surge} = \frac{E_{peak}}{R} = \frac{17 \text{ volts}}{1 \text{ ohm}} = 17 \text{ amps peak}$$

*Basic capacitor principles*

In reality, a transformer with a 12-volt secondary, and a secondary dc resistance of 1 ohm, would not be capable of producing a 17-amp short-circuit current (remember about inductive reactance being additive to resistance), so this gives us a good margin of safety.

After the first half-cycle, C1 is assumed to be fully charged. As long as SW1 is open (turned off), the only current flow in the circuit will be a very small *leakage current* which is considered negligible. The voltage across C1 is pure dc at about +15.6 volts.

A power supply would be of no practical value unless it powered something. The "something" that a power supply powers is called the *load*. The load could be virtually any kind of electrical/electronic circuit imaginable; but in order to operate, it must draw some power from the power supply. In Fig. 5-4, a load is simulated by the resistor $R_{load}$. By closing SW1, the load is placed in the circuit, and the circuit operation will be somewhat changed.

In power supply design, it is important to consider the *source time constant* (calculated previously) as well as the *load time constant*. When C1 is charged, it cannot discharge back through the bridge rectifier and the T1 secondary, because its charged polarity reverse-biases all of the diodes. By closing SW1, a discharge path is provided through $R_{load}$. The load time constant (sometimes called the *discharge time constant*) can be calculated by multiplying the capacitance value of C1, by the resistance value of the load ($R_{load}$):

$TC_{load} = (0.001 \text{ farad})(1000 \text{ ohms}) = 1 \text{ second } [1000 \mu F = 0.001F]$

The importance of the load time constant becomes apparent by examining the "exaggerated" illustration of Fig. 5-5. The waveshape

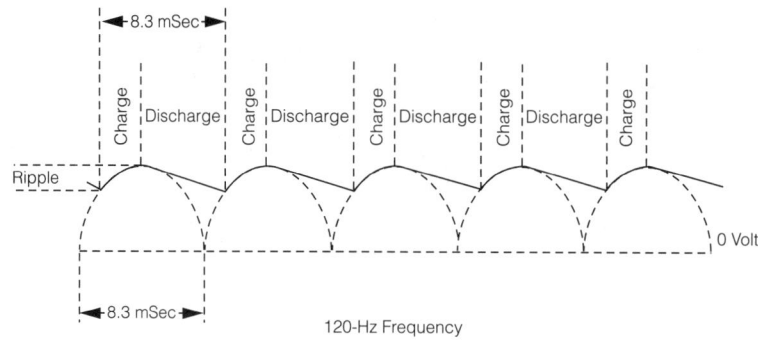

■ **5-5** *Charge/discharge cycle of C1 in Fig. 5-4.*

shown in dotted lines is the 120-Hz, full-wave rectified waveshape that would appear across $R_{load}$ if C1 was not in the circuit. The solid line represents the dc voltage levels across C1 with the charge-discharge amplitudes exaggerated for the sake of illustration. The charge periods represent the source time constant showing C1 receiving a charge (electrical energy) from the bridge rectifier circuit. As calculated earlier, this time constant is very short (only 1 millisecond). So, for illustration purposes, this charge is shown to be concurrent with the applied 120-Hz half-cycles.

The discharge periods represent the load time constant. During these periods, C1 is supplying its stored energy to power the load, causing its voltage level to drop by some percentage until the next charge period replenishes the drained energy. The variations in voltage level between the charge and discharge periods is called *ripple*. In a well-designed power supply, ripple is a very small, undesirable ac component "riding" on a dc level. Ripple can be specified as a peak-to-peak value, an rms value, or a percentage relationship, as compared to the dc level.

As stated earlier, the illustration in Fig. 5-5 is exaggerated; the ripple variations would not be nearly as pronounced, because the load time constant (1 second) is much longer than the peak charging intervals, which occur every 8.3 milliseconds. As a general "rule-of-thumb," the *load time constant should be at least 10 times as long as the charge interval*. In the case of a full-wave rectifier circuit, as in Fig. 5-4, the charge interval is 8.3 milliseconds, so the load time constant should be at least 83 milliseconds. With a half-wave rectifier circuit, the load time constant would have to be twice as long for the same quality of performance.

The power supply illustrated in Fig. 5-4 is referred to as a *raw dc power supply*. This simply means that it is not voltage or current regulated. Regulated power supplies will be discussed in chapter 6.

## Designing raw dc power supplies

Every type of electrical or electronic apparatus needs a source of electrical energy to function. The source of electrical energy is called the *power supply*. The two main classifications of power supplies are *line operated power supplies* (operated from a standard 120-Vac wall outlet) and *battery supplies* (electrical energy is provided through a chemical reaction). It is relatively safe to say that any device capable of functioning properly from a battery power source, can function equally well from a properly designed "raw" dc power supply, receiving its energy from a wall outlet. This is important because you will probably run into many situations where you will want to test or operate a battery powered device from standard household power. As an exercise to test all you have learned thus far, here is a "hypothetical" exercise in designing a raw dc power supply for a practical application.

Assume that you own an automobile CB radio that you would occasionally like to bring into your home and operate as a "base station." In addition to installing an external stationary antenna (which is irrelevant to our present topic of discussion), you would have to provide a substitute for the automobile battery as a power source. The CB radio is specified as needing "+12 to +14 Vdc at 1.5 amps" for proper operation.

The CB radio power supply will have three primary parts; the transformer, a rectifier network, and a filter. You should choose a transformer with a secondary "peak" (not *rms*) voltage rating close to the maximum desired dc output of the power supply. In this case, a 10-Vac secondary would do nicely, and they are commonly available. The peak voltage output of a 10-volt secondary would be:

$$Peak = 1.414(rms) = 1.414(10\text{ volts}) = 14.14\text{ volts}$$

The rectifier network will drop about 1 volt, so that would leave about 13 volts (peak) to apply to the filter capacitor. The current rating of the transformer secondary could be as low as 1.5 amps, but a 2-amp secondary current rating is more common, and the transformer would operate at a lower temperature. A transformer with a 10-Vac @ 2-amp secondary rating can also be specified as a 10-V 20-VA transformer. The *volt-amp* (*VA*) rating is simply the current rating multiplied by the voltage rating (10 volts × 2 amps = 20 VA).

A full-wave bridge rectifier can be constructed using four separate diodes, or it can be purchased in a module form. Bridge rectifier modules are often less expensive, and are easier to mount. The average forward current rating should be at least 2 amps, to match the transformer's secondary rating. The peak reverse voltage rating, or PIV, would have to be at least 15 volts (the peak output voltage of the transformer is 14.14 volts); but it is usually prudent to double the minimum PIV as a safety margin. However, a 30-volt PIV rating is uncommon, so a good choice would be diodes (or a rectifier module) with at least a 2-amp, 50-volt PIV rating.

If you purchase these rectifiers from your local electronic parts store, don't be surprised if they don't have an associated "peak forward surge current" rating. Most modern semiconductor diodes will easily handle the surge current if the average forward current rating has been properly observed. This is especially true of smaller dc power supplies, such as the hypothetical one presently being discussed. If it is desirable to estimate the peak forward surge current, measure the dc resistance of the transformer secondary and follow the procedure given earlier in this chapter. (The secondary dc resistance can be difficult to measure with some DVM's, because of its very low value.)

In order to choose a proper value of filter capacitance, you can equate the CB radio to a resistor. Its power requirement is "+12 to +14 volts at 1.5 amps". Using Ohm's law, you can calculate its *apparent resistance*:

$$R = \frac{E}{I} = \frac{12 \text{ volts}}{1.5 \text{ amps}} = 8 \text{ ohms (worst case)}$$

As far as the power supply is concerned, the CB radio will "look" like an 8-ohm load. Note that the 8-ohm calculation is also the worst case condition. If the upper voltage limit (+14 volts) had been used in the calculation, the answer would have been a little over 9.3 ohms. Eight ohms is a greater current load to a power supply than 9.3 ohms (as the load resistance decreases, the current flow from the power supply must increase).

You now know two variables in the load time constant equation; the *apparent load resistance* ($R_{load}$) and the *desired time constant* (83 milliseconds with a full-wave rectifier). To solve for the *capacitance value*, the time constant equation must be rearranged. Divide both sides by $R$:

$$\frac{Tc}{R} = \frac{R(C)}{R}$$

The $R$s on the right side of the equation cancel each other leaving:

$$\frac{Tc}{R} = C$$

or

$$C = \frac{Tc}{R}$$

By plugging our known variables in the equation, it becomes:

$$C = \frac{83 \text{ milliseconds}}{8 \text{ ohms}} = 0.010375 \text{ farad (or) } 10{,}375 \text{ } \mu F$$

Based on the previous calculation, the filter capacitor needed for the CB radio power supply should be about 10,000 $\mu F$. A capacitor of this size will always be an electrolytic, so polarity must be observed. The voltage rating should be about 20 to 25 WVdc (this provides a little safety margin over the actual dc output voltage), depending on availability. If this calculation had been based on a half-wave rectifier circuit, the required capacitance value for the same performance would have been about 20,000 $\mu F$.

## Assembling and testing the third section of a lab power supply

Materials needed for the completion of this section:

| Quantity | Item description |
|---|---|
| 3 | phenolic type, 2 lug solder strips (see text) |
| 2 | 4400-$\mu F$, 50-WVdc electrolytic capacitors |

The phenolic solder strips are specified only because they are inexpensive and effective. If your local electronic parts store doesn't have these in stock, there are many good alternatives. Any method of providing three chassis mountable, *insulated tie points* that will hold the filter capacitors firmly in place, and allow easy connections to the capacitor leads, will function equally well.

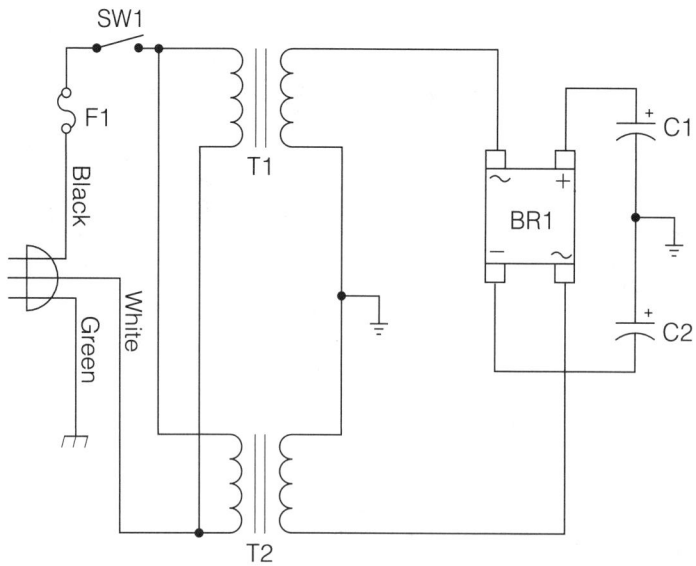

■ **5-6** *Schematic diagram of the first, second, and third sections of the lab power supply.*

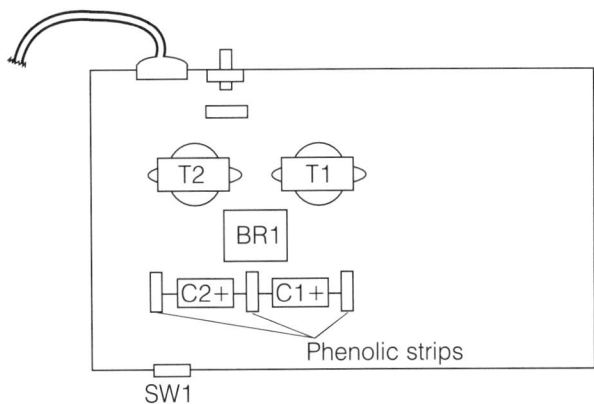

■ **5-7** *Approximate physical layout of the major components for the lab power supply project.*

Referring to Figs. 5-6 and 5-7, mount the solder strips to the chassis, being careful to space them far enough apart to allow room for the capacitors (C1 and C2). Connect the capacitors to the insulated lugs and crimp the leads to hold them in place, until the remaining wiring is completed. Connect a piece of hook-up wire, from the joint connection of C1 and C2, to the circuit common point, on the cardboard, between the secondaries of T1 and T2. Use another piece of hook-up wire to connect the positive side of C1 to the positive terminal of BR1. Connect another piece of hook-up wire from the negative side of C2 to the negative terminal of BR1. **Do not connect circuit common to chassis ground.** At this stage of the project, the only connection to circuit common should be the junction of C1 and C2. Double-check all wiring connections, and be sure all of the voltage polarities are correct. Solder all of the connections.

The power supply you have constructed thus far is called a *dual-polarity, 34-volt raw dc power supply*. It is the same type of power supply as is illustrated in Fig. 5-4 (minus $R_{load}$ and SW1). Power supplies similar to this one are commonly used in audio power amplifiers. This particular power supply could provide the electrical energy that a power amplifier would need to drive an 8-ohm speaker at about a 50-watt rms level. More about audio amplifiers will be covered in chapter 8. Chapter 6 will discuss how to add an adjustable regulator section, to this design, for improved lab performance.

## Testing the power supply

Note: to limit redundancy, it is assumed at this point that you are practicing all of the safety procedures that have been discussed previously. In the successive chapters, I will only mention the special safety considerations which can apply to unique situations. **Review all of the safety recommendations presented thus far, and put them to use all of the time**.

Set your DVM to measure "dc volts" on the 100-volt range (or higher). Plug the power supply into the outlet strip. Set SW1 to the OFF position and turn ON the outlet strip. Briefly, turn SW1 to the ON position, and then back to the OFF position. Turn the outlet strip OFF. Measure the dc voltage across C1 and C2, paying close attention to the polarity (C1 should be positive, and C2 should be negative, in reference to circuit common). The actual amplitude of the voltage is not important at this point in the test.

You have simply "pulsed" the power supply on and off, to verify that the capacitors are charging and in the correct polarity. As you measured the dc voltages, they should have been decreasing in amplitude as the charge was draining off. The draining of the charge is caused by the internal leakage inherent in all electrolytic capacitors (new capacitors can be very leaky until they have the chance to reform during circuit operation). Also, the capacitors will discharge, to some degree, through the internal input impedance of the DVM while you are measuring the voltage.

If you measured some voltage level across C1 and C2, with the correct voltage polarities, re-apply power to the circuit and measure the dc voltages across C1 and C2. The calculated voltage across each capacitor should be about 33 volts (the peak value of the 24-volt secondaries is about 34 volts, minus an estimated 1-volt drop across the rectifier). In reality, you will probably measure about 36 to 38 volts across each capacitor. There are several reasons for this higher level. Transformer manufacturers typically rate transformers based on minimum "worst-case" conditions, so it is common for the secondaries to measure a little high under normal conditions. Also, you are measuring the voltage levels under a no-load condition (often abbreviated N.L. in data books). If you took these same measurements while the power supply was operating under a full-load (abbreviated F.L.), they would be considerably lower.

Leaving power applied to the power supply circuit, set your DVM to measure "ac volts" beginning on the 100-volt (or higher) range. Measure the ac voltage across C1 and C2. If you get a zero indication on the 100 volt range, set the range one setting lower and try again. Continue this procedure until you find the correct range for the ac voltage being measured. (When measuring an unknown voltage or current, always begin with a range setting higher than what you could possibly measure and work your way down. Obviously, if you are using an "autoranging" DVM, you won't have to worry about setting the range.) If the circuit is functioning properly, you should measure an ac component (ripple) of about 5 to 20 millivolts. Turn OFF the circuit.

Many of the more expensive DVM's are specified as measuring true rms. If you are using this type of DVM to measure the ripple content, the indication you'll obtain will be the true rms value. The majority of DVM's, however, will only give an accurate rms voltage measurement of sine wave ac. Referring back to Fig. 5-5, note that the ripple waveshape is not a sine wave; it is more like a "sawtooth" (you'll learn more about differing waveshapes in succeeding chapters). The point here is, there can be some error in the ripple measurement you just performed. High accuracy is not important in this case, but you can experience circumstances in the future, where you must consider the type of ac waveshape that you are measuring with a DVM, and compensate accordingly.

As an additional test, I used a 100-ohm, 25-watt resistor to apply a load to the circuit. If you have a comparable resistor, you might want to try this also, but be careful with the resistor; it gets *hot*.

The resistor is connected across each capacitor and the subsequent ac and dc voltage measurements are taken. The loading effect was practically identical between the positive and negative supplies, which is to be expected. The dc voltage dropped by about 3.1 volts, and the ripple voltage increased to about 135 millivolts. These effects are typical.

## Food for thought

Throughout this chapter, I have followed a more traditional, and commonly accepted, method of teaching and analyzing capacitor theory. I suggest that you continue to comprehend capacitor operation from this perspective. However, in the interest of accuracy, you will find the following story to be of interest.

Michael Faraday, the great English chemist and physicist, had a theory that more closely approaches the way a capacitor really works. His theory stated that the charge is actually contained in the dielectric material—not the capacitor's plates. Inside the dielectric material are tiny *molecular dipoles* arranged in a random fashion. Applying a voltage to the plates of a capacitor stresses these dipoles causing them to line up in rows, storing the energy by their alignment. In many ways, this is similar to the physical change occurring in iron, when it becomes a temporary magnet by being exposed to magnetic flux lines. When a capacitor is discharged, the dipoles flex back like a spring, and their energy is released.

The fact that the stored energy within a capacitor is actually contained within the dielectric explains the reason why different dielectric materials have such a profound effect on the capacity value. Dielectric materials are given a *dielectric constant* rating (usually based on the quality of air as a dielectric; air = 1.0) relative to their overall effect on capacity. A dielectric material with a rating of 5, for example, would increase the *capacitance value* of a capacitor to a value 5 times higher than air, when all other variables remained the same.

# Transistors

BEFORE GETTING INTO TRANSISTOR THEORY, THERE ARE A few definitions that are better discussed in advance.

## Preliminary definitions

*Gain* is a term used to describe a ratio of increase. The most common types of gain are *current gain*, *voltage gain*, and *power gain*. Gain is simply the ratio of the input (voltage, current, or power) to the output (voltage, current, or power). For example, if a 1-volt signal is applied to the input of a circuit and, on the output, the signal amplitude has been increased to 10 volts, you say this circuit has a voltage gain of 10 (10 divided by $1 = 10$). It is possible to have a gain of less than 1. For example, if the output of a circuit is only one-half of the value of the input, it can be said this circuit has a gain of 0.5. However, it is preferable to say the output is being *attenuated* (reduced) by a factor of 2.

The symbol for gain as used in equations and formulas is $A$. Typically, the upper case $A$ (symbolizing gain) will be followed by a small case suffix letter designating the gain type. For example, $A_e$ symbolizes "voltage gain."

*Power gain* is a specific type of gain indicating that more "energy" is being delivered at the output (of a device or circuit) than is fed into the input. A transformer, for instance, is capable of providing voltage gain, if it is a step-up transformer; but the secondary current is reduced by the same factor (turns ratio) as the voltage is increased. Because power is equal to *voltage times current*, the equation seems to balance; equal power in and power out, but this is still not a gain. Additionally, all components have losses. The transformer's efficiency loss is called its *efficiency ratio*. A good unit will have about a 90% ratio, and the primary to secondary power transfer ratio will always be less than 1.

Therefore, a transformer is not capable of producing power gain. Electronic components capable of providing power gain are called *active devices*. These include transistors, vacuum tubes, some integrated circuits, and many other devices. Electronic components that cannot produce power gain are called *passive devices*. Some examples of passive components are resistors, capacitors, transformers, and diodes.

## Introduction to transistors

The development of the transistor was the foundational basis for all modern solid-state electronics. William Shockley, John Bardeen, and Walter Brattian discovered *transistor action* while working at the Bell Telephone Laboratory in 1947. The term *transistor* started out as a combination of the phrase "transferring current across a resistor." The important developmental aspect of the transistor is that it became the first "active" solid-state device and opened a new perspective of design ideology. Its hard to imagine what our lives would be like today without it!

A *transistor* is a solid-state, three-layer semiconductor device. Figure 6-1 shows the basic construction of a transistor, and com-

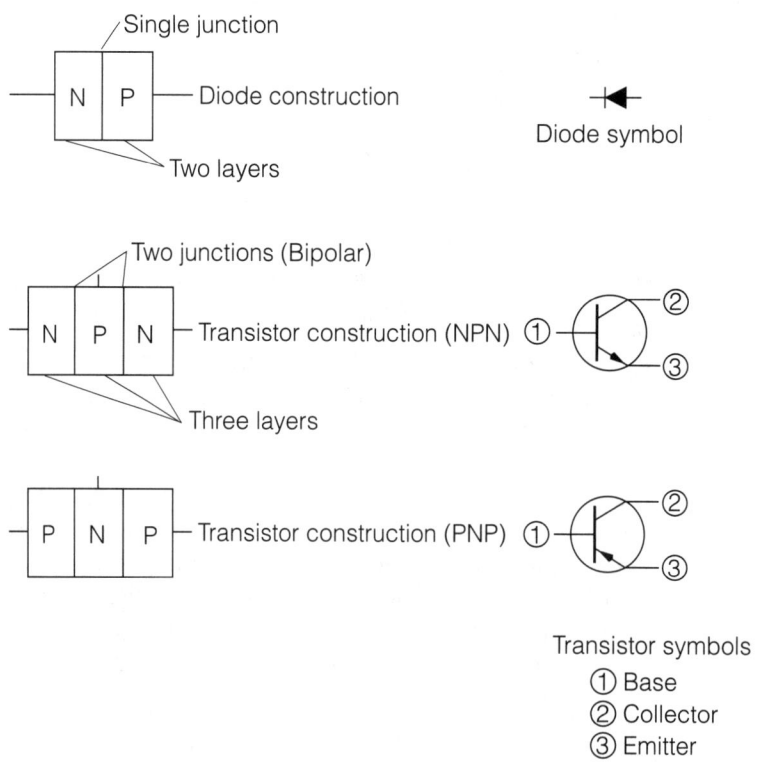

■ **6-1** *Transistor construction and symbols.*

pares transistor construction to diode construction. Note that a *diode* contains only one junction, whereas a transistor contains two junctions. Because a transistor contains two junctions, it is often referred to as a (dual-junction) *bipolar device*. Figure 6-1 also illustrates how bipolar transistors can be constructed in either of two configurations; NPN or PNP.

Bipolar transistors have three connection points, or leads. These are called the *emitter,* the *base,* and the *collector*. The symbols for bipolar transistors are shown in Fig. 6-1. The only difference between the NPN and PNP symbols is the direction of the arrow in the emitter lead. Associate these bipolar types with their descriptions: *N*ot *P*ointing i*N* (the arrow) and the *P*ointing i*N P*oint.

## Transistor principles

Figure 6-2 shows an NPN transistor connected in a simple circuit to illustrate basic transistor operation. A PNP transistor would operate in exactly the same manner, only the voltage polarities would have to be changed. Note that the emitter lead is connected to circuit common (the most negative potential in the circuit) through the *emitter resistor* ($R_e$), the base lead is connected to a potentiometer (P1), and the collector lead is connected to +30 volts, through the *collector resistor* ($R_c$).

A transistor actually consists of two diode junctions; the *base-to-emitter junction* and the *base-to-collector junction*. Assume P1 is adjusted to provide +1.7 volts to the base (note how P1 could provide any voltage to the base from 0 to +30 volts). The +1.7-volt potential applied to the P-material base, in reference to the N-material emitter at 0 volts (circuit common), creates a "forward-biased" diode and causes current to flow from emitter to base. However, because of a phenomenon known as *transistor action*, an additional current will also flow from emitter to collector.

To understand transistor action, we have to consider several conditions occurring simultaneously within the transistor. First, notice that the base-to-collector junction is reverse-biased. The collector is at a much higher positive potential than the base, causing the base to be negative in respect to the collector. Therefore, current will not flow from base to collector. The high positive potential on the collector has the tendency to attract all of the negative charge carriers (electrons) away from the base-collector junction area. This creates a *depletion area* of negative charge carriers close to the base layer. The depletion area seems very positive for two reasons; all of the negative charge carriers have

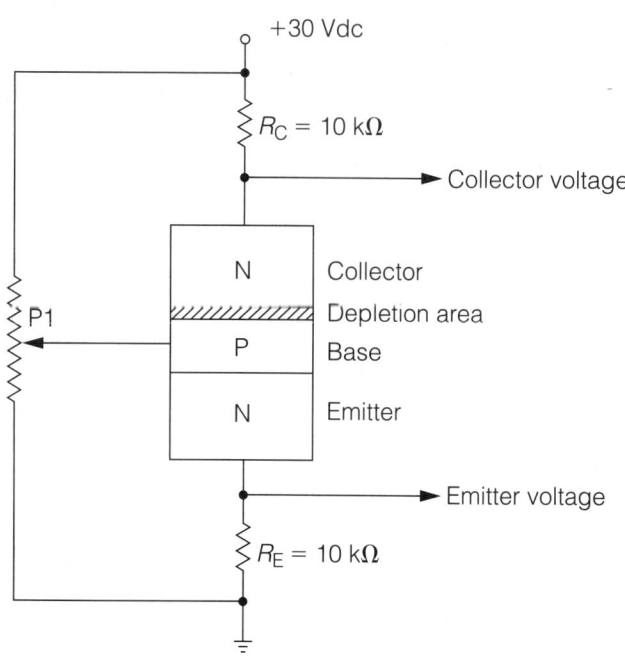

**6-2** *Simple circuit to illustrate basic transistor operation.*

been drawn up close to the collector terminal and, because it is part of the collector layer, it is connected to the highest positive potential. Keeping this condition in mind, turn your attention back to the base-emitter junction.

Referring again to Fig. 6-2, as previously stated, the base-emitter junction is forward biased and current is flowing. However, note how the base layer is much thinner than the emitter layer. The emitter has many more *negative charge carriers* (electrons) than the thin base material has *holes* (absence of electrons) to combine with. This causes an "overcrowded" condition of electrons in the base layer. These crowded electrons have two directions in which to flow (or *combine*, which gives the appearance of flowing); some will continue to flow out to the +1.7-volt base terminal, but the majority will flow toward the very positive "looking" depletion area created in the collector area close to the base layer.

The end result is a much higher current flow through the collector than is flowing through the base. The parameter (component specification) that defines the ratio of the base current to the collector current is called *beta* (abbreviated $B$ or $hfe$). In essence, beta is the maximum possible current gain that can be produced in a given transistor. Typical beta values for small signal transistors are in the range of 100 to 200. In contrast, power transistors can have beta values of 20 to 70. The equation for calculating beta is:

$$Beta = \frac{Ic}{Ib}$$

This equation states that beta is equal to the collector current divided by the base current. The important point to recognize about transistor current gain is that the higher collector current is controlled by the much smaller base current.

Going back to Fig. 6-2, assume the transistor illustrated has a beta of 100. With +1.7 volts applied to the base, about 0.7 volts will be dropped across the base-to-emitter junction (like any other forward-biased silicon diode). The remaining 1 volt (1.7 volts − 0.7 volt = 1 volt) will be dropped across the emitter resistor ($R_e$). Because we know the resistance value of $R_e$ and the voltage across it, you can use ohm's law to calculate the current flow through it:

$$I = \frac{E(R_e)}{R} = \frac{1 \text{ volt}}{1000 \text{ ohms}} = 1 \text{ milliamp}$$

The 1 milliamp of current flow through $R_e$ is the "sum" of the *base current* and the *collector current*. You can think of the emitter as the layer that "emits" the total current flow. The majority of this current is "collected" by the collector, and the overall current flow is controlled by the base. Our assumed beta value tells us that the collector current will be 100 times larger than the base current. Therefore, the base current flow will be about 9.9 microamps, and the collector current will be about 990 microamps (990 microamps +9.9 microamps = 999.9 microamps, or about 1 milliamp). Because the collector current must flow through $R_c$, the voltage drop across $R_c$ can be calculated using ohm's law:

$$E = IR = (990 \text{ microamps})(10{,}000 \text{ ohms}) = 9.9 \text{ volts}$$

There are three individual voltage drops in Fig. 6-2 that must be analyzed to understand the action of the transistor. Two of these have already been calculated: the voltage across the emitter resistor ($R_e$), and the voltage dropped by the collector resistor ($R_c$).

The third important voltage drop occurs across the transistor itself. All three of these voltage drops are in series; the emitter resistor is in series, with the transistor, which is in series with the collector resistor. Going back to our discussion of simple series circuits, you know that the sum of these three voltage drops must equal the source voltage of +30 volts. Because you already know the value of two of the voltage drops, you can simply add these two values, subtract the sum from the source voltage, and the difference must be the voltage drop across the transistor:

$$1 \text{ volt } (R_e) + 9.9 \text{ volts } (R_c) = 10.9 \text{ volts}$$

$$30 \text{ volts source} - 10.9 \text{ volts} = 19.1 \text{ volts}$$

19.1 volts are being dropped across the transistor, but this is not the collector voltage. Unless otherwise noted, all voltage measurements are always made in reference to circuit common (or ground, whichever is applicable). Therefore, to calculate the collector voltage, the voltage drop across the transistor is added to the voltage drop across the emitter resistor ($R_e$). This must be done because, from a circuit common point of reference, the emitter resistor voltage drop is in series with the transistor voltage drop. If this is confusing, look at it this way. If you made the source voltage your point of reference and measured the collector voltage, you would actually be measuring the voltage drop across $R_c$. If you made the transistor's emitter lead your point of reference, and measured the collector voltage, you would be measuring the voltage across the transistor. By making the circuit common your point of reference, you are actually measuring the voltage across the transistor and the voltage drop across $R_e$. Therefore, the collector voltage would be:

$$1 \text{ volt } (R_e) + 19.1 \text{ volts (transistor voltage drop)} = +20.1 \text{ volts}$$

Notice that this is the same value you could have calculated by simply subtracting the voltage drop across the collector resistor ($R_c$) from the source voltage.

Now, observe transistor action by changing the base potential. (Note: to eliminate redundancy, many of the previous calculations and discussions will not be repeated). Assume P1 is adjusted to increase the base bias potential to 2.7 volts. The base-emitter junction will still drop about 0.7 volts, so the voltage across $R_e$ will increase to 2 volts. The emitter current increases to 2 milliamps. With a beta of 100, about 19.8 microamps of the emitter current will flow through the base lead. The remaining 1.98 milliamps of collector current will flow through the collector lead, dropping about 19.8 volts across $R_c$. Subtracting the voltage drop across $R_c$ from the +30−volt source produces a collector voltage of 10.2 volts.

When the base voltage was originally set to 1.7 volts, the collector voltage was +20.1 volts. By increasing the base voltage by 1 volt (up to +2.7 volts), the collector voltage decreased to +10.2 volts. In other words, the one-volt change in the base voltage resulted in a 9.9-volt change in the collector voltage (20.1 volts − 10.2 volts = 9.9 volts). This is an example of voltage gain. In this circuit, we have a voltage gain (abbreviated $A_e$) of 9.9 (9.9 volt change at the collector divided by the 1 volt change at the base = 9.9). There was also an "internal" current gain equal to beta. Obviously, this also results in power gain, because both the voltage and current increased.

If you applied a 1-volt peak-to-peak signal to the base lead of this circuit (Fig. 6-2), it would be increased to a 9.9 volt peak-to-peak signal at the collector lead. However, the amplified output would be inverted, or 180 degrees out-of-phase, with the input signal. As you might have noticed, as the base voltage increased (from 1.7 to 2.7 volts), the collector voltage dropped (from 20.1 to 10.2 volts).

There are three "general" transistor configuration methods; the "common base," "common emitter," and "common collector." In the previous discussion, involving Fig. 6-2, you looked at the operation of a common emitter configuration. In this configuration, the output is taken off of the transistor collector, and the signal is always inverted.

If you haven't had any prior experience with transistors, your head is probably "buzzing" with all of the voltages and currents relating to Fig. 6-2. Fear not. As you progress and gain experience, the haze will clear, and these principles will seem simple. This section is necessary to provide you with a good working knowledge of basic transistor operational principles.

Now, for practical purposes, you can simplify things according to the transistor configuration. However, before proceeding, there is a very important principle to establish. A bipolar transistor is a "current" device. Although there will always be voltages present in an operating circuit, the "effect" produced by a transistor is *current gain*, and the controlling "cause" of a transistor is the input current.

Consider a transistor as being similar to a water valve. A water valve controls the flow of water. Obviously, water pressure must exist to push the water through any type of system, but we never think of a water valve in terms of controlling water pressure (even though pressure changes will occur with differing valve adjustments). A water valve is always used as a device to control the flow of water. Similarly, always think of a bipolar transistor as a device used to control electrical current flow.

# Common transistor configurations

Transistors are "active" devices because they are capable of producing power gain. In actuality, bipolar transistors are current amplifiers, but an increase of current while maintaining the same voltage is an increase of power ($P = IE$). Depending on the circuit configuration a transistor is designed into, the output can produce current gain, voltage gain, or both. The following transistor circuit configurations will differ in their ability to provide voltage, current, and power gain, but to avoid confusion, a parallel analysis will be given at the end of this section.

## Adding some new concepts

Before getting into more circuit analysis, it is helpful to understand a few new terms and concepts that will make the whole process simpler.

The remainder of this text contains a lot of discussion involving "impedance." While examining some of the basic operational fundamentals of inductors and capacitors, you learned they have a certain *frequency-dependent* nature (more about this will be discussed in chapter 15). We call such components "reactive". In a *reactive component*, the opposition to ac current flow is frequency dependent, and these components exhibit a special form of ac resistance called *reactance*. For example, with inductors, as the frequency of an applied voltage rises, the inductor's opposition to that voltage increases. Capacitive action is just the opposite. In other words, a dc voltage applied to a reactive circuit will promote a current flow dependent upon the resistance in the circuit, but an applied ac voltage of the same amplitude might cause a totally different current flow, dependent upon its frequency. For this reason, you need a term to describe the "total" opposition to ac current flow; taking both the resistive and reactive components into consideration. This term is *impedance*. Impedance is defined in ohms, just as resistance; it is composed of dc resistance, plus the inductive reactance and/or the capacitive reactance in a circuit; and its symbol, as used in equations and formulas, is $Z$.

In practical transistor circuits, capacitors are used extensively for a variety of purposes. The following transistor circuits use capacitors for "coupling" (sometimes referred to as "blocking") and "bypass" functions. Remember back to our discussions on basic capacitor operation; you know that current cannot pass through the dielectric (insulating) material under normal operation. However, the "effect" of a changing electrostatic field on one plate can be transferred to the other plate, even though no actual current is passing through the dielectric. Technically speaking, the subatomic distortion occurring on one plate, and in the dielectric, must cause a subsequent distortion in the other plate.

In effect, a capacitor can appear to pass an ac current with virtually no opposition, while totally "blocking" any dc current. A capacitor used to block dc while passing an ac signal current is called a *coupling capacitor*. There are some situations where we want just the opposite to occur; the dc voltage present with no ac component in it. Capacitors used for this function are called *bypass capacitors*. (The previous chapter covered filter capacitors. Filter capacitors are actually a type of bypass capacitor; they maintain the dc voltage, while reducing the ac component, or ripple, by *shunting* the ac current to ground.)

In the following transistor configuration circuits, it would be advantageous for you to start thinking of them as "building blocks" to more complex circuits. The most sophisticated electronic equipment can be broken down into simpler subassemblies and, in turn, these subassemblies can be broken down into basic circuit blocks. In performing electronic design, we start with basic blocks and put them together in such a way that they will collectively produce a desired result. Electronic equipment manufacturers will often include "block diagrams" as part of the overall documentation package for their products. This provides an efficient method of becoming thoroughly familiar with the technical operation of the products, without having to analyze the operation down to a component level. As you gain more experience in the electrical/electronic fields, you'll appreciate the usefulness of a block approach in design, analysis, or troubleshooting ventures.

### The common-emitter configuration

As previously stated, you analyzed the circuit in Figure 6-2 as though it was a common-emitter amplifier for the purpose of demonstrating voltage gain. In actuality, Fig. 6-2 can be either a common-emitter amplifier or a common-collector amplifier depending on whether you use the output from the collector or the emitter, respectively. However, a practical common-emitter transistor amplifier would probably require some improvements, as illustrated in Fig. 6-3.

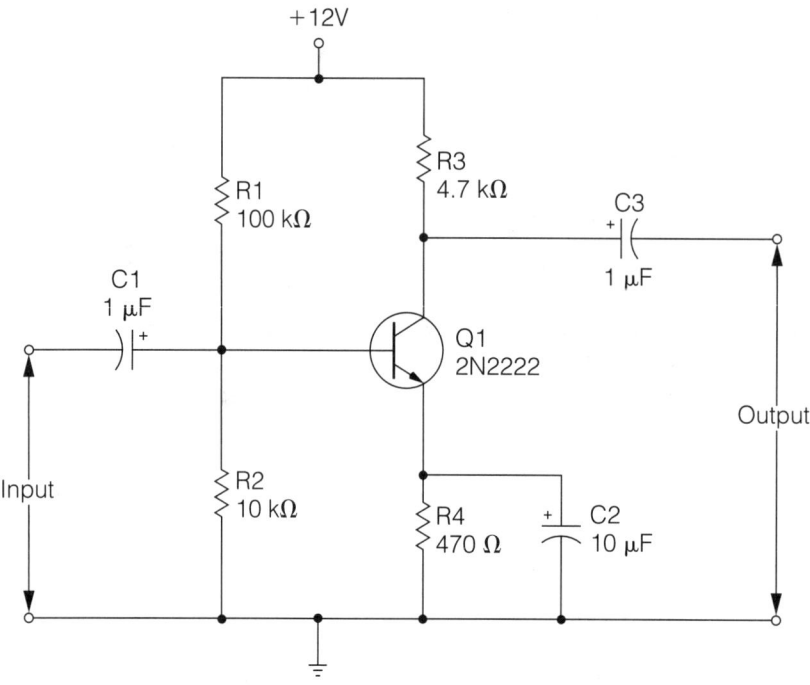

■ **6-3** *Practical example of a common-emitter transistor amplifier.*

Bipolar transistors have a negative temperature coefficient; that is, as a transistor's temperature increases, its internal resistances decrease. Also, temperature increases cause an increase in undesirable "leakage" currents that can further add to temperature build-up. In high-power transistor circuits, this chain-reaction temperature effect can lead to a condition called *thermal runaway*, which renders the circuit inoperative. In small-signal transistor circuits, as we are presently examining, varying temperatures will cause shifts in operating points.

In addition to temperature considerations, typical bipolar transistors do not have precise beta values. Manufacturers specify beta values within minimum and maximum ranges for each transistor type. Consequently, two transistors with the same exact part number can have dramatically differing beta values.

A third problematic variable to consider is the *source voltage*. Transistor circuits intended to be powered from a battery power supply must be capable of operating with a relatively broad range of supply variance. Even 120-Vac line-powered transistor circuits might experience voltage variations if the power supply is not well regulated. A "perfect" transistor amplifier circuit would be self-correcting with temperature variations, unaffected by the transistor's beta value, and totally immune to source voltage variations. Although you can't achieve perfection, you can come close to it with the circuit in Fig. 6-3.

Notice that P1 (Fig. 6-2) has been replaced with two fixed resistors to set the *base bias*. C1 is a coupling (or blocking) capacitor to keep the dc base bias voltage from being applied to the input source. Similarly, C3 serves the same function of keeping the dc collector voltage from being applied to the output. C2 is a bypass capacitor which effectively "shorts" the ac emitter (input signal) voltage to circuit common, while leaving the dc emitter voltage unaffected.

The *bias voltage divider*, consisting of R1 and R2, keeps the correct "percentage" of source voltage applied as a base bias, regardless of the "actual" value of the source voltage. In other words, the actual resistor values in Fig. 6-3 have been chosen so that about 9% of the source voltage will be dropped across R2 (without considering the parallel base impedance). Regardless of the actual value of the source voltage, about 9% of it will be applied as a base bias. This has the effect of keeping the bias voltage optimized, even with wide variations in source voltage. Temperature stability is also improved by this method. However, R2 is not absolutely necessary and it has the undesirable effect (for most applications) of lowering the input impedance. For these reasons, R2 is not incorporated into all common-emitter designs.

Although R4 is commonly called the *emitter resistor*, from an operational perspective, R4 is also a "negative feedback" resistor. *Negative feedback* is the term given for applying a percentage of an amplifier's output back into the input. This improves amplifier stability, but it also reduces gain. In Fig. 6-3, the negative feedback provided by R4 makes the circuit less dependent upon the individual transistor current gain (beta), and aids in increasing temperature stability.

As stated earlier, C2 "couples" (shorts) the ac signal to circuit common, but it has little effect on the dc emitter voltage. In reality, this causes two individual gain responses within the circuit. From a dc perspective, C2 doesn't exist. The negative feedback produced by R4 causes the dc voltage gain to be a ratio of the value of the collector resistor (R3), divided by the value of the emitter resistor (R4). With the circuit illustrated, the dc voltage gain is 10. From an ac perspective, R4 doesn't exist, because C2 looks like a short tying the transistor emitter directly to circuit common. Consequently, the ac voltage gain becomes the ratio of the collector resistor value divided by the internal base-emitter junction resistance. The junction resistance of a forward-biased semiconductor junction is very low, so the ac voltage gain is reasonably high; typically in the range of 200, using the 2N2222 transistor as illustrated. This high gain is advantageous, but is highly dependent upon individual transistor characteristics. If C2 were removed from the circuit, the ac voltage gain would become the same as the dc voltage gain; or about 10.

The dc input impedance of this circuit (Fig. 6-3) can be considered infinite, because C1 blocks any dc current flow. The ac input impedance consists of three parallel resistive elements: R1, R2, and the forward-biased base-emitter junction resistance multiplied by beta. The forward-biased junction resistance is a low value, typically only a few ohms. Even after multiplying this value by a high beta value, the product would still be much lower than a typical R2 value. Therefore, for practical analysis purposes, you can usually eliminate R1 and R2 from consideration and estimate the input impedance to be the base-emitter junction resistance times the beta value. If C2 is removed from the circuit, the ac input impedance then becomes the parallel resistance value of R1, R2, and the value of R4 multiplied by the beta value. Because R4 times beta is typically a high resistance value, and R1 is usually a high resistance value, a reasonably close estimate of the input impedance with C2 removed is the value of R2. The output impedance of this circuit can be considered to be equal to the value of the collector resistor (R3).

Figure 6-3 is a good, stable design with reasonably high voltage gain and wide-range immunity from beta, temperature, and source voltage variations. You might use this circuit as a good building block for most voltage amplification applications, or you can tailor the component values according to specific needs.

### The common-collector configuration

Figure 6-4 is a practical example of a common-collector transistor amplifier. Notice the output is taken off of the emitter instead of the collector (as in the common-emitter configuration). A common-collector amplifier is not capable of voltage gain. In fact, there is a very slight loss of voltage amplitude between the input and output. However, for all practical purposes, we can consider the voltage gain at unity. Common-collector amplifiers are *noninverting*, meaning the output signal is in-phase with the input signal. Essentially, the output signal is an exact duplicate of the input signal. For this reason, common-collector amplifiers are often called *emitter-follower amplifiers*, because the emitter voltage follows the base voltage.

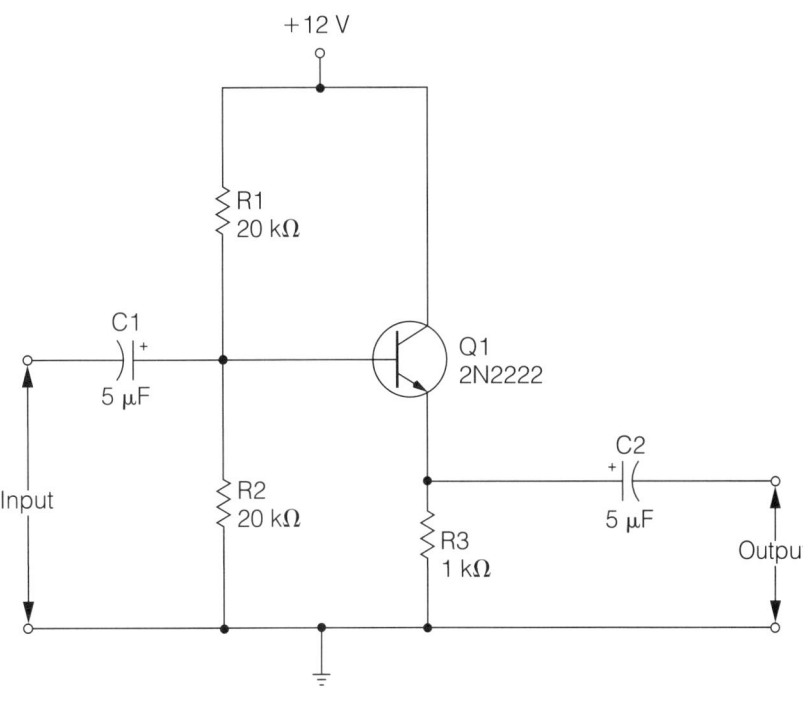

■ **6-4** *Practical example of a common-collector (emitter-follower) transistor amplifier.*

Common-collector amplifiers are current amplifiers. The current gain for the circuit illustrated in Fig. 6-4 is the parallel resistance value of R1 and R2, divided by the resistance value of R3. R1 and R2 are both 20 kΩ in value, so their parallel resistance value is 10 kΩ. 10 kΩ divided by 1 kΩ (the value of R3) gives us a current gain of 10 for this circuit. Because the voltage gain is considered to be unity (1), the *power gain* for a common-collector amplifier is considered equal to the current gain (10, in this particular case).

The input impedance of common-collector amplifiers is typically higher than the other transistor configurations. It is the parallel resistive effect of R1, R2, and the product of the value of R3 times the beta value. Because beta times the R3 value is usually much higher than that of R1 or R2, you can closely estimate the input impedance by simply considering it to be the parallel resistance of R1 and R2. In this case, the input impedance would be about 10 kΩ. The traditional method of calculating the output impedance of common-collector amplifiers is to divide the value of R3 by the transistor's beta value. Although this method is still appropriate, a closer estimate can probably be obtained by considering the output impedance of most transistors to be about 80 ohms. This 80-ohm output impedance should be viewed as being in parallel with R3, giving us a calculated output impedance of about 74 ohms (80 ohms in parallel with 1000 ohms).

R1 and R2 have the same function within a common-collector amplifier, as previously discussed with common-emitter amplifiers. The high negative feedback produced by R3 provides excellent temperature stability and immunity from transistor variables. As in the case of Fig. 6-3, the circuit illustrated in 6-4 can be a valuable building block toward future projects.

## The common-base configuration

I have included the common-base transistor amplifier configuration in this text for the sake of completeness, but the applications for it are few. They are often used as the first RF amplifier stage, amplifying signals from radio antennas, but are seldom seen otherwise.

Figure 6-5 illustrates a practical example of a *common-base amplifier*. Common-base amplifiers have the unique characteristic of a variable input impedance dependent upon the emitter current flow. The equation for calculating the input impedance is:

$$Z_{in} = \frac{26}{I_e}$$

where: $I_e$ is the emitter current in milliamps.

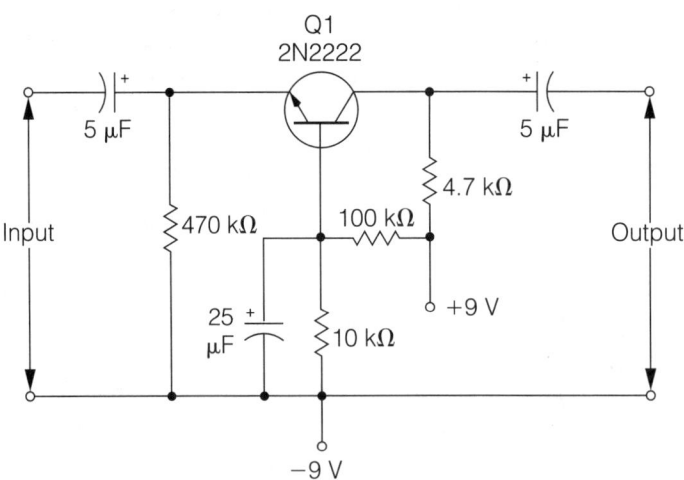

**6-5** *Practical example of a common-base transistor amplifier.*

As can be seen from the previous equation, the input impedance is low. Furthermore, common-base amplifiers have a *high output impedance*, and a *power gain* slightly higher than common-emitter amplifiers.

## Transistor amplifier comparisons

The common-emitter configuration is used in applications requiring reasonably high voltage and power gains. The output is inverted. Common-emitter amplifiers have low input impedances, and high output impedances.

The common-collector configuration in used for impedance matching applications. They have high input impedances with low output impedances. Voltage gain is considered at unity, and the output is non-inverted.

The common-base configuration is seldom used because of its very low input impedance and high output impedance. They also have a very unstable nature at high gain values.

## Impedance matching

Fig. 6-6 illustrates the importance of correctly matching input and output impedances between circuit stages. The ac signal source illustrated in the left half of Fig. 6-6 could be the output of a common-emitter amplifier, a laboratory signal generator output, the "line" output of a FM radio receiver, or thousands of other possible sources. The important point is that all signal sources will have an internal impedance; shown as $Z_S$ in the illustration. Internal source impedances can range from fractions of an ohm to millions of ohms in value. In this example, assume the internal impedance to be 1000 ohms. The signal level you want to apply to $R_L$ is 10 volts

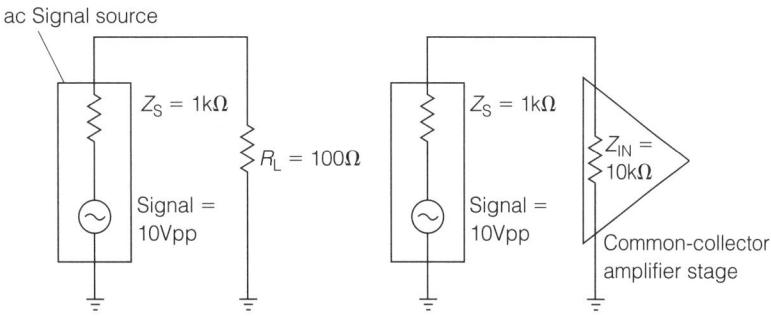

■ **6-6** *Demonstration of impedance matching.*

peak-to-peak. Notice that $Z_s$ is in series with $R_L$. You know that, in a series circuit, the voltage drop will be proportional to the resistance values. Therefore, in this circuit, about 9.1 volts P-P will be dropped across the internal source impedance, and only about 0.9 volts P-P will actually be applied to $R_L$. This is undesirable because over 90% of the signal has been lost.

The circuit illustrated in the right half of Fig. 6-6 shows the same signal source connected to the input of a common-collector amplifier stage. (Note the triangular symbol for the common collector amplifier. This is the symbol used for amplifiers in most block diagrams.) Because the amplifier has an input impedance of 10,000 ohms, only 0.9 volts P-P will be dropped across the internal source impedance, and 9.1 volts P-P will be applied to the amplifier. This is much better because the signal loss is only about 9%.

Here are a few general rules to remember regarding impedance matching. For the *maximum transfer of voltage* (as discussed in the previous example), the output impedance should be as low as possible and the input impedance should be as high as possible. For the *maximum transfer of power*, the output impedance should be the same value as the input impedance. For the *maximum transfer of current*, the output impedance and input impedance should be as low as possible.

## Assembling and testing of the last section of a lab power supply

Materials Needed for the Completion of this Section

| Quantity | Schematic ref. | Item description |
|---|---|---|
| 6 | D1 through D6 | 1N4001 (NTE 116) general purpose silicon diode |
| 2 | R1 and R2 | 1000-ohm, 2-watt resistor |
| 2 | R3 and R4 | 2.2- k$\Omega$, 2-watt resistor |
| 2 | R7 and R8 | 470-ohm, 1/2-watt resistor |
| 2 | R5 and R6 | 0.4-ohm, 5-watt resistor |
| 2 | P1 and P2 | 1000-ohm, linear taper 2-watt potentiometer |
| 2 | C3 and C4 | 0.1-$\mu$F ceramic disc capacitor |
| 2 | Q3 and Q5 | TIP 31C (NTE 291) transistor |

| Quantity | Schematic ref. | Item description |
|---|---|---|
| 2 | Q4 and Q6 | TIP 32C (NTE 292) transistor |
| 1 | Q1 | 2SC3280 (NTE 2328) transistor |
| 1 | Q2 | 2SA1301 (NTE 2329) transistor |
| 2 | (not illustrated) | 3AG Size (1/4″ × 1 1/4″) fuse blocks (optional) |
| 2 | (not illustrated) | 2-amp slow-blow fuses (optional) |
| 3 | See Fig. 6-8 | Insulated banana jack binding posts (Radio Shack #274-661A or #274-662A or equivalent) |
| 1 | See Fig. 6-7 | Universal grid board (Radio Shack #276-168A or #276-150A or #276-159A or equivalent) |

■ **6-7** *Examples of universal type circuit boards that can be used to permanently construct many projects within this book.*

The fuse blocks and fuses are listed if you want to fuse-protect the positive and negative outputs. This is a good idea, but it is not mandatory. The fuse blocks can be mounted in any convenient location prior to the binding posts. The fuse/fuse block assemblies are simply placed in series with the output wires going to the positive and negative binding posts (see Fig. 6-10).

*Assembling and testing of the last section of a lab power supply*

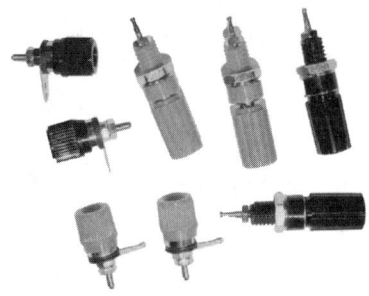

■ **6-8** *Examples of the most popular styles of insulated binding posts.*

It might be a little difficult to obtain the *series pass transistors* (Q1 and Q2) on a local basis. However, they are readily available through catalog suppliers.

### Assembling the circuit board

The specified grid boards are universal types, designed for building one-of-a-kind circuits or prototypes. They consist of "three-point" solder pads and long strips, or rails, called "buses." Circuits are assembled using the three-point solder pads as connection points and the buss lines are used for common connections to power supply voltages or circuit common.

To begin laying out the circuit board, refer to Fig. 6-9. The first step is to recognize that all of the components shown in the schematic will not be placed on the circuit board. C1 and C2 are already installed on solder strips connected to the chassis (this was performed in the last chapter). Q1 and Q2 will be externally mounted to the chassis bottom for heatsinking. P1 and P2 will be mounted to the front panel, as your voltage adjustment controls.

As the second step, notice how the *positive regulator* is almost a "mirror-image" of the *negative regulator* (the circuitry above the circuit common line is the positive regulator; the negative regulator is below the circuit common line). Divide the grid board in half and concentrate on the positive regulator section first. Upon completion, the negative regulator is simply a duplication on the first half of the grid board.

By the time you have completed the prior steps, you are left with four resistors, two transistors, three diodes, and one capacitor (the circuit board mountable components for the positive regulator). It is not difficult to arrange these few components for proper connections (Fig. 6-9). There is not a critical nature to these parts placements, so you might install them in any way that is easy and organized.

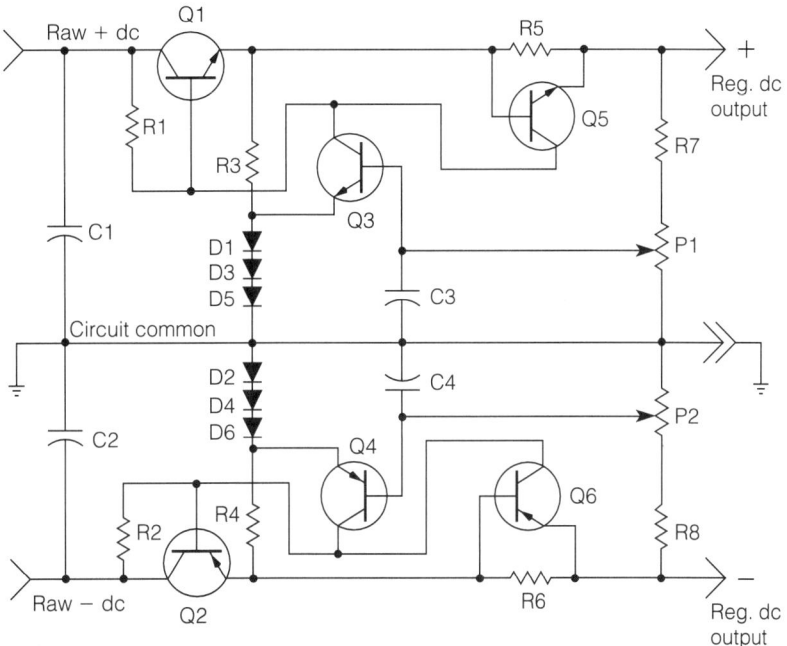

■ **6-9** *Regulator section of the lab power supply project.*

When you finish with the positive regulator section, copy the parts layout for the negative regulator. It is usually easier to wait until the entire circuit board is "populated" and doublechecked, before permanently soldering any components into the board. Simply insert the components and bend the leads back to hold them in place temporarily. Don't forget to leave some empty holes in the pads for connecting the external components (Q1, Q2, P1, and P2) and the external wiring (raw dc inputs, regulated dc outputs, and the circuit common).

Figure 6-10 shows a suggested method of physically mounting all of the components together with the circuit board. You might have to move things around a little to accommodate your enclosure. If you choose to fuse the outputs, the fuse blocks should be wired in series between the regulated dc outputs and the binding posts.

Regarding the external wiring to the circuit board, it should be stranded, 18-to 22-gauge insulated hook-up wire. An itemized list of the external wiring is as follows:

☐ 3 wires from the raw dc power supply (raw +dc, raw -dc, and circuit common)

☐ 3 wires from Q1 (emitter, base, collector)

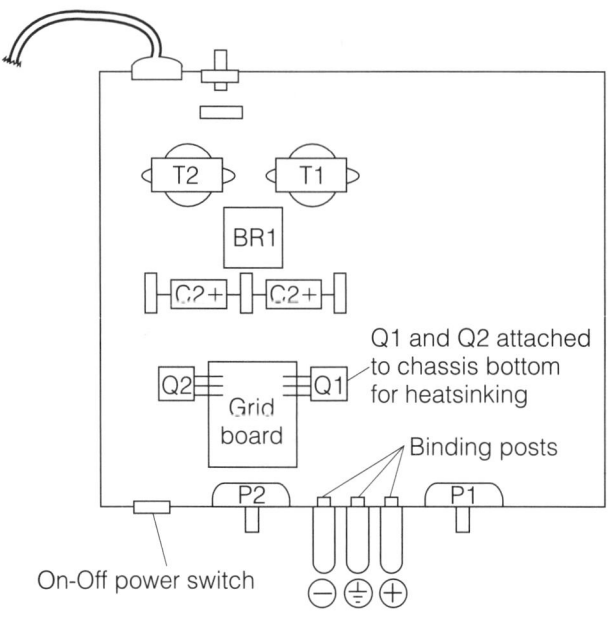

■ **6-10** *Suggested physical layout for the finished lab power supply.*

☐ 3 wires from Q2 (emitter, base, collector)
☐ 6 wires from P1 and P2 (3 wires each)
☐ 3 wires from the three binding posts (regulated +dc, regulated −dc, and circuit common)

### Mounting Q1 and Q2

Under worst-case conditions, Q1 and Q2 might have to dissipate close to 60 watts of power each. To keep them from overheating, some form of *heatsinking* must be provided. The chassis's metal bottom will work nicely for this purpose; unless you try to operate the power supply in the current limit mode at low voltage, or if short-circuit conditions exist for extended periods of time. This "worst-case" condition could cause the series-pass transistors (Q1 and Q2) to overheat, even with the chassis heatsinking. Specially designed aluminum heatsinks can be purchased for the mounting of Q1 and Q2 to eliminate the problem, if desired.

When mounting Q1 and Q2 to the chassis, you will have to use an insulator (mica insulators are the most common for this purpose) to keep the transistors *electrically isolated* from the chassis. To aid in heat conductivity, the mounting surface of the transistor and the insulator is coated with a "thin" layer of thermal joint compound, commonly called "heatsink grease." *Thermal joint compound* is a white, silicon oil-based grease used to fill the minute air gaps between the transistor, insulator, and heatsink material. Some power transistors require the use of an *insulating sleeve* to keep the mounting bolt from making electrical contact with the transistor collector. After mounting Q1 and Q2, use your DVM (in the ohm's setting) to check for any continuity between the transistor's collector lead and the chassis. You should obtain an infinite resistance.

You might be wondering how you could check Q1 and Q2 during the power supply operation to know if they are getting too hot. Without the aid of expensive temperature measurement equipment, there is a very "unscientific" way to "guesstimate" possible temperature problems with semiconductors. **Be sure that the power is off**. Then, touch the transistors immediately after a period of operation. If they are too hot to comfortably hold your finger on them for more than 1 second, you should probably acquire some better heatsinking.

## Circuit description

The negative regulator functions identically to the positive regulator, except for the difference in voltage polarity. Therefore, I will only detail the operation of the positive regulator.

When power is first applied to the circuit, C1 charges to about +37 volts (as discussed in chapter 5). R1 supplies sufficient base current to allow uninhibited current flow through Q1. As the Q1 emitter voltage approaches the "raw" dc level of C1, diodes D1, D3, and D5 all go into forward conduction through resistor R3. Once conducting, these diodes maintain a relatively constant *reference voltage* of about 2.1 volts, which is the sum of their individual forward threshold voltages.

As the Q1 emitter voltage continues to rise, it is applied to the R7 and P1 voltage divider. The P1 wiper is connected directly to the Q3 base. When the wiper voltage reaches about 2.8 volts, Q3 will begin to conduct. This occurs because the Q3 emitter is being held at 2.1 volts by the diode assembly, and its base has to be about 0.7 volts more positive than the emitter to overcome the forward threshold voltage requirement of the base to emitter junction. When Q3 starts to conduct, it begins to divert some of the current, being supplied by R1, away from the Q1 base. Consequently, Q1 begins to restrict some of its emitter-collector current flow. This *leveling effect* continues until the output is stabilized at a voltage output relative to the P1 setting.

When a "load" of some kind is placed on the output, the *loading effect* will "try" to decrease the output voltage amplitude. As the voltage starts to decrease, this decrease is seen at the P1 wiper and transferred to the Q3 base. Consequently, Q3 decreases its current conduction by an amount relative to the output voltage decrease. This allows more current to flow through the Q1 base causing Q1 to become more "conductive" than it was previously. The end result is that the load is compensated for, and the output voltage remains stable (at the same amplitude as before the load was applied).

If a load is placed on the output, causing the current flow to exceed about 1.5 amps, the resultant current flow will cause the voltage drop across R5 to exceed the forward threshold voltage of the base-emitter junction of Q5. When Q5 begins to conduct, it diverts current flow away from the Q1 base causing Q1 to decrease its current flow, and thus reduce the output voltage. The Q1 and Q5 combination will not allow the output current to exceed about 1.5 amps, even under a *short-circuit condition*.

### Testing the power supply

Before applying power to the power supply, set your DVM to measure ohms, and check for continuity of virtually every component to chassis ground. There is not a single part of this circuit that should be electrically connected to the chassis. If you measure resistance from any component to chassis ground, re-check your wiring and connections. Remember that you have a *"floating" circuit common* that is the power supply ground, but not connected to the chassis ground.

If you have double-checked your wiring, and are certain that everything is correct; set P1 and P2 to their approximate center positions, and "pulse" the line power to the power supply ON and OFF quickly. If there were no blown fuses, or other signs of an obvious problem (smoke, the smell of burning plastic insulation, or the "pop" of exploding plastic semiconductor cases), re-apply the line power and measure the output voltages at the binding posts with your DVM. The positive and negative output voltages should be measured in reference to circuit common (the center binding post as illustrated in Fig. 6-10), not chassis ground. You should be able to smoothly adjust the positive output voltage by rotating P1 from a minimum of about +3.8 volts, to a maximum of about +36 volts. The negative regulated output should perform identically by rotating P2, except for the difference in voltage polarity.

## In case of difficulty

If the power supply is blowing fuses upon application of line power, you probably have a *short-circuit condition* on one of the raw dc supplies. **Be sure that the line power is off**, and disconnect the raw dc wires from the circuit board. Replace any blown fuses and apply line power. If you still blow fuses, the problem is most likely in the bridge rectifier (BR1). You learned how to check a bridge rectifier in chapter 4.

It is most likely that the fuses won't blow after disconnecting the raw dc lines because the complete raw dc power supply was previously tested in chapter 5. If this is the case, **be sure that the line power is off**, and use your DVM to measure the resistance from the collector of Q1 to circuit common. This should be a very high indication (my circuit measures over 2 M$\Omega$). Make the same measurement from the Q2 collector. If either one of these resistance measurements are low, you have a wiring error or a defective component.

The most likely components to be defective in this circuit are the transistors and diodes. You have already learned how to test diodes with a DVM set to the "diode test" function. Transistors can be tested in this manner also.

## Checking transistors with a DVM

If a need arises to functionally test a bipolar transistor, it can be thought of as two back-to-back diodes. With your DVM set to the "diode test" mode, check the base to emitter junction just like any diode junction. The DVM should indicate infinity with one test lead orientation while reversing the test leads will cause a low indication, or vice versa. Because a transistor is a two junction device, you must perform the same test on the base to collector junction also. If both junctions appear to be functioning normally, you should measure the resistance from the emitter to collector. This resistance should be high with both orientations of the DVM test leads.

To check transistors that have already been installed in a circuit, it is usually necessary to consider the parallel effects of other associated components that might influence the measurement. For example, in Fig. 6-9, the base to emitter measurement of Q5 will be predominantly influenced by R5. Because R5 is only 0.4 ohms, it will not be possible to measure the forward or reverse resistance of the base-emitter junction because 0.4 ohms is much less than either. In this case, you would need to temporarily disconnect the base lead before checking the transistor with a DVM. In some cases, it might be necessary to disconnect two of the transistor's leads to obtain a reliable test.

If you discover a defective transistor, there is a good chance it was destroyed by an error in the wiring. After replacing the defective transistor, doublecheck its associated wiring for mistakes.

# Special-purpose diodes and optoelectronic devices

TO THIS POINT, YOU HAVE EXAMINED THE APPLICATION OF diodes in only two general areas; rectification and voltage referencing (the three-diode voltage reference in the lab power supply project). There are many other applications, and numerous types of special purpose diodes. *Optoelectronic devices* are electronic devices used in applications involving visual indicators, visible light, infrared radiation, and laser technology. Special purpose diodes make up a large part of this family.

## Zener diodes

*Zener diodes* are used primarily as voltage regulator devices. They are specially manufactured diodes designed to be operated in the reverse-breakdown region. Every zener diode is manufactured for a specific reverse-breakdown voltage called the *zener voltage* (abbreviated $V_z$ in most data books).

To understand the operational aspects of zener diodes, refer to Figure 7-1. Note the symbol used to represent a zener diode. For this illustration, a 5.6-volt zener has been chosen. Assume you can apply a variable dc voltage to the positive and negative terminals of the circuit, ranging from 0 to 10 volts. Beginning at 0 volts, as the voltage is increased up to 5.6 volts, the zener diode behaves like any other reverse-biased diode. It totally blocks any significant current flow. And, because it represents an almost infinite resistance, the entire applied voltage will be dropped across it.

This all holds true up to the point when the applied voltage exceeds the rated zener voltage of the diode. When this *avalanche voltage* occurs, the zener diode will abruptly start to "freely" conduct current. The point at which this abrupt operational change occurs is called the *avalanche point*. The minimum amount of current flow through the zener diode required to keep it in an

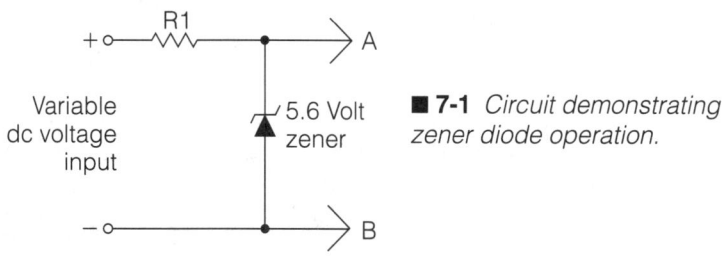

**7-1** *Circuit demonstrating zener diode operation.*

avalanche mode of operation is called the *holding current*. The voltage across the zener diode does not decrease when the avalanche point is obtained, but it does not increase by a very significant amount as the applied circuit voltage is increased substantially. Hence, the voltage across the zener diode is *regulated*, or held constant.

To clarify the previous statements, consider the circuit operation in Fig. 7-1 at specific input voltage levels. If 5 volts is applied to the input (observing the polarity as illustrated), the zener diode will block any significant current flow, because its avalanche point will not occur until the voltage across it reaches a level of at least 5.6 volts. If 6 volts is applied to the input, about 5.6 volts will be dropped across the zener, and about 0.4 volts will be dropped across R1. Increasing the applied input voltage to 7 volts causes the voltage across R1 to increase to about 1.4 volts, but the voltage across the zener diode will remain at about 5.6 volts. If the applied input voltage is increased all the way up to 10 volts, about 4.4 volts will be dropped by R1, but the zener diode will continue to maintain its zener voltage of about 5.6 volts. In other words, as the applied circuit voltage is increased "above" the rated voltage of the zener diode, the voltage across the zener diode will remain relatively constant and the excess voltage will be dropped by its associated series resistor, R1.

If a load of some kind were to be placed in parallel with the zener diode of Fig. 7-1, the zener diode would hold the voltage applied to the load at a relatively constant level, as long as the applied circuit voltage did not drop below the rated zener voltage of the zener diode.

The two most important parameters relating to zener diodes are the zener voltage and the rated power dissipation. Zener diodes are commonly available in voltages ranging from about 3 volts to over 50 volts. If a higher zener voltage is needed, two or more zener diodes can simply be placed in series. For example, if an application required the use of a 90-volt zener, this could be accomplished by placing a 51-volt zener in series with a 39-volt zener.

Unusual zener voltages can be obtained in the same manner. Another method of obtaining an odd (non-standardized) zener voltage value is to incorporate the 0.6-volt or 0.7-volt "forward threshold" voltage drop of a general-purpose silicon diode. When using this method, the general-purpose diode is placed in series with the zener diode, but it is oriented in the forward-biased direction, and the zener diode is reverse-biased.

The standardized *power dissipation ratings* for zener diodes are 1/2, 1, 5, 10, and 50 watts. Zeners rated at 10 and 50 watts are manufactured in stud-mount casings, and must be mounted into appropriately sized heat sinks for maximum power dissipation.

## Designing simple zener-regulated power supplies

Referring to Fig. 7-2, your design problem here is to build a 12-volt, 500-mA zener-regulated power supply to operate the load, designated as $R_L$ in the schematic diagram. Much of the "front-end" of the circuit should be rather familiar to you by now. T1, the bridge rectifier, and C1 comprise a raw dc power supply. (The methods of calculating the values and characteristics of these components will not be discussed in this context because this was covered in previous chapters.)

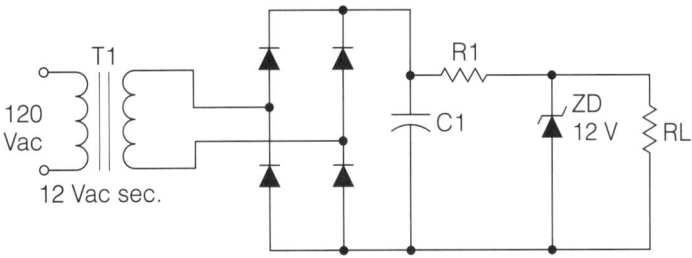

■ **7-2** *Simple zener regulated power supply.*

The design of a well-functioning zener-regulated power supply is a little tricky because of variations in the raw dc power supply. The output voltage of a raw dc power supply may decrease by as much as 25% when placed under a full load. To roughly estimate the no-load voltage that C1 would charge to, the T1 secondary voltage should be multiplied by 1.414 to calculate the peak secondary voltage. In this case, the peak secondary voltage is about 17 volts. After subtracting about 1 volt to compensate for the loss in the bridge rectifier, you are left with about 16 Vdc across C1. Unfortunately, this "no-load" calculation will be in error as soon as the power supply is loaded.

The primary component affecting the full-load voltage decrease of a raw dc power supply is the transformer. The percentage of full-load voltage decrease will depend on how close the full-load requirement comes to the maximum secondary current rating of the transformer. For example, our hypothetical load in Fig. 7-2, as previously stated, will require up to 500 mA. If T1's secondary rating is 600 mA, the secondary voltage will decrease substantially when fully loaded. However, if T1's secondary rating is 2 amps, the 500-mA full-load requirement of $R_L$ will have much less effect. In addition, even transformers with similar ratings can behave somewhat differently, depending on certain manufacturing techniques.

A further complication, although not as dramatic, relates to the value of C1. When a load is placed on a raw dc power supply, the ripple content increases. A high ripple content has the effect of reducing the usable dc level.

The easiest solution to overcoming all of these unknown variables is to simply build the raw dc power supply, and place a dummy load across C1 that will closely approximate the full load requirement of $R_L$. In this case, you would start with the no-load voltage across C1, which is about 16 volts. Knowing that $R_L$ might require as much as 500 mA, the resistance value of the dummy load can be calculated using ohm's law:

$$R = \frac{E}{I} = \frac{16 \text{ volts}}{0.500 \text{ A}} = 32 \text{ ohms}$$

A 33-ohm resistor would be close enough for calculation purposes. But don't forget the power rating! This dummy resistor must be capable of dissipating about 8 watts.

Assume you built the raw dc power supply, placed the 33-ohm dummy load across C1, and measured the "loaded" dc voltage to be 14 volts. This gives you all the information you need to design the rest of the power supply. (You might have realized that when the raw dc voltage decreased under load, the 33-ohm dummy load no longer represented a full-load condition. Experience has shown that this "secondary" error, which is the difference between the "almost fully loaded" voltage and the "fully loaded" voltage, is not significant in the vast majority of design situations.)

There are three variables you must calculate to complete your design problem; the power rating of the zener, the resistance value of R1, and the power rating of R1.

To calculate these variables, you need to understand how the circuit should function under extreme variations of $R_L$. When $R_L$ requires the full load of 500 mA, the current flow through the zener diode should be as close to the minimum holding current as possible. Assuming the holding current is about 2 mA, that means about 502 mA must flow through R1; 2 mA through ZD, plus 500 mA through $R_L$. R1 is in series with the parallel network of ZD and $R_L$. The applied voltage to the series-parallel circuit of R1, ZD, and $R_L$ is the voltage developed across C1, which you are assuming to be 14 volts, under loaded conditions. Because 12 volts is being dropped across the parallel network of ZD and $R_L$, the remaining 2 volts must be dropped by R1 (the sum of all of the series voltage drops in a circuit must equal the source voltage). You now know the voltage across R1, and the current flow through it. Therefore, ohm's law can be used to calculate the resistance value:

$$R = \frac{E}{I} = \frac{2 \text{ volts}}{0.502 \text{ A}} = 3.98 \text{ ohms}$$

Of course, 3.98 ohms is not a standard resistance value. You don't want to go to the nearest standard value above 3.98 ohms because this would risk "starving" the zener diode from its holding current when the current flow through $R_L$ was maximum. The nearest standard value below 3.98 ohms is 3.9 ohms, which is the best choice. By using any of the familiar power equations, the power dissipated by R1 comes out to be about 1 watt. A 2-watt resistor should be used to provide a good safety margin.

The "worst-case" power dissipation condition for ZD occurs when there is no current flow through $R_L$. If all current flow through $R_L$ ceases, the full 502 mA must flow through ZD. Actually, the maximum current flow through ZD could be as high as 513 mA because you chose a 3.9-ohm resistor for R1 instead of the calculated 3.98 ohms. The power dissipated by ZD is the voltage across it (12 volts), multiplied by the current flow through it (the worst case is 513 mA). This is the familiar power equation $P = IE$. The answer is 6.15 watts. Therefore, ZD would need to be a 12-volt, 10-watt zener with an appropriate heat sink. Another option would be to use two 5-watt, 6-volt zeners in series. The latter option eliminates the need for a heat sink, but care must be exercised to assure plenty of "air space" around the zener diodes for adequate convection cooling.

As the previous design example illustrates, zener regulated power supplies are not extremely efficient because the zener diode wastes a significant amount of power when the current flow through the load is small. For this reason, zener regulated power supplies are typically restricted to low-power applications. However, zener diodes are commonly used as voltage references in high-power circuits, as is illustrated later in this chapter.

## Varactor diodes

Going back to diode fundamentals, you might recall that when a diode is reverse-biased, a "depletion region" of current carriers is formed around the junction area. This depletion region acts as an insulator resulting in the restriction of any appreciable current flow. A "side-effect" of this depletion region is to "look like" the dielectric of a capacitor, with the anode and cathode ends of the diode acting like capacitor plates. As the reverse bias voltage across a capacitor is varied, the depletion region will also vary in size. This gives the effect of varying the distance between the plates of a capacitor, which varies the capacitance value. A diode that is specifically designed to take advantage of this capacitive effect is called a *varactor diode*. In essence, a varactor is a *voltage-controlled capacitor*.

Varactor diodes are manufactured to exhibit up to 450 picofarads of capacitance for AM (MW) radio tuning applications; but they are more commonly found in VHF (very high frequency) and UHF (ultra high frequency) applications, with capacitance values ranging from 2 to 6 picofarads. Virtually every modern television and radio receiver incorporates varactor diodes for tuning purposes, to reduce costs and to improve long-term performance by eliminating mechanical wear problems.

The schematic symbol used to represent varactor diodes is the same symbol used for general-purpose diodes, but with the addition of a small capacitor symbol placed beside it. Figure 7-3 illustrates the commonly used electronic symbols for special-purpose diodes and optoelectronic devices.

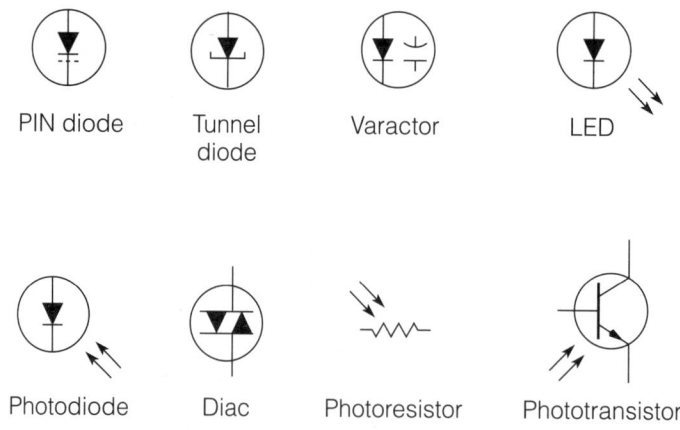

**7-3** *Commonly used electronic symbols for special purpose diodes and photoelectronic devices.*

*Special-purpose diodes and optoelectronic devices*

## Schottky diodes

*Schottky diodes* are sometimes called "four-layer" diodes, because their construction includes two layers of each type of semiconductor material. The result is a NPNP device. As with all other types of diodes, they are two-lead devices.

The schottky family of diodes includes the *PIN diode*, sometimes referred to as the *silicon hot-carrier diode*.

These are *breakdown devices*, meaning that their useful function is to become highly conductive when the reverse voltage across them exceeds an inherent *trigger voltage*. Unlike zener diodes, they do not maintain their avalanche voltage across them after the trigger voltage has been reached. In contrast, their internal resistance drops to an extremely low value of only a few ohms, and remains there as long as a *minimum holding current* is maintained.

Schottky diodes are commonly used in high frequency switching, detecting, and oscillator circuits.

## Tunnel diodes

*Tunnel diodes* are constructed similarly to ordinary diodes, with the exception of heavier impurity doping in the semiconductor material. This results in an extremely thin depletion region under reverse-bias conditions and causes a tunnel diode to be a reasonably good conductor when reverse-biased. The unique characteristic of tunnel diode behavior, however, is in the forward-biased mode.

As the forward-biased voltage across a tunnel diode is increased, there will be three specific voltage levels where the tunnel diode will exhibit *negative resistance* characteristics. This means the current through the tunnel diode will decrease as the voltage across it increases. As the forward voltage across a tunnel diode is smoothly increased from minimum to maximum, the current response will show a series of three peaks and valleys as the tunnel diode vacillates between positive and negative resistance responses.

Tunnel diodes are used in oscillators and high-speed switching applications for digital circuitry.

## Diacs

*Diacs* are 3-layer bilateral trigger diodes. Like PIN diodes, diacs are breakdown devices. However, unlike PIN diodes, diacs are triggered from a blocking-to-conduction state in either polarity of applied voltage. Consequently, there is not a band encircling the device body to indicate the cathode end, because the orientation of the device is irrelevant.

For example, if the rated breakover voltage (the breakdown voltage, or avalanche point) for a specific diac is 30 volts, it will present an extremely high resistance until the voltage drop across it equals about 30 volts. At that point, it will become highly conductive; and it will remain in this state until the voltage across it reaches a minimum level. At that point, it becomes highly resistive again. It will react this way regardless of the voltage polarity, hence, it is "bilateral" in operation.

Diacs are most commonly used in high-power control circuitry to provide the "turn-on" pulses for SCRs and TRIACs. (SCR and TRIAC operation will be discussed in a later chapter.) Diacs were designed to be a solid-state replacement for "neon tubes".

## Fast recovery diodes

All diodes possess a characteristic called their *recovery time*, which is the amount of time required for the diode to turn off after being in the forward conduction state. This is usually a very short time period; and for 60-Hz rectification applications, it is negligible. However, in high-frequency rectification applications, the recovery time becomes critical. For these applications, specially manufactured diodes with very fast recovery times are implemented. Logically enough, they are called *fast recovery diodes*.

## Noise and transient suppression diodes

Common household and commercial ac power is fine for powering motors, heaters, and most electro-mechanical devices. However, in applications where ac power is used to provide the operational power for sensitive and high-speed solid-state circuitry, it can create many problems resulting from noise, voltage spikes, lighting surges, and other undesirable interference signals, which might be conducted into the home or industrial facility by the power lines. To help in eliminating these problems, a whole family of noise and transient suppression diodes have been developed.

Common names for such devices are *unidirectional surge clamping diodes, varistor diodes, unidirectional transient suppression diodes, bidirectional transient suppression diodes, transorbers*, and many others. All of these devices utilize the nonlinear resistive effect or the avalanche effect of semiconductor materials to reduce voltage spikes or over-voltage surges. Their uses are primarily in solid-state power supplies and ac line filters.

## A basic course in quantum physics

Light is a form of radiated energy. As such, it makes up a small part of the total range of radiated energies called the *electromagnetic spectrum*. Radiated energy is comprised of extremely small particles of wavelike energy called *quantums* (technically speaking, the plural form of quantum is "quanta", but the effect of science fiction and media inaccuracy has brought "quantums" into the colloquial language). In reference to the visible and near-visible light frequencies, the older term "quantum" has been replaced with the newer term *photon*.

In the early days of solid-state electronics, it was discovered, quite by accident, that a solid-state diode would emit a small quantity of light as a side-effect of the *recombination process* occurring in the PN junction, while forward-biased. This led to the development of the modern *light-emitting diode*, or LED. Further research led to the discovery that if an "outside" light source was focused on the junction area of a solid-state diode, the light photons had the tendency to "dislodge" some electrons from their atomic shell positions resulting in an increase of "minority" carriers. In other words, the "leakage" current in a reverse-biased diode would increase proportionally to the light intensity falling on the junction. This photoconductive property resulted in the development of the photodiode, the phototransistor, and the photoconductive cell. It is also possible to directly convert the energy of photons (light energy) into electrical energy. Devices capable of performing this energy conversion are called *photovoltaic cells*, or, more commonly, *solar cells*.

Further developments, involving a wider diversity of materials and manufacturing processes, led to the more recent member in the optoelectronic field; the laser diode.

The acronym *LASER* stands for "Light Amplification through Stimulated Emission of Radiation." As stated earlier, normal light consists of small "packets" of energy called *photons*. The typical light emitted all around us consists of photons all traveling in random fashion and random frequencies. Laser light differs from normal light in several ways. First, it is "coherent," meaning the photons are all traveling in the same direction. To understand this difference, consider a typical flashlight. As you shine a flashlight beam into a distance, the diameter of the beam of light will increase with distance, becoming very broad after only a hundred feet or so. In contrast, a laser light beam will not broaden with distance because all of the photons comprising the beam are going in the same direction. A high-coherency laser beam can easily be bounced off of the moon!

The second radical difference, between laser light and standard "white" light, relates to frequency. Solid-state semiconductor light-emitting devices, such as LEDs and laser diodes, typically emit only a narrow wavelength of light. Therefore, the emitted light consists of only one "pure" color. Common white light, on the other hand, contains all of the colors (meaning all of the frequencies) in the visible light spectrum.

Now that most of the basic principles relating to optoelectronics have been defined, it is appropriate to discuss these devices individually, in more detail.

## LEDs (light emitting diodes)

An *LED* (*light emitting diode*) is a specially manufactured diode that is designed to glow, or emit light, when forward-biased. When reverse-biased, it will act like any common diode; it will neither emit light or allow substantial current flow. LEDs can be manufactured to emit any color of the visible light spectrum desired, including "white" light. Red is, by far, the most common color. For certain physical reasons, semiconductor material is especially efficient and sensitive to "near-visible" light in the infrared region. Consequently, many *photoelectric eyes* used for presence detection and industrial control functions operate in the infrared region.

LEDs are primarily used as indicator devices. The brightness, or intensity, of an LED is relative to the forward current flow through it. Most LEDs are low-voltage, low-current devices; but more recent developments in optoelectronics have led to a family of high-intensity LEDs that approach the light intensity levels of incandescent bulbs.

Most commonly available LEDs operate in the 5- to 50-milliamp range and drop about 1.4 to 1.6 volts in the forward-biased mode. In most applications, LEDs require the use of a series resistor to limit the maximum current flow.

LEDs have far too many available case styles, shapes, and colors to describe in detail within this context, but each type will use some physical method to indicate the cathode lead. You will learn many of the indication methods through experience, but when in doubt, simply use your DVM in the "diode test" mode to check lead identification. (Most DVMs, in diode test mode, will cause an LED to glow very faintly when checked in forward-biased orientation.)

A common "alpha-numeric" type of LED indicator device is the seven-segment display. (The term *alpha-numeric* refers to display devices capable of displaying some, or all, of the characters of the alphabet, as well as numbers.) Seven-segment LED displays are actually seven individual elongated LEDs arranged in a "block 8" pattern. Seven-segment LEDs will have a common connection point to all seven diodes. This common connection might connect all of the cathodes together (making a common cathode display), or all of the anodes together (making a common anode display). The choice of using a common cathode, or a common anode display, is simply a convenience choice, depending on the circuit configuration and the polarity of voltages used. In addition, many types of "decoder" integrated circuits (integrated circuits designed to convert logic signals into seven-segment outputs), will specify the use of either common-cathode or common-anode displays.

A seven-segment LED will have eight connection pins to the case. One pin is the common connection point to all of the cathodes or anodes. The other seven pins connect to each individual diode within the package. Thus, by connecting the common pin to the appropriate polarity, and forward-biasing various combinations of the LEDs with the remaining seven pins, any seven-segment alpha-numeric character can be displayed.

## Optoisolators, optocouplers, and photoeyes

Regarding their principles of operation, optoisolators, optocouplers, and photoeyes are equivalent; but their sizes, construction, and intended applications can vary dramatically. In essence, all of these devices consist of a light emitter (LED) and a light receiver (photodiode, phototransistor, photoresistor, photo-SCR, or photo-TRIAC).

The intensity of the light emitter can be varied proportionally to an electrical signal. The light receiver can convert the varying light intensity back into the original electrical signal. This process completely eliminates any electrical connection between emitter and receiver resulting in total isolation between the two. Total electrical isolation is very desirable in circuits that could malfunction from electrical noise, or other interference signals, "feeding back" to the more susceptible areas. Light emitter-receiver pairs used in this manner are called *optical isolators* (*optoisolators*) or *optical* couplers (*optocouplers*). The electrical signal being transmitted to the receiver might be either analog (linear) or digital (pulses).

These same basic components are often used in a photoeye mode. When used in this manner, the light emitter is held constant and sends a continuous beam of light to the receiver. The intended application requires an external object to come between the emitter and receiver, breaking the beam, and thereby causing the receiver to produce a "loss-of-light" signal. This type of "presence detection" is used extensively in VCRs, industrial control applications, and security systems.

Optoisolators and optocouplers utilizing a SCR or TRIAC as the receiver are designed for ac power-control applications. The primary advantage in this configuration is the complete isolation from any noise or voltage spikes present on the ac line.

### Photodiodes, phototransistors, and photoresistive cells

*Photodiodes* are manufactured with a clear window in the case to allow external light to reach the junction area. When photodiodes are reverse-biased in a circuit, the amount of "leakage" current allowed to flow through the diode will be proportional to the light intensity reaching the junction. In effect, it becomes a light-controlled variable resistor.

*Photoresistors* function much like photodiodes, but with a few differences. Unlike photodiodes, photoresistors are junctionless devices. Therefore, like resistors, they do not have fixed orientation in respect to voltage polarity. Also, photoresistors react to light intensity with a very broad resistance range; typically 10,000 to 1. The typical "dark" resistance value of a photoresistor is about 1 Megohm; this resistance then decreases proportionally with exposure to increasing light intensity.

*Phototransistors*, like photodiodes, incorporate a clear window in the casing to allow ambient light to reach the junction area. The external light affects the transistor operation much like a base signal voltage, so in most cases, the base lead is left unconnected (some phototransistors don't even have base leads). Phototransistors are especially useful in some applications, because they can be used as amplifiers with external light either substituting for, or adding to (modulating), the base signal.

### Laser diodes

The widespread common use of coherent light in the average home has been made a reality by the solid-state laser diode. Every compact disc player or compact disc ROM (Read-Only Memory) system utilizes a laser diode as the light source for reading the disc data. The commonly seen "laser pointers," familiar to office environments, are little more than a laser diode and a couple of batteries enclosed in a case.

Laser diodes are actually a type of LED. Their operation is similar, with the primary difference being in the type of light emitted. Laser diodes emit coherent light.

Laser diodes are available in power ranges from about 0.5 milliwatts to 5 milliwatts. They are also available as visible red or infrared emitters.

Note: Please use caution if you plan to use or experiment with laser diodes. Laser light is dangerous to the eyes. Always follow the manufacturer's recommended safety precautions.

## LCDs (liquid crystal displays)

*Liquid crystal displays* have rapidly replaced LED systems in many indicator applications due to several advantages. First, LCDs require much less operational power than comparable LED systems. This is because LCDs do not actually produce any light of their own; LCD operation depends on ambient light for character display. The second LCD advantage relates to the first. Because LCDs depend on ambient light for operation, they are the most visible in the strongest light where LEDs often appear faint.

An LCD is an optically transparent sandwich, often including an opaque backing. The inner surfaces of the panels making up the sandwich have a thin metallic film deposited on them. On one of the panels, this film is deposited in the form of the desired characters or symbols to be displayed. The space between the two panels contains a fluid called *nematic liquid*. This liquid is normally transparent. When an electrical field is placed between the back panel and the desired character to be displayed, the liquid turns black and is displayed, providing that the ambient light is strong enough to see it. This is really no different than using a black "magic marker" to write a character on a piece of white paper; such a character is clearly visible in normal light, but you couldn't see it in the dark.

Although research is continuing in the LCD field, to date, there are some severe disadvantages. For one, LCDs are much slower than LEDs, making their use in high-speed display applications (such as television) limited. Their speed of operation is greatly affected by temperature; operation becomes visibly slow in cold temperatures. Another disadvantage is versatility. LCD displays must be manufactured for specific applications. For example, an LCD intended for use as a clock display could not be used as a counter display because the colon, which normally appears between the hour and minute characters, places an undesired space between the numerals. After an LCD is manufactured, its character display cannot be modified for another application. A third disadvantage, relating to the home hobbyist or experimenter, is the decoding required for correctly displaying the characters. It could be very complex, requiring ICs that might not be readily available. If you plan on ordering LCD displays from any of the surplus electronic suppliers, be sure that it includes all of the necessary interface documentation.

### CCDs (charge-coupled devices)

*Charge-coupled devices* are actually digital circuits used primarily to replace the older "vidicon" tubes in video cameras. They require less power to operate and provide a much sharper and clearer picture.

Although CCDs are currently used exclusively for video "reception," research toward using CCD technology for solid-state display applications is very promising.

## Circuit potpourri!

As an electrical engineer, instructor, and amateur scientist, I have come to fully appreciate the electronic field as being one of infinite creativity and infinite possibilities. As I write the text for this book, I am surrounded by devices that would have been considered incomprehensible "miracles" only a few decades ago. Currently, scientists are taking the first infantile steps toward uncovering the mysteries of subatomic structure and quantum physics. These areas could lead to gravity and antigravitational generators, total annihilation fusion reactors, and deep space travel by means of bending the time/space continuum! Virtual reality systems are available today that can positively "knock your socks off!" Does all this sound a little more than "mildly interesting?"

Although I can't show you how to build a time machine, this section, together with the concurrent "circuit potpourri" sections, contains a collection of projects and circuit building blocks that can be practical, fascinating, and fun (with the emphasis on "fun"). This particular section allows you to start taking the first steps toward discovering all of the creative and ingenious facets within your own self. I'm hoping someday you'll be able to show me how to build a time machine.

I suggest you read through the description of each circuit even if you do not intend to build or experiment with it. The practical aspects of much of the previous theory is illustrated within them.

## Preliminary steps to project building

At this point, I am assuming that you have a lab power supply, DVM, electronic data books, soldering iron, hand tools, suppliers for electronic parts, and miscellaneous supplies needed for project building. In addition, I highly recommend that you purchase a *solderless breadboard* for testing the following projects before permanently building them. A solderless breadboard is a plastic rectangular block with hundreds, or thousands, of contact points internally mounted. Electronic components and interconnection wiring (ordinary #22 solid-conductor insulated wire) is simply inserted into the breadboard (without soldering) and the completed circuit can be tested in a matter of minutes.

The circuit can then be modified by simply unplugging the original components and inserting new ones, until the operation is satisfactory. At this point, the user can then remove the components of the perfected circuit, and permanently install them in a universal perfboard or PC board. The solderless breadboard is not damaged in this process, and it can be used repeatedly for designing thousands of additional circuits. An illustration of some excellent-quality solderless breadboards is given in Fig. 7-4. You can also buy prestripped, prebent hook-up wire intended to be used with solderless breadboards. For the modest cost involved, I believe this is a good investment.

## Flashing lights, anyone?

Figure 7-5 illustrates a good basic circuit to cause two LEDs to spontaneously blink on and off. The frequency will be about 1 Hz, depending on component tolerances and the type of transistors used.

The basic circuit is called an *astable multivibrator*. Multivibrators are covered in more detail in successive chapters, but for now you can think of it as a free-running oscillator.

When power is first applied to this circuit, one transistor will *saturate* (the state of being turned-on fully) before the other because of slight component variations. For discussion's sake, assume Q1

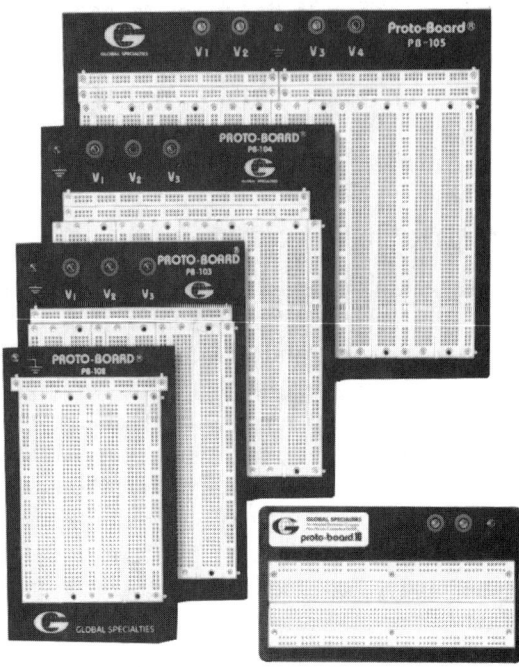

■ **7-4** *Examples of solderless breadboards. Photograph courtesy of Interplex Electronics Inc., New Haven, CT.*

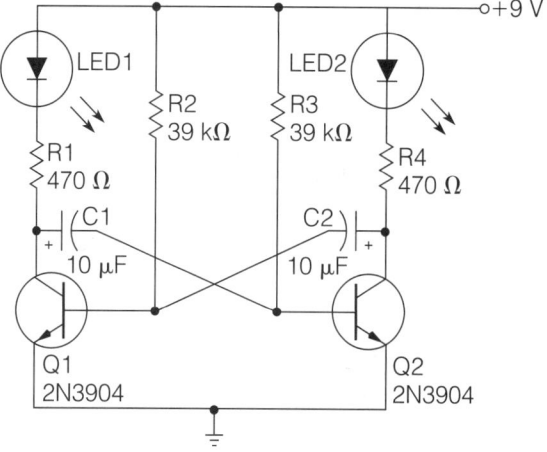

■ **7-5** *A dual LED flasher.*

*Special-purpose diodes and optoelectronic devices*

saturates first. In the saturation state, Q1 conducts the maximum collector current lighting LED1. At the same time, this condition makes the positive side of C1 appear to be connected to ground, and it begins to charge to the supply potential through R3. When the charge across C1, which is also the base voltage of Q2, charges to a high enough potential, it causes Q2 to turn on and lights LED2.

At the same time, the positive side of C2 is now placed at ground potential causing the base voltage of Q1 to go low forcing Q1 into cutoff (the state of being fully turned off). With Q1 at cutoff, LED1 is dark and C1 begins to discharge back through R3. In the meantime, C2 is now charging and applying a rising voltage to the base of Q1. This continues until Q1 saturates again and the whole process starts over again. (You might have to slowly re-read this functional description and study Fig. 7-5 several times to fully grasp the operation.)

The astable multivibrator shown in Fig. 7-5 is certainly a basic building block for many future applications. Here are a few examples of how this circuit could be modified for a variety of projects. The frequency and on-time/off-time relationship (called the *duty cycle*) can be changed by changing the values of C1, C2, R2, and R3. Experiment with changing the value of each of these components, one at a time, and observe the results. R2 and R3 can be replaced with rheostats (potentiometers connected as variable resistors) for continuously variable frequencies and duty cycles. C1 and C2 can be replaced with smaller values of capacitance making the circuit useful as a simple *square-wave frequency generator*. With the correct choice of transistors and capacitors, this circuit is usable well into the megahertz region. If you wanted to flash brighter lights, LED1 and LED2 could be replaced with small 6-volt relay coils. The relay contacts, in turn, could be connected to the line voltage (120 Vac), and incandescent lamps for high-brightness flashing (be careful not to exceed the contact current and voltage ratings of the relays). There are many more applications. The fun is in using your imagination.

### Three lights are better than two!

Figure 7-6 illustrates a variation of the same circuit illustrated in Fig. 7-5. The primary differences are: the LEDs will light when their associated transistor is in cutoff, rather than saturation, and an extra transistor circuit has been added for a sequential *three-light effect*.

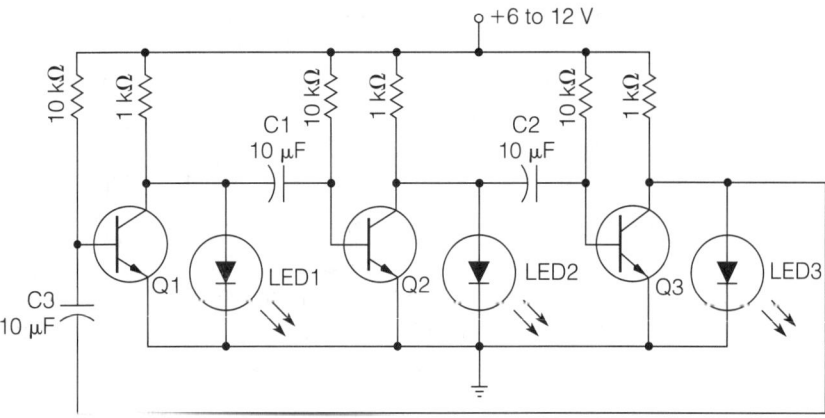

All capacitors rated at 16 WVdc or higher

■ **7-6** *A triple LED flasher circuit.*

Only one transistor will be in the cutoff state at any one time. Assume transistors Q1 and Q3 are saturated, and that Q2 is cutoff. In this state, LED2 is bright from the current flow through Q2's 1-kΩ collector resistor. Meanwhile, C1 is charging through Q2's 10-kΩ base resistor. When the voltage across C1 reaches a high enough potential, it turns on Q2 and causes LED2 to become dark (the saturation of Q2 effectively "shorts" the voltage drop across LED2).

Here is where the circuit operation becomes a little tricky. Going back to the prior condition when Q2 was in cutoff, the voltage on the negative side of C2 was actually a little more positive than the voltage on the positive side. This is because the voltage drop across LED2 was a little higher than the base to emitter voltage drop of Q3. Therefore, C2 actually takes on a slight reverse voltage charge. When Q2 saturates, this has the effect of forcing Q3 into "hard" cutoff; because a slight negative voltage is applied to its base, before C2 has the chance to start charging in the positive direction through Q3's 10-kΩ base resistor. With Q3 in the cutoff state, C3 begins to take on a small reverse charge, while C2 begins to charge toward the point where it will drive Q3 back into saturation. The cycle continues to progress in a sequential manner, with LED1 lighting next, and so on. Even though C1, C2, and C3 are electrolytic capacitors, the small reverse charge is not damaging because the charging current and voltage are very low.

The frequency of operation is a function of the time constant of the capacitors and base resistors. Increasing the value of either component will slow down the sequence. Any general purpose NPN transistor should operate satisfactorily in this circuit. The LED type is not critical either, although you would have to adjust the resistor values somewhat to accommodate the newer "high brightness" LEDs.

Many of the applications that applied to Fig. 7-5 will also apply to this circuit. One of the advantages of this circuit design is that additional transistor stages can be added on for a longer sequential flashing string.

## A mouse in the house

Figure 7-7 is definitely a "fun" circuit. It consists of two astable multivibrator circuits, very similar to the circuit in Fig. 7-5. The first multivibrator circuit will oscillate very slowly, because of the component values chosen. However, the second multivibrator circuit, consisting of Q3 and Q4, will not oscillate until Q2 saturates, which will occur every few seconds. The component values for the second multivibrator are chosen so it will oscillate very rapidly and produce a "chirping" or "squeaking" sound if a small speaker is connected to it.

This circuit can be assembled on a very small universal perfboard and enclosed in a small plastic project box together with the

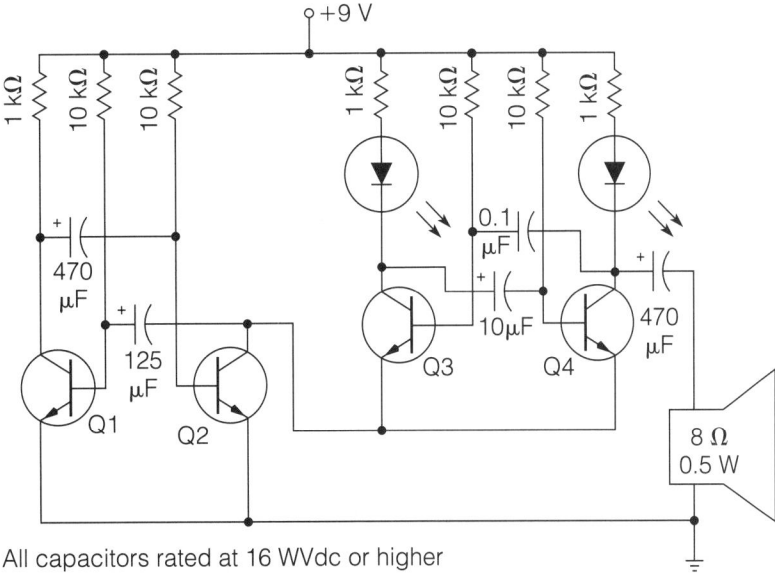

All capacitors rated at 16 WVdc or higher

■ **7-7** *An electronic chirping circuit.*

speaker, an on-off switch, and a 9-volt "transistor" battery (which powers it nicely). The two LEDs are optional, but their effect is dramatic.

Once completed, is should be about the size of a pack of cigarettes. It can easily be put in your pocket and carried to a friend's house. When the opportunity arises, turn it on and hide it somewhere inconspicuous. Then, wait for the fun to start! The chirping sounds like a mouse, or some type of large insect. It is not loud enough to cause instant attention, but everyone in the room will notice it in a few minutes. The frequencies and harmonics produced by the multivibrator have the effect of making the sound omnidirectional, so it will be difficult to locate. In the meantime, everyone who is a little squeamish toward mice or large insects will get "seriously" nervous.

It is important to use a very small speaker for this project to achieve the desired effect. Virtually any type of general-purpose NPN transistors will perform well.

If you would like to try a variation on this circuit to produce some really weird sounds, try replacing Q1 and Q3 with a couple of "3-lead" phototransistors. Connect the phototransistors into the circuit exactly like the original transistors. This causes the changes in ambient light to "sum" with the original base voltages. Various capacitor and resistor combinations will produce some remarkable sounds in conjunction with changing light intensities.

To carry this idea one step further, you can mount this circuit in the center of a "bulls-eye" target and convert a laser pointer into a "gun" (put a dummy handle on it, and fabricate the on-off switch into a trigger). Using various component values, the target can be made to produce any number of strange sounds, when the laser beam hits the bulls eye. Including a small power amplifier into the circuit, to boost the output volume, will improve the effect. If you built multiple circuits, adjusted them for individual sound effects, and mounted them in a variety of targets, you could have a "high-tech" shooting gallery in your own home!

### A sound improvement

Figure 7-8 is a Hi-Z audio amplifier circuit that will greatly increase the volume level of a high-impedance headphone (two of these circuits will be needed for stereo headphones). This circuit can come in handy if you want to use your headphones to listen (loudly!) to some of the sounds that you can create with these

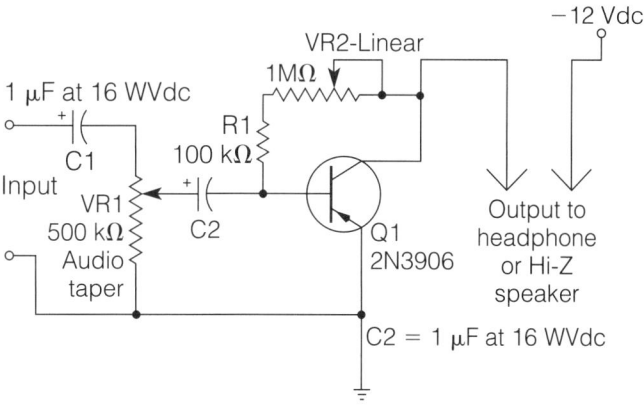

■ **7-8** *High-impedance headphone amplifier.*

multivibrator circuits. The input impedance is high enough to keep it from loading down most circuits. You can also use this circuit with most types of speakers that have an impedance-matching transformer connected to the speaker frame. Don't try to use a standard 8-ohm or 4-ohm speaker; you'll destroy the transistor, or the speaker, or both!

This circuit is a modified form of the common-emitter transistor amplifier discussed in chapter 6. VR2 should be adjusted for the best quality of sound, and VR1 is the volume control.

## A delay is sometimes beneficial

In the field of electronics, there are many control applications that require a *TDR* (*time delay relay*). Commercial TDRs are very expensive. Figure 7-9 is a *time-off TDR*, which is both useful and inexpensive. It can also be modified for a variety of functions.

Q1 and Q2 are connected in a configuration called a darlington pair. The *darlington pair configuration* is essentially a *beta multiplier*, causing the beta value of Q1 to be multiplied by the beta value of Q2. For example, if both transistors had a beta value of 100, the overall beta value for the pair would be 10,000. The high beta value is particularly useful in this circuit, because only an extremely small Q1 base current is needed to saturate the pair. A discussion of the circuit operation will illustrate why this is important.

When the momentary switch is closed, C1 will appear to charge instantly, because there is no significant series resistance to limit the charge rate. At the same time, Q1 is supplied with more than

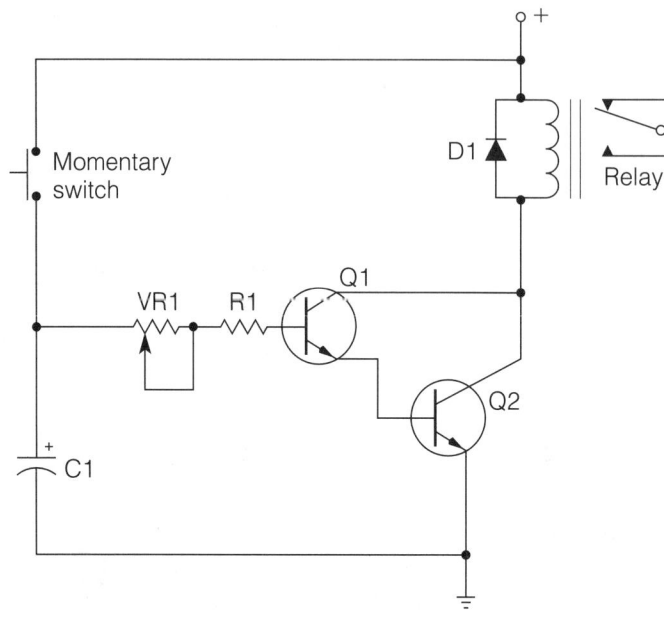

**7-9** *Time-off delay relay driver.*

enough base current to saturate the transistor pair, and the relay is energized. When the switch is released, opening the charge path to C1, C1 begins to "slowly" discharge through the Q1-Q2 base-emitter circuits. Because very little base current is needed to keep the transistor pair saturated, VR1 and R1 can be of a high resistance value, causing a very slow discharge of C1. The majority of C1's discharge cycle will maintain the saturated condition of Q1 and Q2, causing the relay to remain energized for a substantial time period after the switch is released. Theoretically speaking, if you tried to perform this same function with only a single transistor, the resistance values of VR1 and R1 would have to be about 100 times smaller to maintain a base current adequate for saturation (assuming both transistors have a beta value of 100), and the discharge rate of C1 would be very rapid. You could accomplish the same operation if you increased the value of C1 by a factor of 100, but large electrolytic capacitors are both expensive and bulky.

Figure 7-9 is a "time-off" TDR; meaning that after the control action is instigated (closing and releasing the momentary switch), there is a time delay before the relay de-energizes. The length of the time delay depends on the setting of VR1, which largely controls the discharge rate of C1.

This circuit can be easily modified to provide a time-on, time-off delay. Remove VR1 and connect the opened end of R1 to the positive side of C1. Then connect VR1 into the C1 charge path, between the momentary switch and the applied power source. In this configuration, when the momentary switch is closed, C1 must charge through VR1 causing a "time-on" delay, until C1 charges to a high enough potential to cause the transistor pair to saturate. The length of this delay would depend on the setting of VR1. Upon releasing the momentary switch, a "time-off" delay would occur, while C1 discharged through R1 and the base-emitter junctions of Q1 and Q2. This delay would be largely controlled by the value of R1.

If you wanted a "time-on" TDR (without the time-off function), a reasonably good facsimile can be made by simply removing R1, and by connecting the base of Q1 directly to the positive side of C1. Connect R1 in series with VR1 in the C1 charge path. By experimenting with different values of VR1 and C1, a significant "time-on" delay can be achieved with a fairly rapid turn-off.

Q1 and Q2 are general purpose NPN transistors. For best results, use "low-leakage" type transistors. The voltage amplitude of the circuit power source should be about equal with the relay coil voltage. For experimentation purposes, start with a C1 value of about 100 $\mu$F, and a R1 and VR1 value of 100 k$\Omega$ and 1 M$\Omega$, respectively. These values can then be adjusted to meet your requirements. Whenever relay coils are incorporated into dc powered solid-state circuitry, they should always be paralleled with a reverse-biased general-purpose diode. Note the orientation and connection of D1 in Fig. 7-9. The purpose of D1 is to suppress the inductive kickback, transient voltage spike, which will occur when the relay is de-energized. This kickback voltage spike can easily damage solid-state devices. It is generated by the stored energy in the electromagnetic field surrounding the relay coil. Fortunately, these voltage spikes will always be in the opposite polarity of the applied power source. Therefore, D1 will short out the spike, and render it harmless.

## A long-running series

Earlier in this chapter, during the discussion of zener diode regulators, it was shown why zeners are not very power efficient as high-current regulators. A circuit to greatly improve the efficiency and operation of *voltage regulation* is illustrated in Fig. 7-10.

Much of this circuit should already be familiar to you. D1 is only needed if you plan on using this regulator circuit with a battery as

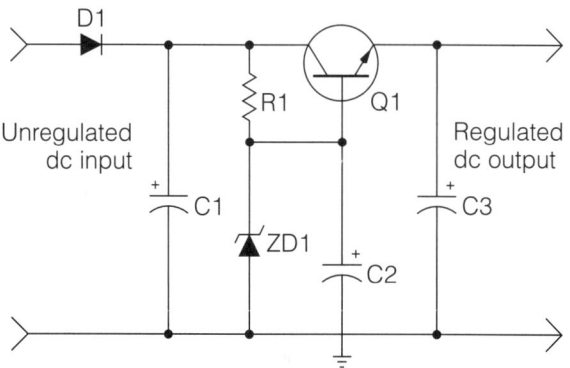

■ **7-10** *A series-pass regulator using a zener diode as a voltage reference.*

the unregulated power source. C1 is the filter capacitor(s) for the raw dc power supply. This raw dc power supply can be of any design you choose. R1 and ZD1 form a simple zener regulator, as discussed previously in this chapter. However, in this circuit, the zener serves as a *voltage reference* for the "series pass" transistor Q1. Transistor Q1 serves as a "current multiplier" for the zener. For example, if Q1 has a *beta* value of 100, and the current requirement for the load is 1 amp, the zener would only have to supply 10 milliamps of current to the base of Q1 for a 1-amp output. This means the value of R1 can be chosen so that the current flow through ZD1 is only slightly above its minimum holding current. Therefore, ZD1 is only required to dissipate a small quantity of power, and a much higher load can be regulated (remember, a "high" load means that the load resistance is "small," and vice versa). C3 serves as an additional filter for smoothing the regulated dc.

For regulating most low-voltage loads requiring up to about 1 amp, ZD can be a 1-watt zener. Its zener voltage value should be 0.6 volts above the desired regulated output voltage. Q1 will drop this 0.6-volt excess across the base to emitter junction. For example, if you wanted a 5-volt regulated output, ZD should be a 5.6-volt zener. The value of R1 should be chosen to place the zener diode at about 15 to 20 milliamps above its rated holding current. Commonly used transistors for Q1 are the TIP31, TIP3055, and 2N3055 types.

Capacitor C2 serves a unique purpose in this circuit. Connected as shown, transistor Q1 serves as a *capacitor multiplier*, multiplying the filtering effect of C2 by its beta value. If capacitor C2 had a value of 1000 $\mu F$ and Q1 had a beta of 100, the regulated output voltage would be filtered as if a 100,000 $\mu F$ capacitor had been placed in parallel with the output.

## Keep it steady

You will probably run into many situations where you will want to use an LED as an indicator for a variable voltage circuit. Trying to use a single resistor for current limiting will prove ineffective for this application because the current flow through the LED will vary proportionally to the voltage, and it is likely to go too high, or too low, for good results. Figure 7-11 is a quick and easy solution to the problem. A low-power zener (ZD1) will maintain the voltage across the resistor-LED combination at a relatively constant level, regulating the current flow through the LED, and the voltage variations will be dropped across R1.

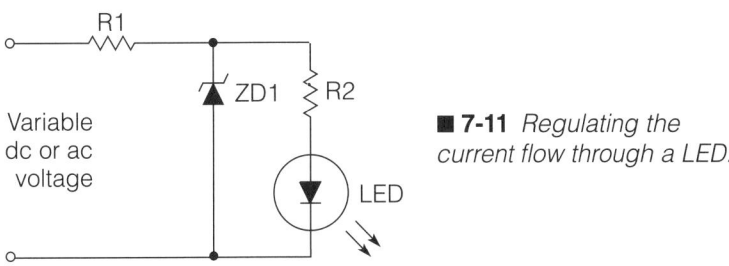

■ **7-11** *Regulating the current flow through a LED.*

## Double your pleasure

Do you have a transformer in your junkbox that you would like to use for a circuit application, but the secondary voltage is too low? If so, you can use the voltage doubler circuit illustrated in Fig. 7-12 to approximately double the secondary output during the rectification process.

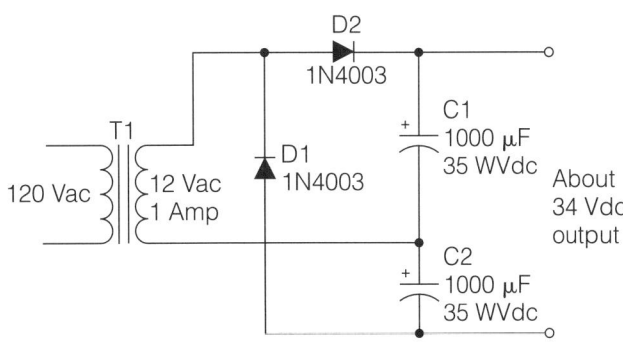

■ **7-12** *A voltage-doubler rectification circuit.*

In Fig. 7-12, a transformer with a 12-volt, 1-amp secondary is used to illustrate the principle. D1 and D2 are configured as two half-wave rectifiers, with C2 and C1 being their associated filters. C1 and C2 are simply connected so that their voltages are additive.

C1 and C2 filter a half-wave rectified voltage, so they must have a much higher capacity than comparable capacitors used for full-wave filtering. D1 and D2 are common, general-purpose diodes. Also notice that neither secondary transformer lead is used as the common reference.

### Show the blow

As my last entry into this section of "circuit potpourri", I submit the blown fuse alarm circuit illustrated in Fig. 7-13. In most homes, there are situations where it is critical to maintain electrical power to certain devices. For example, a chest freezer or a sump pump located in the basement have critical needs for constant power. A blown fuse (or tripped circuit breaker) to either of these appliances could result in a flooded basement, or the loss of hundreds of dollars worth of food. Unfortunately, it is likely that the blown fuse will not be discovered until the damage has already occurred. The circuit shown in Fig. 7-13 solves that problem by providing an alarm when a blown-fuse condition occurs.

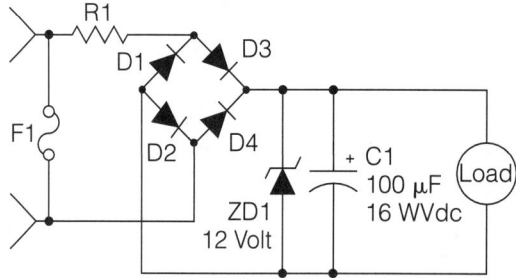

■ **7-13** *An electronic fuse monitor circuit.*

When a fuse blows, it represents an infinite resistance (like an open switch) within an electrical circuit. Therefore, the entire source voltage for the circuit will be dropped across it. This is how the fuse monitoring circuit obtains its operational power.

Assume the fuse (F1) is protecting a 120-Vac circuit (I do not recommend this monitoring circuit for ac voltages higher than 120 Vac). Upon blowing, it will apply 120 Vac to the monitor circuit. R1 limits the current, and drops most of the applied voltage. Diodes D1 through D4 rectify the voltage and apply pulsating dc to the zener diode (ZD1). C1 filters the pulsating dc to apply smooth dc to the load. The load can be a piezo buzzer (such as used in smoke detectors) or any other type of low-power, low-voltage visible or audible indicator.

The resistance value and power rating of R1 will depend on the load requirement. Build the circuit as illustrated using a 1-amp, 200-volt PIV bridge rectifier (or comparable diodes), a 1-watt zener diode, and a 100-k$\Omega$, 1/2-watt resistor for R1. If the load will not operate when 120 Vac is applied to the circuit, start decreasing the value of R1 a little at a time until you reach the point of reliable operation. If reliable operation requires going below 12 k$\Omega$, I suggest you try using an indicator requiring less operational power. At R1 values lower than 68 k$\Omega$, the power rating should be increased to 2 watts.

This circuit can also be used to monitor fuses used in dc circuits. For these applications, the bridge rectifier is not required, but be sure to observe the correct polarity.

The enclosure for this circuit will depend upon the intended application. 120-Vac applications require the use of an "approved" metal enclosure, properly grounded, and with wiring and conduit meeting national and local safety standards. It might also be necessary to "fuse protect" the monitor circuit.

A final word of caution: **please don't try to connect this circuit into a fuse box or breaker box**, unless you're fully qualified to do so. Mistakes resulting from a lack of knowledge or experience in this area can result in property damage, fire hazard, and electrocution (especially if you're working on a damp basement floor)!

# Audio amplification systems

IN GENERAL, AUDIO AMPLIFICATION CIRCUITS ARE probably the most common type of circuits to be found in the field of electronics. Radios, television sets, stereo systems, musical instruments, public address systems, and even "multimedia" computer systems, require the use of a high-quality audio amplifier. Audio amplification systems were comparatively simple in the vacuum tube days, but modern audio systems can be as complex and challenging as any other field of electronics.

## Transistor biasing and load considerations

The circuit illustrated in Figure 8-1 should already be familiar to you from the previous discussions of transistor amplifiers. It is a common-emitter amplifier because the output (to the speaker) is taken off from the collector, and the input signal to be amplified is coupled to the base. C1 is a coupling capacitor (blocking the dc bias voltage, but passing the ac audio signal). R1 and R2 form a voltage-divider network to apply the proper dc bias to the base. The emitter resistor ($R_E$) increases the input impedance, and it improves temperature and voltage stability.

Transformer T1 is an audio transformer. It serves two important functions in this circuit. First, it isolates the dc quiescent (steady-state) current flow from the speaker coil (speaker coils can be damaged by even relatively small dc currents). Secondly, it provides a more appropriate load impedance for a transistor collector than would a low-impedance 8-ohm speaker. A transistor amplifier of this configuration could not operate very well with an extremely low collector impedance. A typical audio transformer might have a primary impedance of 100 ohms, for connection into the transistor circuit, and a secondary impedance of 8 ohms for connection to the speaker. Generally speaking, an audio amplifier of this type performs satisfactorily for

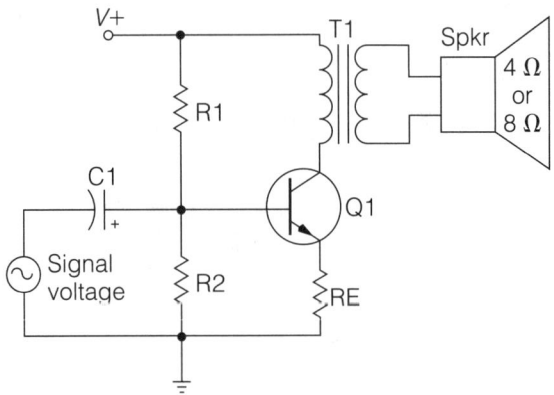

**8-1** *Basic audio amplifier circuit.*

low-power applications. However, based on modern standards, it has severe problems and limitations.

To begin, examine the "real-life" problems relating to efficiency and biasing considerations. Choosing some simple numbers for discussion purposes, assume the source voltage in Fig. 8-1 is 35 volts, T1's primary impedance is 100 ohms, and $R_E$ is 10 ohms. As you might recall, the voltage gain of this circuit is approximately equal to the collector resistor (or impedance) divided by the emitter resistor. Therefore, the 100-ohm collector impedance (T1) divided by the 10-ohm emitter resistor (RE) places the voltage gain ($A_e$) at 10.

R1 and R2 are chosen so that the base voltage is about 2.1 volts. If Q1 drops about 0.6 volt across the base-emitter junction, this leaves 1.5 volts across RE. The 1.5-volt drop across the 10-ohm emitter resistor (RE) indicates the emitter current is at 150 milliamps. Because the collector current is approximately equal to the emitter current, the collector current is also about 150 milliamps. (The 150 milliamps is the "quiescent" collector/emitter current flow. The term *quiescent* refers to a steady-state voltage or current established by a bias.) 150 milliamps of current flow through the 100-ohm T1 primary causes it to drop 15 volts. If 15 volts is being dropped across T1's primary, the collector voltage must be 20 volts (in reference to ground). 15 volts plus 20 volts adds up to the source voltage of 35 volts.

If a 500-millivolt rms signal voltage was applied to the input of this amplifier, a 5-volt rms voltage would be applied to the primary of T1 ($A_e$ = 10). If T1 happened to be a "perfect" transformer, it would transfer the total ac power of the primary to the secondary

*Audio amplification systems*

load. In this example, the total power being supplied to the primary is 250 milliwatts rms. Even with no T1 losses, 250 milliwatts of power would not produce a very loud sound out of the speaker.

In contrast, examine the power being dissipated by Q1. As stated earlier, in its quiescent state, the collector voltage of this circuit is 20 volts. The emitter resistor is dropping 1.5 volts; therefore, 18.5 volts is being dropped across the transistor (the emitter to collector voltage). With a 250-milliamp collector/emitter current flow, that comes out to 4.625 watts of power dissipation by Q1. In other words, about 4.6 watts of power is being wasted (in the form of heat) to supply 250 milliwatts of power to the speaker. That translates to an efficiency of about 5%.

With a better choice of component values, and a more optimum bias setting, the efficiency of this amplifier design could be improved. However, about the best "real-life" efficiency that can be hoped for is around 20% at maximum output, in a class-A amplifier.

There are actually two purposes to this efficiency discussion. The first, of course, is to demonstrate why a simple common-emitter amplifier makes a poor high-power amplifier. Secondly, this is a refresher course in transistor amplifier basics. If you had some trouble understanding the circuit description, you might want to review chapter 6 before proceeding.

## Amplifier classes

Although some audio "purists" still insist on wasting enormous quantities of power to obtain the high linearity characteristics of amplifiers, such as the one shown in Fig. 8-1; most people who specialize in audio electronics recognize the impracticability of such circuits. For this reason, audio power amplifiers have been developed, using different modes of operation, that are much more efficient. These differing operational techniques are arranged into general groups, or "classes." The class categorization is based on the way the output "drivers" (transistors, FETs, or vacuum tubes) are *biased*.

The amplifier circuit illustrated in Fig. 8-1 is a *class-A audio amplifier* because the output driver (Q1) is biased to amplify the full, peak-to-peak audio signal. This is also referred to as biasing in the linear mode.

Continuing to refer to Fig. 8-1, assume that the bias to Q1 was modified to provide only 0.6 volt to the base. Assuming Q1 will drop about 0.6 volt across the base-emitter junction, this leaves zero voltage across $R_E$. In other words, Q1 is biased just below the point of conduction. In this quiescent state, Q1 would not dissipate any significant power (a little power would be dissipated because of leakage current) because there essentially is no current flow through it. If an audio signal voltage were applied to the input of the circuit in this bias condition, the positive half-cycles would be amplified (because the positive voltage cycles from the audio signal on the base would "push" Q1 into the conductive region, above 0.7 volts); but the negative half-cycles would only drive Q1 further into the cutoff region and would not be amplified. Naturally, this results in severe distortion of the original audio signal, but the efficiency of the circuit, in reference to transferring power to the speaker, would be greatly improved. This mode of amplification is referred to as *class B*, where conduction occurs for about 50% of the cycle.

Of course, the circuit shown in Fig. 8-1 (biased for class-B operation) is not very practical for amplifying audio signals because of the *high distortion* that the more efficient classes of bias promote. But, if a second transistor were incorporated in the output stage, also biased for class-B operation, but configured to amplify only the "negative" half-cycles of the audio signal, it would be possible to re-create the complete original amplified audio signal at the output. This is the basic principle behind the operation of a *class-B push-pull audio amplifier*.

There is still one drawback with class-B amplification. At the point where one transistor goes into cutoff and the other transistor begins to conduct (the *zero reference point* of the ac audio signal), a little distortion will occur. This is referred to as *crossover distortion*. Crossover distortion can be eliminated by biasing both output transistors just "slightly" into the conductive region in the quiescent state. Consequently, each output transistor will begin to conduct at a point slightly in advance of the other transistor going into cutoff. By this method, crossover distortion is essentially eliminated, without degrading the amplifier's efficiency by a significant factor. This mode of operation is referred to as *class AB*. An efficiency factor of about 40% usually applies to class-AB amplifiers.

Figure 8-2 is an example of a "theoretical" class-AB audio amplifier. C1 is the input coupling capacitor, R1 and R2 form the familiar voltage divider bias network for biasing Q1, RE is Q1's emitter resistor, and RC is Q1's collector resistor. These components comprise a typical common-emitter transistor amplifier. Q2 and poten-

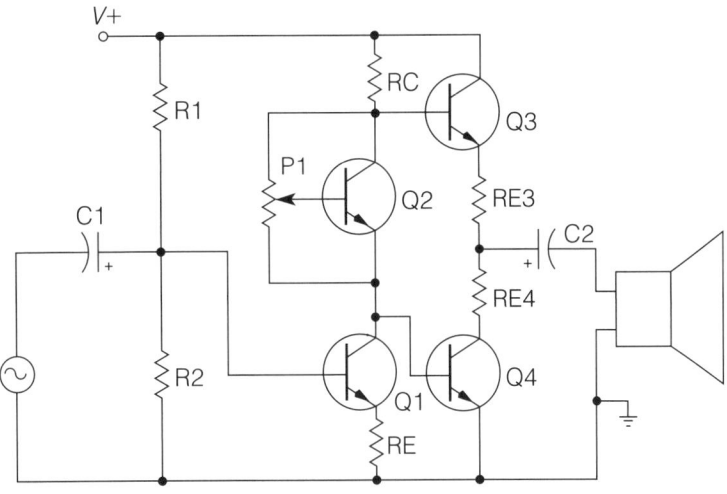

**8-2** A "theoretical" class-AB audio amplifier.

tiometer P1 are configured in a circuit arrangement called an amplified diode. The purpose of this circuit is to provide the slight forward bias required on both output driver transistors to eliminate crossover distortion. Q3 and Q4 are the output drivers; with Q3 amplifying the positive half-cycles of the audio signal, and Q4 amplifying the negative half-cycles. C2 is an output coupling capacitor; it serves to block the dc quiescent voltage from reaching the speaker, while allowing the amplified ac output voltage to pass.

The amplified diode circuit of Q2 and P1 could be replaced with two forward-biased diodes. In theory, each diode would drop about the same voltage as the forward biased base-emitter junction of each output transistor. The problem with this method is a lack of adjustment. If the forward threshold voltage of each diode is not exactly equal to the base-emitter junction voltage of each transistor, some crossover distortion can occur. If three diodes are used, the quiescent conduction current of each output transistor might be too high, resulting in excessive heating of the output transistors.

The amplified diode circuit could also be replaced with an adjustable biasing resistor for biasing purposes. Although this system will function well and eliminate crossover distortion, the adjustable resistor will not thermally "track" with the output transistors. As you might recall, bipolar transistors have a negative temperature coefficient, meaning they exhibit a decrease in resistance with an increase in temperature. In reference to transistors, a decrease in resistance actually means an increase in leakage current. In other words, bipolar transistors become more "leaky" when they get hot.

In bipolar transistor amplifiers, this is a major problem. As output transistors begin to heat up, the leakage current also increases, causing an increase in heat, causing an increase in leakage current, causing an additional increase in heat, causing an additional increase in leakage current, and so forth. This condition will continue to degrade until the output transistors breakdown. A breakdown of this nature is called *thermal runaway*.

A means of automatic thermal compensation is needed to correct the problem. An adjustable resistor cannot do this (most resistors have a positive temperature coefficient), but that is the beauty of an amplified diode circuit. Referring to Fig. 8-2, if Q2 is placed on the same heatsink as the output transistors, its temperature rise will closely approximate that of the output drivers. As the leakage current increases with a temperature rise in the output transistors, the leakage current through Q2 also increases. The increase of leakage current through Q2 causes the voltage drop across it to "decrease," resulting in a decrease of forward bias to the output transistors. The decrease in forward bias compensates for the increase in leakage current, thus resulting in good temperature stabilization.

## Additional amplifier classification

There are additional classes of amplifier operation, but they are not typically used for audio amplifiers. *Class-C amplifiers* are biased to amplify only a small portion of a half-cycle. They are used primarily in RF (radio frequency) applications, and their efficiency factors are usually about 80%.

*Class-D amplifiers* are designed to amplify "pulses," or square waves. A class D amplifier is strictly a "switching device," amplifying no part of an input signal in a linear fashion. Strangely enough, class-D amplifiers are available (although rarely) as audio amplifiers through a technique called *pulse width modulation (PWM)*. A *PWM audio amplifier* outputs a high frequency (about 100 kHz to 200 kHz) square wave to the audio speaker. Because this is well above human hearing, the speaker cannot respond. But the *duty cycle* (on-time/off-time ratio) of the square-wave output is varied according to the audio input signal. In effect, this creates a proportional "power signal," which the speaker does respond to and the audio input signal is amplified. Class-D audio amplifiers boast extremely high efficiencies, but they are expensive, and they have drawbacks in other areas. Class-D audio amplifiers are sometimes called *digital audio amplifiers*. Most class-D amplifiers are more commonly used for high-power switching and power conversion applications.

## Audio amplifier output configurations

The circuit illustrated in Fig. 8-2 has a complementary symmetry output stage. This term means that the output drivers are opposite types (one is NPN, and the other is PNP) but have symmetrical characteristics (same beta value, base-emitter forward voltage drop, voltage ratings, etc.). Generally speaking, most audiophiles consider this to be the best type of output driver design. Transistor manufacturers offer a large variety of "matched pair," or "complementary pair," transistor sets designed for this purpose.

Another common type of output design is called the quasi-complementary symmetry configuration. It requires a complementary symmetry "pre-driver" set, but the actual output transistors are of the same type (either both NPN, or both PNP, with NPN outputs being vastly more popular). This type of output design used to be a lot more popular than it is now. The current availability of a large variety of high-power, high-quality complementary transistor pairs has overshadowed this older design. However, it produces good-quality sound with only slightly higher-distortion characteristics than complementary symmetry.

## Audio amplifier definitions

The field of audio electronics is an "entertainment oriented" field. The close association between audio systems and the "arts" has led to a kind of semi-artistic aura surrounding the electronic and electromechanical systems themselves. As with any art form, personal preference and taste plays a major role. This is the reason why there are so many disputes among audiophiles regarding amplifier and speaker design. My advice is to simply accept what sounds good to you, without falling prey to current trends and fads.

Unfortunately, there have been many "scams" and sly "stigmas" perpetrated by unethical, get-rich-quick manufacturers over the years. This has led to much misunderstanding and confusion regarding the various terms used to define audio amplifier performance.

The most heavily abused characteristic of amplifier performance is "power." *Power*, of course, is measured in watts. The only standardized method of designating ac wattage, for comparison purposes, is by using the RMS value. Any other method of rating an amplifier's power output should be subject to suspicion.

Output power is also rated according to the speaker load. For example, an amplifier specification might rate the output power as being 120-watts rms into a 4-ohm load, and 80-watts rms into an 8-ohm load. You might expect the power output to double when going from an 8-ohm to a 4-ohm load, but there are certain physical reasons why this will not happen. However, when comparing amplifiers, be sure you compare "apples with apples;" an amplifier rated at 100-watts rms into an 8-ohm load is more powerful than an amplifier rated at 120-watts rms into a 4-ohm load.

The human ear does not respond in a linear fashion to differing amplitudes of sound. It is very fortunate for you that you are made this way, because the nonlinear ear response allows you to hear a full range of sounds; from the soft rustling of leaves to a jack-hammer pounding on the pavement. For example, a loud sound that is right on the threshold of causing pain to a normal ear is about 1,000,000,000,000 times louder than the softest sound that can be heard. Our ears tend to "compress" louder sounds, and amplify smaller ones. In this way, we are able hear the extremely broad spectrum of audible sound levels.

When trying to express differing sound levels, power ratios, noise content, and various other audio parameters, the non-linear characteristic of human hearing presents a problem. It was necessary to develop a term to relate linear mathematical ratios with non-linear hearing response. That term is the *decibel*. The prefix "deci" means 1/10 so that the term decibel actually means "one-tenth of a bel."

The bel is based on a "logarithmic" scale. Although I can't thoroughly explain the concepts of logarithms within this context, I can give a basic "feel" for how they operate. *Logarithms* are trigonometric functions, and are based on the number of decimal "columns" contained within a number, rather than the decimal values themselves. Another way of putting this is to say that a *logarithmic scale* is linearized according to "powers of ten." For example, the log of 10 is 1; the log of 100 is 2; the log of 1000 is 3. Notice, in each case, the log of a number is actually the number of weighted columns within the number minus the "units" column.

The *bel* is a ratio of a "reference" value, to an "expressed" value, stated logarithmically. A decibel is simply the bel value multiplied by 10 (bels are a little too large to conveniently work with).

In this case, I believe a good example is worth a thousand words. Assume you have a small radio with a power output of 100-milliwatts rms. During a party, you connect the speaker output of this radio into a power amplifier which boosts the output to 100-watts rms. You would like to express, in decibels, the power increase. The power level that you started with, 100 milliwatts (0.1 watt), is your reference value. Dividing this number into 100 watts gives you your ratio, which is 1000. The log of 1000 is 3 (bel value). Finally, multiply 3 by 10 (to convert bels to decibels), and the answer is 30 decibels.

Each 3-dB increase means a doubling of power. 6 dB gives four times the power [3 dB + 3 dB equates to: $2x$ power times $2x$ power; $2x(2x) = 4x$]. A 9-dB power increase converts to an $8x$ power increase [6 dB + 3 dB, or $4x(2x) = 8x$].

Each 10-dB increase equals a 10-fold increase. 13 dB yields $20x$ [10 dB + 3 dB, or $10x(2x) = 20x$]. 16 dB yields $40x$ [13 dB + 3 dB, or $20x(2x) = 40x$]. 20 dB yields $100x$ [10 dB + 10 dB, or $10x$ times $10x$]. And finally, as per the example above: a 30 dB power gain means a 1000-fold increase [10 dB + 10 dB + 10 dB = $10x^3 = 1000x$].

Also, please be aware that negative values of dBs represent negative gain, or *attenuation*. A $-3$-dB gain means that the power has been halved. Similarly, a $-10$-dB gain represents a 10-fold attenuation, or a $0.1x$ change in power output.

These figures represent *power logs*. *Voltage* and *current decibel logs* are somewhat different; the square root of the power logs. This is because power is voltage times current, $P = IE$. 30-dB volts is 31.620. 30-dB amps is also 31.620. Thus, $_{\log}P = {_{\log}}I \times {_{\log}}E = 31.62 \times 31.62 = 999.8x$. Most electronic reference books have dB log tables for easy reference. Just be aware that there is a difference between the power logs, and the voltage or current logs.

I recognize that if you have not been exposed to the concept of logarithms, or exponential numbering systems, this entire discussion of decibels is probably rather abstract. If you would like to research it further, most good electronics math books should be able to help you understand it in more detail.

*Dynamic range* is a term used to describe the difference (in decibels) between the softest and loudest passages in audio program material. In a practical sense, it means that if you are listening to an audio system at a 10-watt rms level, then for optimum performance, you would probably want about a 50-watt rms amplifier to handle the instantaneous high-volume passages that might be contained within the program material (a cymbal crash, for example). Compact disc and "hi-fi" video tape recorders offer the widest dynamic range commonly available in today's market.

*Frequency response* defines the frequency spectrum that an amplifier can reproduce. The normal range of human hearing is from 20 to 20,000 Hz (if you're a new-born baby, and had Superman as a father). In theory, there are situations occurring in music where ultrasonic frequencies are produced which are not audible, but without them, the audible frequencies are "colored" to some degree causing a variance from the original sound. For this reason, many high-quality power amplifiers have frequency responses up to 100,000 hertz. The high-end and low-end *frequency response limits* are specified from the point where the amplifier output drops to 50% ($-3$ decibels) of its rated output.

*Distortion* is a specification defining how much an amplifier changes, or "colors," the original sound. A perfect amplifier would be perfectly "linear," meaning the output would be "exactly" like the input, only amplified. However, all amplifiers distort the original signal by some percentage. Two decades ago, it was a commonly accepted fact that the human ear could not distinguish distortion levels below 1%. That has since been proven wrong. It is a commonly accepted rule-of-thumb today that even a trained ear has difficulty detecting distortion below 0.1%, although this figure is often disputed as being too high among many audiophiles. In any case, the lower the distortion specifications, the better.

Distortion is subdivided down into two more specific categories in modern audio amplifiers: harmonic distortion and intermodulation distortion. *Harmonic distortion* describes the nonlinear qualities of an amplifier. In contrast, *intermodulation distortion* defines how well an amplifier can amplify two specific frequencies simultaneously, without the frequencies interfering with each other in a nonlinear fashion. Typical ratings for both of these these distortion types is 0.1% or lower in modern high-quality audio amplifiers.

*Load impedance* defines the recommended speaker system impedance for use with the amplifier. For example, if the amplifier specification indicates the load impedance as 4 ohms or 8 ohms, you might use either a 4-ohm or 8-ohm speaker system (or any impedance in-between) as the output load for the amplifier.

*Input impedance* describes the impedance "seen" by the audio input signal. This value should be moderately high; 20 k$\Omega$ or higher.

*Sensitivity* defines the *rms* voltage level of the input audio signal required to drive the amplifier to full output power. Typical values for this specification are 1 to 2 volts rms.

*Signal-to-noise ratio* is a specification given to compare the inherent noise level of the amplifier with the amplified output signal. Random noise is produced in semiconductor devices by the recombination process occurring in the junction areas, as well as other sources. High-quality audio amplifiers incorporate various noise reduction techniques to reduce this undesirable effect, but a certain amount of noise will still exist and be amplified right along with the audio signal. Typical signal-to-noise ratios are $-70$ to $-92$ decibels, meaning that the noise level is 70 to 92 decibels below the maximum amplifier output. $-70$ dB$_p$ is $10^{-7}$!

There are additional specifications that might or might not be given in conjunction with audio amplifiers, but the previous terms are the most common and the most important.

Before proceeding, here is a note of caution. Research has proven that continued exposure to high-volume noise (meaning music or any other audio program material) causes degradation of human hearing response. It saddens me to hear young people driving by in their cars with expensive audio systems blasting out internal sound pressure levels at 120 decibels. Even relatively short exposures to this level of sound can cause them to develop serious nerve-deafness problems by the time they're middle-aged. Of course, exposure to high-volume levels at any age is destructive. It isn't worth it. **Keep the volume down** to reasonable levels for your ear's sake.

## Power amplifier operational basics

Now that some of the basic audio terms have been established, this section will concentrate on the "front end" of modern audio amplifier design. In addition, this section establishes some of the basics relating to integrated circuit "operational amplifiers," which will be discussed later in this book.

Fig. 8-3 is a kind of semi-block diagram illustrating the input stage of most high-power audio amplifiers. Q1, D1, D2, and P1 form a circuit called a *constant current source*. For discussion's sake, assume D1 drops the same voltage as the base-emitter junction of

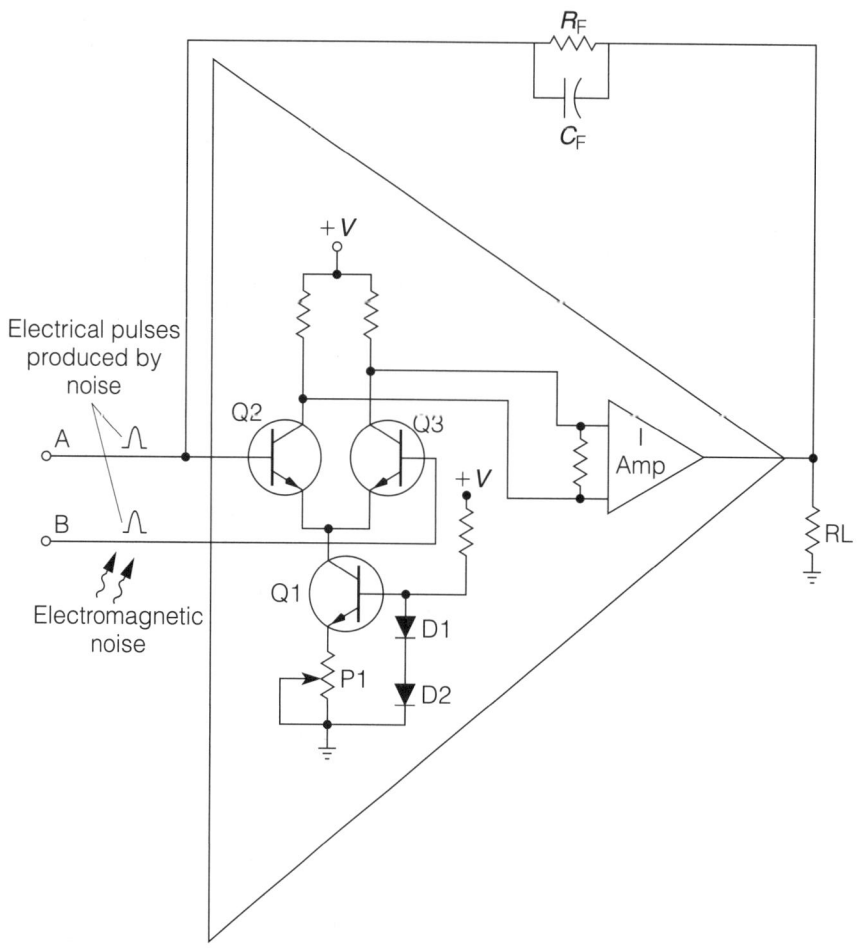

■ 8-3 *Power amplifier operational basics.*

Q1 (which should be a close assumption). That would mean that the voltage drop across D2 would also be the voltage drop across P1. If the voltage drop across D2/P1 is 0.6 volt, and P1 is adjusted to be 600 ohms, the emitter current flow will be 1 milliamp. Because the collector current of Q1 will approximately equal the emitter current, the collector current is also "held" at 1 milliamp. The important point to note here is that the collector current is regulated; it is not dependent upon the collector load or the amplitude of the source voltage. The only variables controlling the collector current are D2's forward threshold voltage and the setting of P1. Therefore, it is appropriately named a *constant current source*.

Q2 and Q3 form a differential amplifier. Think of a differential amplifier as being like a "see-saw" in a school playground. As long as everything is balanced on a see-saw, it stays in a horizontal position. If something unbalances it, it tilts, causing one end to go up proportionally, as the other end goes down. This is exactly how a differential amplifier operates with the current flow. As discussed previously, it is assumed that the constant current source will provide a regulated 1 milliamp of current flow to the emitters of Q2 and Q3. If Q2 and Q3 are in a balanced condition, the 1 milliamp of current will divide evenly between each transistor, providing 0.5 milliamp of current flow through each collector. If an input voltage is applied between the two base inputs (A and B) so that point A is at a different potential than point B, the balance will be upset. But as the collector current rises through one transistor, it must decrease by the same amount through the other, because the constant current source will not allow a varying "total" current. For example, if the differential voltage between the inputs caused the collector current through Q2 to rise to 0.6 milliamp, the collector current through Q3 will fall to 0.4 milliamp. The total current through both transistors still adds up to 1 milliamp.

Notice that the output of the differential amplifier is not taken off of one transistor in reference to ground; the output of a differential amplifier is the difference between the two collectors. Now let's discuss the advantages of such a circuit.

Assume the source voltage (supplied externally) increases. In a common transistor amplifier, an increase in the source voltage will cause a corresponding change throughout the entire transistor circuit. In Fig. 8-3, an increase in the source voltage (+V) does not cause an increase in current flow from the constant current source, because it is regulated by the forward voltage drop across D2 which doesn't change (by practical amounts) with an increase in current. Q2 and Q3 would still have a combined total current flow of 1 milliamp. The collector voltages of Q2 and Q3 would increase, but they would increase by the same amount, even if the circuit was in an unbalanced state.

Therefore, the voltage differential between the two collectors would not change. For example, assume Q2's collector voltage is 6 volts, and Q3's collector voltage is 4 volts. If you used a voltmeter to measure the difference in voltage between the two collectors, it would measure 2 volts (6 − 4 = 2). Now assume the source voltage increased by an amount sufficient to cause the collector voltages of Q2 and Q3 to increase by 1 volt. Q2's collector voltage would rise to 7 volts, and Q3's would rise to 5 volts. This didn't change the voltage differential between the two collectors at all; it still remained at 2 volts. In other words, the output of a differential amplifier is immune to power supply fluctuations. This not only applies to gradual changes in dc levels, but the effect works just as well with power supply ripple and other sources of undesirable noise signals that might enter through the power supply.

One of the most common problems with high-gain amplifiers is noise and *interference signals* being applied to the input through the input wires. Input wires and cables can pick up a variety of unwanted signals, just as an antenna is receptive to radio waves. If you have ever touched an uninsulated input to an amplifier, you undoubtably heard a loud 60-Hz roar (called "hum") through the speaker. This is because your body picks up electromagnetically radiated 60-Hz signals from power lines all around you. Fluorescent lights are especially bad electromagnetic radiators. Figure 8-3 illustrates an example of some electromagnetic radiation causing noise pulses on the A and B inputs to the differential amplifier. Because electromagnetic radiation travels at the speed of light (186,000 miles per second), the noise pulses will occur at the same time, and in the same polarity. This is called *common mode interference*. A very desirable attribute of differential amplifiers is that they exhibit *common mode rejection*. The noise pulses illustrated in Fig. 8-3 would not be amplified.

To understand the principle behind common mode rejection, assume that the positive going noise pulse on the A input is of sufficient amplitude to "try" to cause a 1 volt decrease in Q2's collector voltage. Because the noise is common mode (as is all externally generated noise), an identical noise pulse on the B input is "trying" to cause Q3's collector voltage to decrease by 1 volt also. If both collectors decreased in voltage at the same time, it would require an increase in the combined total current flow through both transistors. This can't happen because the constant current source is maintaining that value at 1 milliamp.

Therefore, neither transistor can react to the noise pulse, and it is totally rejected. (Even if both transistors did react slightly, they would react by the same amount. Because the output of the pair is the difference across their collectors, a slight reaction by both at the same time would not affect their differential output.) This goes back to the analogy of the see-saw I made earlier. If a see-saw is balanced and you placed equal weights on both ends at the same time, it would simply remain stationary. In contrast, the desired signal voltage to be amplified is not common mode. For example, the B input might be at signal ground while the A input is at 1-volt

RMS. Differential amplifiers respond very well to differential signals. That is why they are called *differential amplifiers*.

One final consideration of Fig. 8-3 is in reference to $R_F$ and $C_F$. Notice that this resistor/capacitor combination is connected from the output back to one of the inputs. The process of applying a percentage of the output back into the input is called *feedback*. In high-gain amplifiers, this feedback is almost always in the form of *negative feedback*, meaning the feedback is 180 degrees out-of-phase with the input. This has the effect of canceling out part of the input signal, and reducing the overall gain. Negative feedback is necessary to temperature stabilize the amplifier, flatten out the gain, increase the frequency response, and eliminate oscillations. Various combinations of resistors and capacitors are chosen to tailor the frequency response, and to provide the best overall performance. More about feedback will be discussed in later chapters.

## Building high-quality audio systems

You don't have to be an electrical engineer to build much of your own high-quality audio equipment. Even if your interests don't lie in the audio field, you are almost certain to need a lab-quality audio amplifier for many related fields. In this section, I have provided a selection of audio circuits that are time-proven, and which provide excellent performance.

Figure 8-4 is a block diagram of a typical audio amplification system. It is mostly self-explanatory, with the exception of the volume control and the two 10 µF capacitors. The volume control potentiometer should have an audio taper (logarithmic response). 100 kΩ is a typical value. For stereo applications, this is usually a "two-ganged" pot, with one pot controlling the right channel and one pot controlling the left. The two 10-µF capacitors are used to block unwanted dc shifts that might occur if the volume control is rotated too fast.

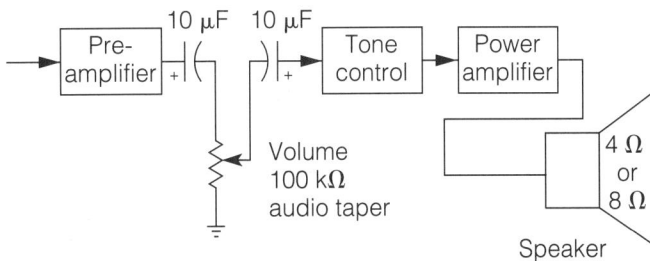

■ **8-4** *Block diagram of a typical audio amplification system.*

Figure 8-5 is a simple preamplifier circuit for use with high-impedance signal sources, such as crystal or ceramic microphones. It is merely a common-collector amplifier with a few refinements. R5 and C4 are used to "decouple" the circuit's power source. The simple RC filter formed by R5 and C4 serves to isolate this circuit from any effects of other circuits sharing the same power supply source. R3 and C2 provide some positive feedback (in-phase with the input) called "bootstrapping." *Bootstrapping* has the effect of raising the input impedance of this circuit to several megohms.

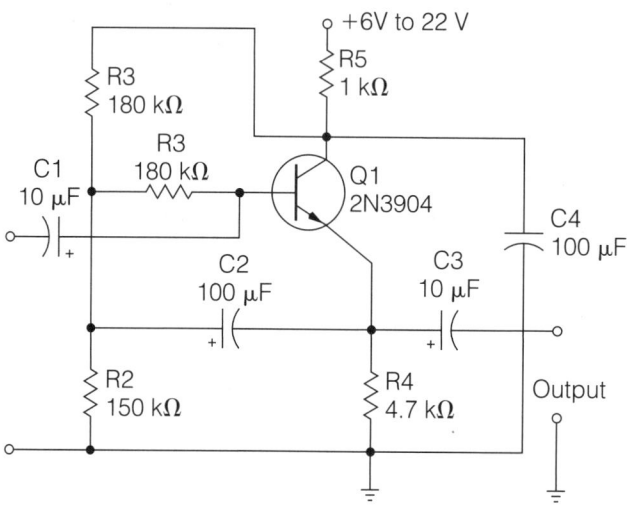

■ **8-5** *Preamplifier for use with high-Z signal sources.*

Figure 8-6 is a *high-gain preamplifier circuit* for use with very low input signal sources. Dynamic microphones, and some types of musical instruments (such as electric guitars), work well with this type of circuit. R1 provides negative feedback for stabilization and temperature compensation purposes. Notice that this circuit is also decoupled by R5 and C2.

Figure 8-7 is an *active tone control circuit* for use with the outputs of the previous preamplifier circuits, or any "line-level" output. Active tone controls incorporate the use of an active device (transistor, FET, operational amplifier, etc.), and can provide better overall response with gain. *Passive tone controls*, in contrast, do not use any active devices within their circuits, and will always "attenuate" (reduce) the input signal. *Line-level outputs* are signal voltages that have already been preamplified. The audio out-

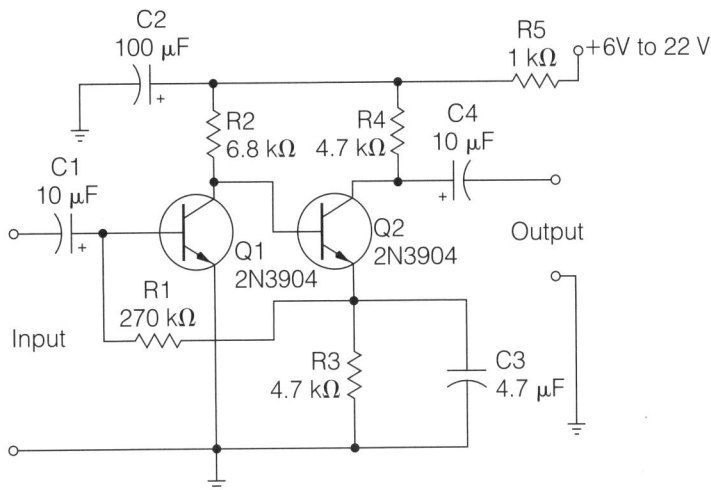

**8-6** High-gain preamplifier.

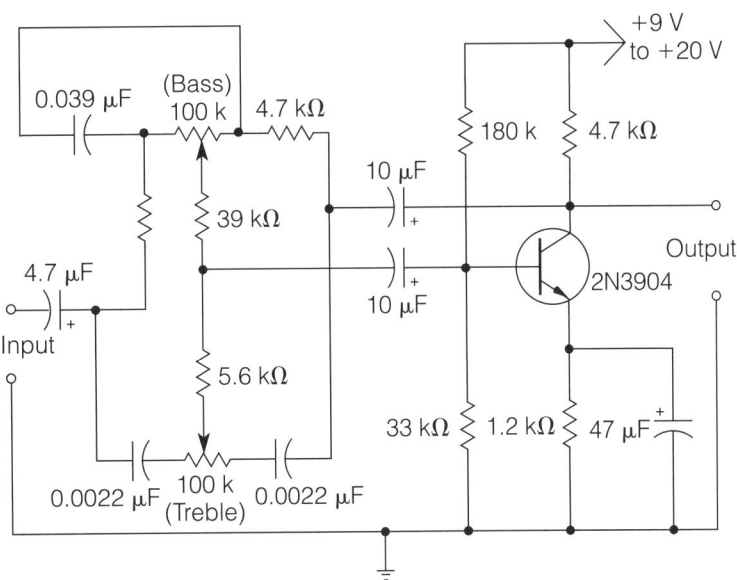

**8-7** Active tone control circuit.

puts from CD players, VCRs, tape players, and other types of consumer electronic equipment are usually line outputs. Line outputs are connected to the *line inputs* of audio amplifiers, bypassing any preamplification circuitry.

Figure 8-8 is a *12-watt rms audio amplifier*. It is relatively easy to build, provides good linearity characteristics, and it operates from a single dc supply (many audio amplifiers require a dual-polarity power supply). The complete circuit can be assembled on a small universal perfboard; with the exception of Q2, Q5, and Q6, which should be mounted to a small heatsink (be sure to use the appropriate transistor insulators, heatsink compound, and mounting hardware to electrically isolate the transistors). The parts list for this amplifier is given in Table 8-1.

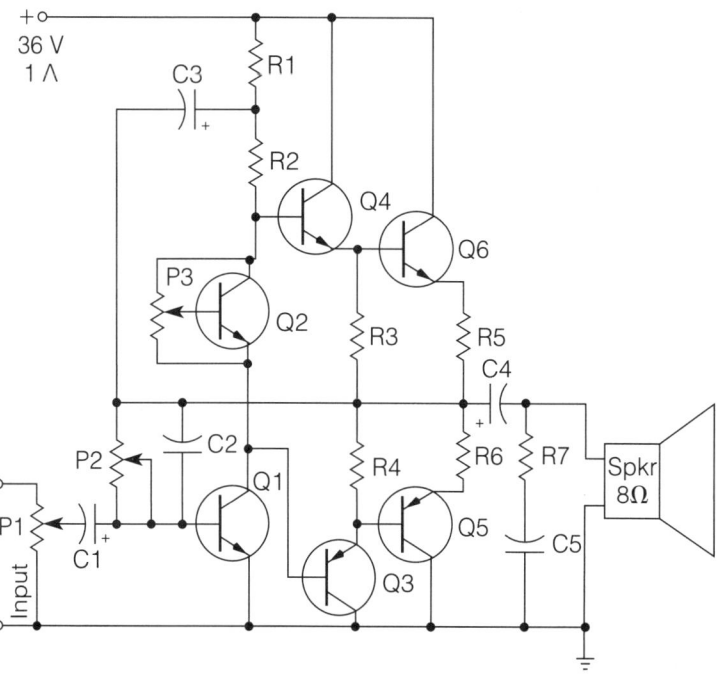

■ **8-8** *12-watt rms audio power amplifier.*

The *initial set-up* for the amplifier is as follows. Before applying power, set P2 and P3 to mid-position. Apply 36 Vdc. Adjust P2 to obtain 1/2 of the power supply voltage (about 18 Vdc) between the positive side of C4 and ground. Connect a speaker and a line-level input signal source from a radio, tape player, etc. Turn P1 down until you can hardly hear any sound coming from the speaker. At low volume levels, one extreme position of P3 should produce some audible crossover distortion (crossover distortion is more easily heard at low volume levels). Starting at this extreme position of P3, begin to slowly rotate it in the opposite direction until you reach the point where the distortion is no longer audible.

■ **Table 8-1 Parts list for 12 watt RMS audio amplifier**

### 12 WATT AUDIO AMPLIFIER PARTS LIST

| Part designation | Definition | Description |
|---|---|---|
| Q1,Q2 | NPN transistor | 2N2904 or most general-purpose NPN transistors |
| Q4,Q6 | NPN transistor | TIP31C |
| Q3,Q5 | PNP transistor | TIP32C |
| P1 | Potentiometer | 10 k$\Omega$, audio taper |
| P2 | Trimpot | 1 M$\Omega$, linear |
| P3 | Trimpot | 1 k$\Omega$, linear |
| C1 | Electrolytic cap. | 50 $\mu$F, 16 WVdc |
| C2 | Capacitor | 10 pF |
| C3 | Electrolytic cap. | 20 $\mu$F, 50 WVdc |
| C4 | Electrolytic cap. | 2000 $\mu$F, 50 WVdc |
| C5 | Capacitor | 0.05 $\mu$F, 50 V |
| R1 | Resistor | 1 k$\Omega$, 1/2 W |
| R2 | Resistor | 3.9 k$\Omega$, 1/2 W |
| R3,R4 | Resistor | 680 $\Omega$, 1/2 W |
| R5,R6 | Resistor | 0.47 $\Omega$, 2 W (can use two 1 $\Omega$, 1 W resistors in parallel) |
| R7 | Resistor | 10 $\Omega$, 1/2 W |
| Miscellaneous | – – – | Transistor insulators, mounting hardware, heatsink compound |

After the initial set-up, turn the volume up and allow the amplifier to operate "loudly" for a while. Keep checking the temperature of the output transistors. If they should become too hot, re-check your initial set-up. If this doesn't help the problem, you might have to use a larger heatsink. (If you have mounted this amplifier in a small aluminum project box, the project box itself should make an adequate heatsink.)

The circuit description for this amplifier is almost identical to the description given in Fig. 8-2, with a few minor changes. Q4 and Q3 are added to increase the overall current gain of the circuit. They are referred to as the *pre-drivers*. P2 provides dc feedback to set the correct bias for Q1. C2 limits the maximum frequency response to eliminate oscillation, or "ringing" (decaying oscillation), problems. C3 is a bootstrap capacitor to improve ac gain characteristics. R7 and C5 form an RC network that improves stability under loaded conditions.

This amplifier, when enclosed in a suitable housing, makes a good lab audio amplifier for circuit testing purposes.

## Building a professional-quality audio amplifier

A professional-quality audio amplifier is considerably more complex than the 12-watt amplifier in Fig. 8-8. Even if you decide not to build this amplifier, it is worth your time and effort to understand it.

Figure 8-9 is a schematic diagram of a 50-watt professional quality audio power amplifier. Q1, Q2, and Q3 are "junction field-effect transistors," or JFETs. JFETs will be discussed in a later chapter, but for now, simply consider their operation to be similar to bipolar transistors. Q1's circuitry makes up a constant current source supplying current to the high-gain differential amplifier consisting of Q2, Q3, Q4, and Q5 and their associated circuitry. The output of this first differential stage is directly coupled to a second differential amplifier stage consisting of Q7 and Q8. These two differential amplifiers provide high-voltage gain, and almost complete immunity to power-supply variations and power supply noise. They also

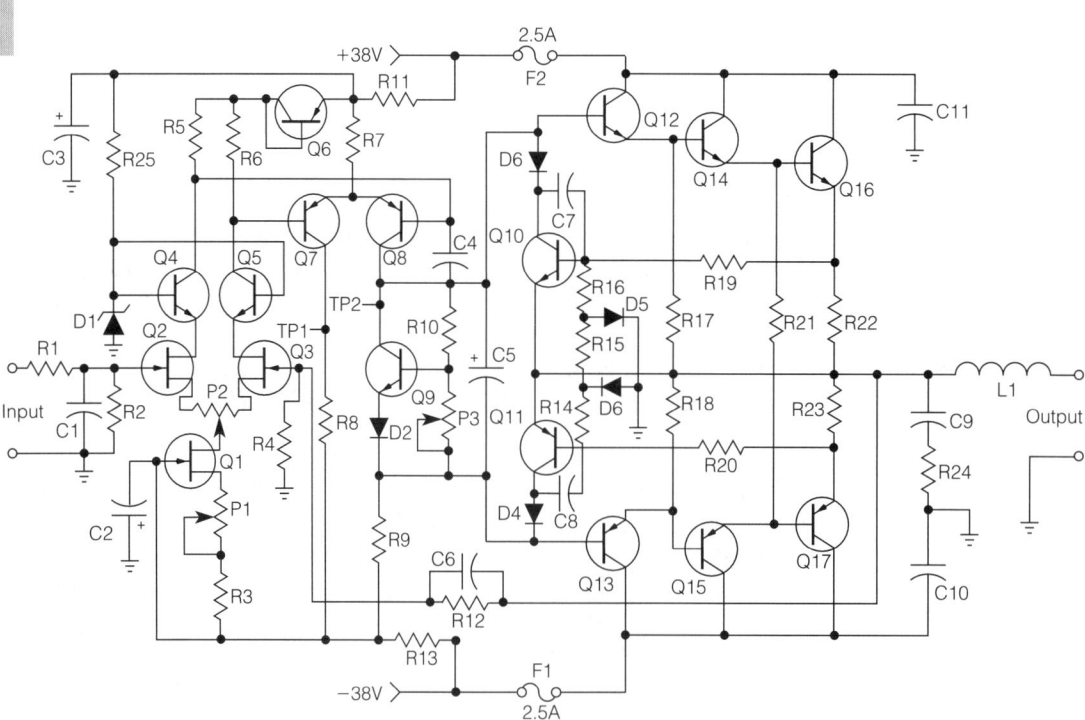

■ **8-9** *Schematic diagram of a professional-quality audio power amplifier.*

reduce inherent noise levels. Notice that these stages are *direct coupled* (no coupling capacitors from stage to stage) which is desirable to increase frequency response and reduce undesirable phase-shift problems.

Q9, D2, R10, and P3 form an amplified diode circuit. Three complementary stages are utilized to increase the current gain to the desired level. The pre-driver transistors are Q14 and Q15. The output driver transistors, Q16 and Q17, are the Toshiba 2SC3280 and 2SA1301 complementary pair. You might recognize these transistors as being the same type recommended for the lab power supply circuit. I respect these transistors as being some of the best available for this type of application.

Q10 and Q11 make up the heart of a circuit that protects the output transistors against short-circuit conditions. They are connected to monitor the output voltage across R22 and R23. Notice that all of the output current to the speaker system must flow through R22 and R23. In the event of a short-circuit on the output, the output current will try to rise to a high level, causing the voltage drop across R22 and R23 to rise proportionally. When this voltage reaches a level high enough to bias Q10 and Q11 into the conductive state, they reduce the drive signal to Q12 and Q13, thus limiting the output current to a safe level. As soon as the output short-circuit is removed, the amplifier returns to normal operation.

Feedback for dc voltage stabilization is provided by R12. C6 provides negative ac feedback to limit the frequency response. When I used a 150-pF capacitor for C6, the frequency response began to drop off at 60 kHz. However, by going as low as 40 picofarads, I achieved a frequency response well over 100 kHz, with no apparent stability problems.

A final point to notice about this design is that the speaker is directly coupled to the output. This eliminates the need for a very large output coupling capacitor and improves the low-end frequency response.

I have seen this same exact design used in virtually hundreds of top-quality amplification systems; both for home entertainment and commercial applications. Naturally, different manufacturers will use different transistor pairs and various component values, but the basic design remains the same. However, there is one difference between this design and most commercially available amplifiers. Because I didn't design this amplifier to be mass produced, I didn't worry about a cost savings of a few dollars per unit. Therefore, the critical parts have been over-rated, increasing the long-term reliability. After building the prototype of this amplifier, I tested it in excess of 110-watts rms (by reducing the value of R22 and R23 down to 0.22 ohms) with no overheating problems or loss of stability. There are some advantages to be had by building your own stuff!

### Building a printed circuit board

If you would like to build this amplifier but are a little bit "shaky" about etching a circuit board and finding all of the needed parts, it is available in kit form from Seal Electronics, listed in the appendix of this book.

Etching a printed circuit board is not that difficult for this type of project. This section describes the easiest way to accomplish this if you are a novice. To begin, you will need to purchase some materials. I recommend an inexpensive "resist ink" type kit that should contain several unetched "single-sided" PC boards, a bottle of ferric chloride etchant solution, a "resist ink" pen, and instructions. In addition, you will need a #61 drill bit (about the same diameter as the leads on a 5-watt carbon resistor), a hand drill, some fine 600 grit emery paper, a small pin punch, a tack hammer, and a small glass tray.

Start by making a "good" copy of the full-size layout diagram illustrated in Fig. 8-10. Cut a piece of PC board material to the same size as the illustration. Cut out the illustration from the copy and tape it securely on the foil side of the PC board. Using a small pin punch and tack hammer, make a dimple in the PC board copper at each spot where a hole is to be drilled (be sure not to miss the holes needed in the wide foil areas). When finished, you should be able to remove the copy and find a dimple in the copper corresponding to every hole shown in the diagram. Next, drill a #61 hole through the PC board at each dimple position. When finished, hold the PC board up to a light, with the diagram placed over the foil side, to make certain you haven't missed any holes and that all of the holes are drilled in the right spot. If everything looks good, lightly sand the entire surface of the copper foil with 600 grit emery paper to remove any burrs and surface corrosion.

Using the resist ink pen, draw a "pad" area around every hole. Make these pads very small for now; you can always go back and make selected ones larger, if needed. Using single lines, connect the pads as shown in the diagram. This last step is easy if you do it right. Being sure you have the board turned correctly to match the diagram, start at one end and connect the simplest points first. Us-

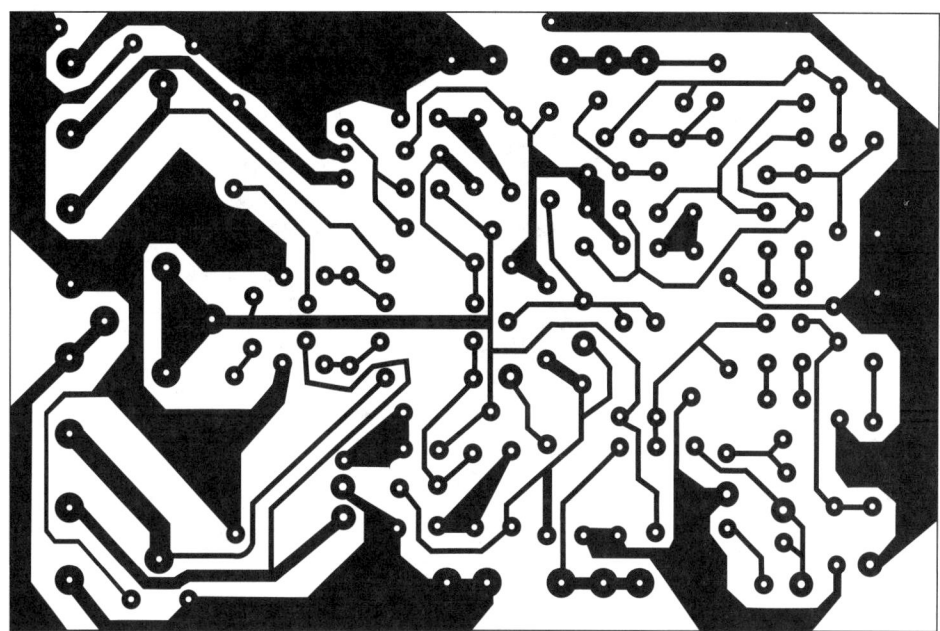

■ **8-10** *Full-size artwork diagram of audio amplifier.*

ing these first points as a reference, eventually proceed on to the more difficult connections. When finished, you'll have a diagram that looks like a "connect the dots" picture in a coloring book. Finally, go back and "color" in the wide foil areas and fill in the wider tracks. The process is actually easier than it looks. If you happen to make a major mistake, just remove all of the ink with ink solvent or a steel wool pad and start over again.

When you're satisfied that the ink pattern on the PC board corresponds "electrically" with the diagram in Fig. 8-10, place it in a small glass or plastic tray, and pour about an inch of etchant solution over it (Fig. 8-11). **Be careful with the etchant solution**; it permanently stains everything it comes in contact with. Wear goggles to protect your eyes, and don't breathe the fumes. After about 15 to 20 minutes, check the board using a pair of tongs to lift it out of the etchant solution. Continue checking it every few minutes until all of the unwanted copper has been removed. When this is accomplished, wash the board under cold water, and remove the ink with solvent or steel wool.

## Amplifier assembly

The parts needed to build this professional quality amplifier are listed in Table 8-2. All of these components should be relatively

■ **8-11** *Illustration of the same PC board kit the author used to etch the PC board for the professional quality audio amp.*

easy to obtain from a variety of sources, and the list is self-explanatory.

When I designed the PC board, my intentions were to use an NTE 287 for Q12 and an NTE 288 for Q13. While inserting the parts into the PC board, I ran out of these NTE transistors. My supplier didn't have them readily available in stock. So, in my impatience to test the design, I substituted a TIP31C and TIP32C pair in their place. These substitutions worked well, although they were a little difficult to stick into the smaller space on the board. You can make the same substitution also, if desired.

If you do not have a signal generator and an oscilloscope for performing waveshape analysis on the finished amplifier, I suggest using a 150-picofarad capacitor for C6. Reducing the value of C6 will increase the upper frequency response.

If you have trouble finding an 8-ohm, 1-watt resistor for R24, you might parallel a 10-ohm, 1-watt resistor and a 33-ohm, 1/2-watt resistor in its place.

L1 is a homemade air core coil. Find something round that is 1/2 inch in diameter to use as a form. Wind 12 loops around this form using 20-gauge insulated "magnet wire." *Magnet wire*, or "coil wire," is solid copper wire with a coating of varnish on the outside serving as the insulator. Remove the form, and push the loops tightly together. Bend the ends of the coil straight down for insertion into the PC board, and scrape the varnish off of the ends to allow for soldering.

■ Table 8-2 Parts list for 50 watt audio amplifier

## LAB QUALITY AUDIO AMPLIFIER PARTS LIST

| Part designation | Definition | Description |
|---|---|---|
| Q1,Q2,Q3 | N-channel JFET | MPF102, Radio Shack #276-2062, NTE 133 |
| Q4,Q5 | NPN transistor | 2N3904, NTE 123AP |
| Q6,Q7,Q8 | PNP transistor | 2N3906, Radio Shack #276-1604, NTE 159 |
| Q9,Q14 | NPN power transistor | TIP31C, NTE 291 |
| Q10,Q12 | NPN transistor | NTE 287 (See text) |
| Q11,Q13 | PNP transistor | NTE 288 (See text) |
| Q15 | PNP power transistor | TIP32C, NTE 292 |
| Q16 | NPN power transistor | 2SC3280, NTE 2328 |
| Q17 | PNP power transistor | 2SA1301 NTE 2329 |
| D1 | Zener diode | 24 V, 1 W |
| D2-D6 | Switching diode | NTE 519 or equivalent |
| C1 | Capacitor | 100 pF |
| C2,C3 | Electrolytic cap. | 100 $\mu$F @ 50 WVdc |
| C4 | Capacitor | 5 pF to 10 pF |
| C5 | Electrolytic cap. | 10 $\mu$F @ 16 WVdc |
| C6 | Capacitor | 40 pF to 150 pF (see text) |
| C7,C8,C9 | Capacitor | 0.047 $\mu$F |
| C10,C11 | Capacitor | 0.1 $\mu$F |
| P1 | Trimpot | 5 k$\Omega$, single turn |
| P2 | Trimpot | 200 $\Omega$, single turn |
| P3 | Trimpot | 1 k$\Omega$, single turn |
| R1,R7,R21 | 1/2-W resistor | 150 $\Omega$ |
| R2 | 1/2-W resistor | 100 k$\Omega$ |
| R3,R11,R13,R19,R20 | 1/2-W resistor | 100 $\Omega$ |
| R4 | 1/2-W resistor | 470 $\Omega$ |
| R5,R6,R25 | 1/2-W resistor | 3.3 k$\Omega$ |
| R8,R9 | 1-W resistor | 10 k$\Omega$ |
| R10 | 1/2-W resistor | 1.2 k$\Omega$ |
| R12 | 1/2-W resistor | 12 k$\Omega$ |
| R14, R16 | 1/2-W resistor | 1 k$\Omega$ |
| R15 | 1/2-W resistor | 6.8 k$\Omega$ |
| R17,R18 | 1/2-W resistor | 820 $\Omega$ |
| R22,R23 | 5-W resistor | 0.47 $\Omega$, Radio Shack #271–130 |
| R24 | 1-W resistor | 8 $\Omega$ (See text) |
| L1 | Coil | 1 $\mu$H (See text) |
| F1,F2 | Fuse | 2.5 A |
| (Not Illustrated) | Fuse blocks | 2 needed |
| (Not Illustrated) | Large heatsink | 1 needed (See text) |
| (Not Illustrated) | Small heatsink | 2 needed (See text) |
| (Not Illustrated) | Transistor insulator | 3 needed (See text) |

*Building high-quality audio systems*

F1 and F2 are intended to be located in close proximity to the PC board, so two fuse blocks will be needed. A small TO220-type heatsink should be mounted to the back of Q14 and Q15; space has been made on the PC board for this purpose.

A large, "flat-on-one-side" heatsink must be used to help dissipate the considerable heat produced by Q16 and Q17. I used a heatsink measuring 9" × 4" × 3/4" for this application. Q9, Q16, and Q17 are mounted to this heatsink (using insulators and heatsink compound), and the PC board can then be mounted over the top of the output transistors on the same heatsink. Spacers should be used; giving at least 1/2" clearance between the top of the output transistors and the bottom of the PC board. The fuse blocks (for F1 and F2) can also be mounted to this heatsink.

Figure 8-12 illustrates a top-side view of how the components are to be mounted on the PC board. The jumper wire is standard 22-gauge insulated hook-up wire. Take your time installing these components (Figs. 8-13 and 8-14), and continually doublecheck your work, using data books for lead identification and the schematic diagram (Fig. 8-9) as an electrical reference. Pay particular attention to capacitor and diode polarities.

There are two test point holes in the PC board. Using a small length of scrap component lead wire, form a test point contact by bending a tiny portion of the bottom of the scrap wire to a 90-degree angle. Insert the top of the wire through the PC board and

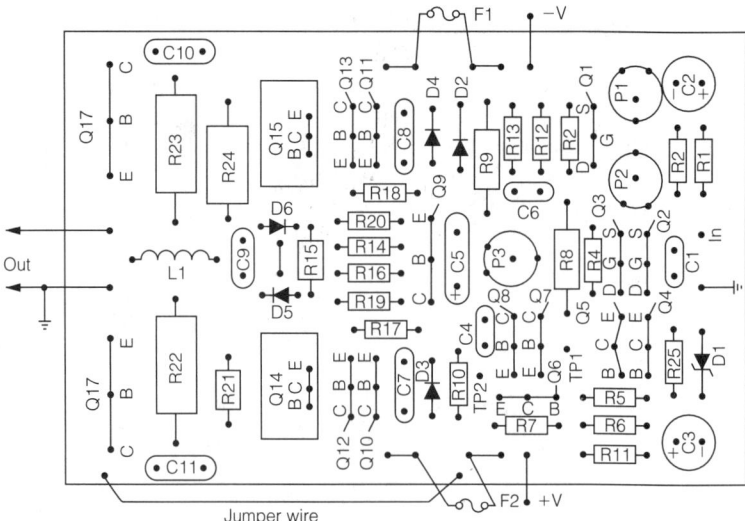

■ **8-12** *Top-side component placement.*

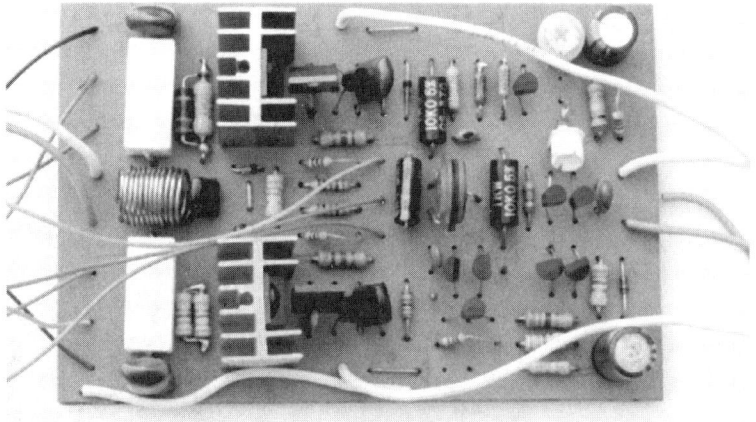

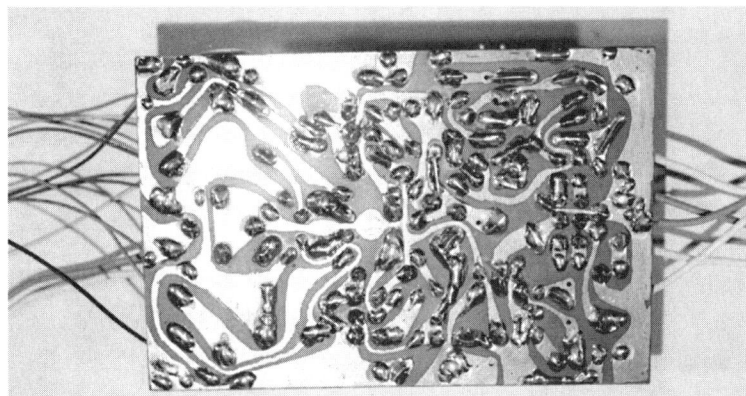

■ **8-13** *Top and bottom view of the finished PC board for the professional quality audio amp.*

solder the bent section to the foil. These two test points will be used for initial set-up.

## Initial set-up

This amplifier requires a dual-polarity power supply, such as the recommended power supply illustrated in Fig. 8-15. The power supply common should be connected to the output common point on the PC board, with the positive and negative connections made through their associated fuses to their designated points on the top of the board.

P1, P2, and P3 should be set to their approximate center positions. Apply dc power, and check F1 and F2. If either of these fuses blow, a wiring error exists, or one of the output drivers is shorted. If

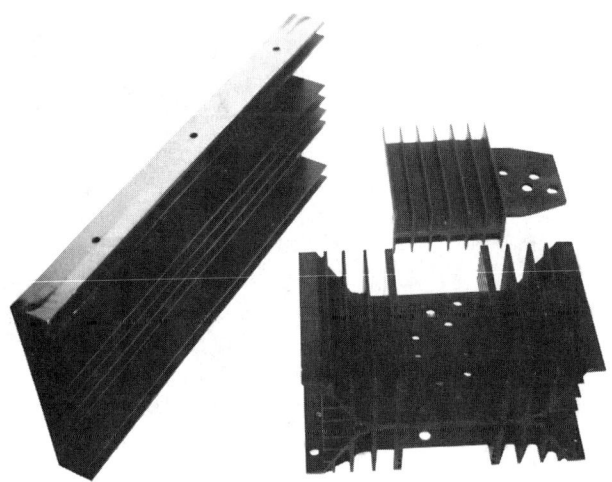

■ **8-14** *Some examples of larger heatsink styles. For the professional-quality audio amplifier, the author used the largest one, shown on the left.*

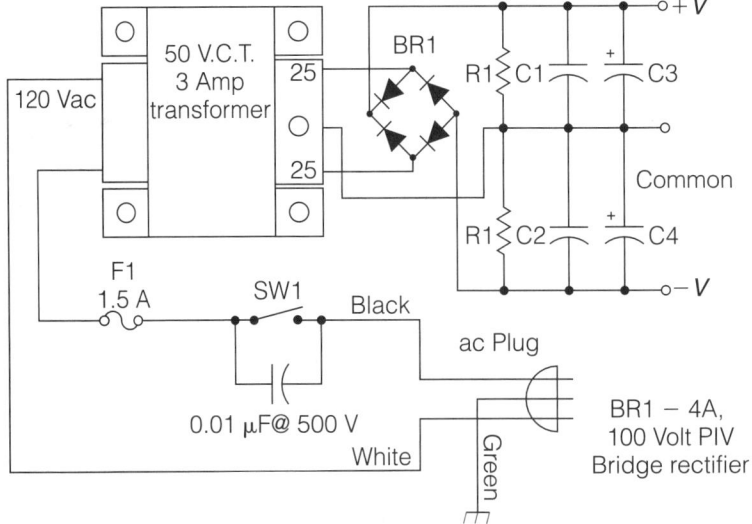

■ **8-15** *Recommended power supply for 50-watt audio amplifier.*

everything looks good, don't leave the power applied for very long until the balance adjustments have been made.

Connect one lead of your DVM (set to measure dc volts on the 200-volt range) to test point one (TP1), and the other lead to test point two (TP2). Apply dc power, and adjust P1 until the dc voltage reading is around 0.1 volt (you should be able to vary this voltage in either polarity by adjusting P1). Next, connect the DVM between circuit common and either side of L1. Adjust P2 for the lowest possible dc voltage (0.1 Vdc or lower). The adjustment of P1 and P2 is somewhat interactive, so you will need to repeat these two steps several times until both dc voltage readings are around 0.1 volt, or lower. When this is accomplished, the amplifier is electrically balanced. Turn off the dc power.

The adjustment of P3 will require a line-level signal source and a speaker. Connect the speaker to the output of the amplifier, and the signal source to the input (be sure there is no dc voltage present on the signal source; if in doubt, connect a 10-$\mu$F capacitor between the signal source and the input to the amplifier). Turn the signal source down as low as it will go, and apply power to the amplifier. Using your DVM, measure the dc voltage across R21; it should only be a couple of volts. Adjust P3 for the minimum voltage across R21 (about 1 volt). Turn the signal source up until you can just hear it through the speaker. There should be some audible crossover distortion detected. Slowly increase P3 until the distortion is gone (do not adjust P3 any higher than this point). When this is completed, the amplifier is ready to use.

## Amplifier power supply

The recommended power supply for the professional quality audio amplifier is illustrated in Fig. 8-15. The power transformer is a 50-volt c.t., 3-amp transformer. BR1 is a 4-amp, 100-volt PIV bridge rectifier. R1 and R2 are 4.7-k$\Omega$, 1/2-watt resistors. C1 and C2 should be about 0.1-$\mu$F, 100-volt ceramic disc capacitors, but this value is not critical. C3 and C4 are 4700-$\mu$F, 50-WVdc electrolytic capacitors.

## Speaker protection circuit

Most high-quality, high-power audio amplifiers have a common problem; the potential risk of applying a dc voltage to the speaker. The use of a coupling capacitor on the output of a modern direct-coupled amplifier defeats many of the high-quality attributes inherent with this amplification system (primarily low frequency response). Certain types of component failures or long-term component value shifts can place destructively high dc potentials directly across the speaker. Another less-serious problem with direct-coupled amplifiers is called *turn-on thumps*. When power is first applied to an amplifier of this type, it might experience a short period of instability while the capacitors (in the power supply) charge and the amplifier itself reaches the point of balance.

This short-term state of imbalance can apply a transient dc pulse to the speaker causing an audible "thump." Typically, this is more annoying than destructive.

Figure 8-16 is a speaker protection circuit. I included this circuit for two reasons. First, it is a very useful and practical circuit to incorporate within many audio systems. Secondly, it is an extremely good example of how many of the circuit "building blocks" that have been discussed thus far can be put together to form more complex circuits.

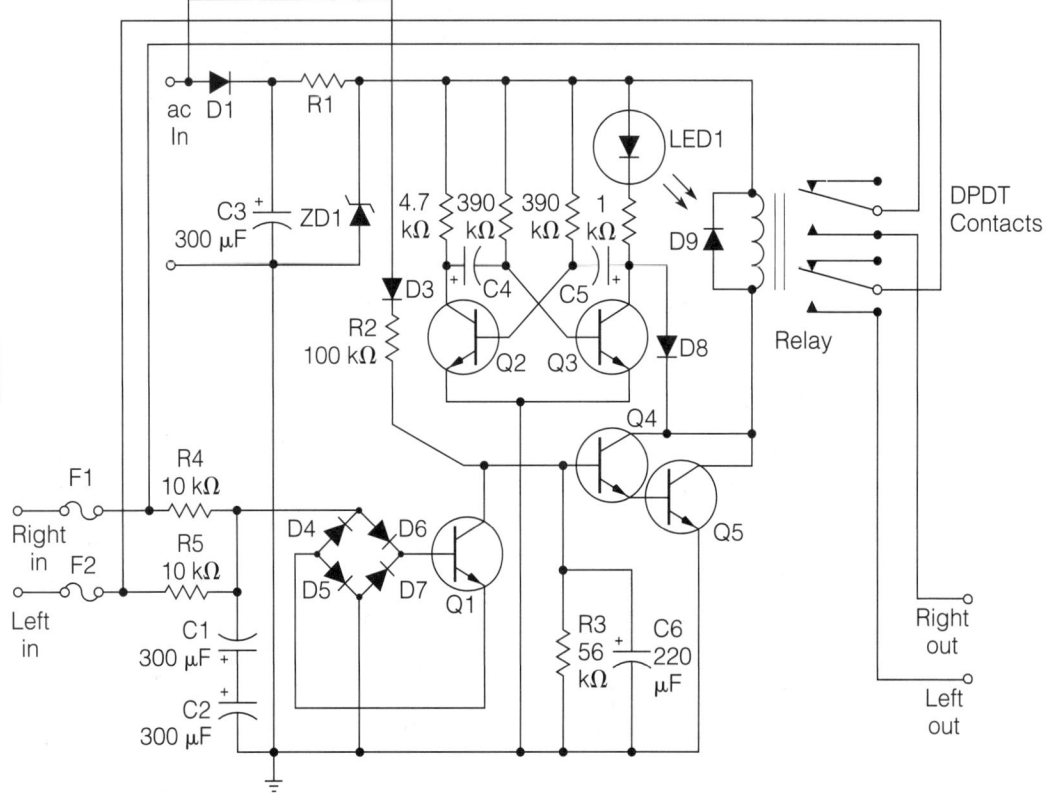

■ **8-16** *Speaker protection circuit.*

D1, C3, R1, and ZD1 form a simple half-wave rectified, zener-regulated power supply. Notice that the negative side of the supply goes to circuit common, and the positive side is the applied circuit power for the remainder of the overall circuit.

D3, R2, R3, C6, Q4, Q5, D9, and the relay form a time-delay relay (TDR). When ac power is first applied to the ac input, D3 is only forward-biased during the positive half-cycles. The positive pulses are applied to the RC circuit of R2 and C6. Because of the RC time constant, it takes several seconds for C6 to charge to a high enough potential to turn on the darlington pair (Q4 and Q5), and energize the relay. D9 is used to protect the circuit against inductive kickback spikes when the relay coil is de-energized. R3 is incorporated to "bleed" off C6's charge when the circuit power is turned off.

Q2, Q3, and their associated circuitry form the familiar astable multivibrator. The values of C4 and C5 (about 0.5 $\mu$F) should be chosen to cause the circuit to oscillate at around 2 Hz.

Now, examine how these circuits are put together to protect a speaker. The ac input voltage can be taken from the power transformer powering the audio amplifier, or from a separate transformer. In either case, the wiring is connected so that ac power is applied to this circuit at the same time it is applied to the power amplifier. The speaker outputs coming from the power amplifier(s) are connected to the "right in" and "left in" connections on the protection circuit (assuming you are using two power amplifiers for stereo).

Although many audiophiles disagree with using fuses in the speaker connection line, I adamantly disagree with them. However, F1 and F2 are optional. (Fuses will not protect a speaker against damage caused by dc voltages, but they will provide some protection against overload; that is, trying to drive a 30-watt speaker at a 100-watt power level.) From the fuses, the amplifier outputs are connected directly to the two poles of the DPDT relay.

When power is first applied to the amplifiers and speaker protection circuit, the speakers will not be connected to the power amplifiers because the relay has not been energized. The relay will not energize until C6 reaches a potential high enough to turn on Q4 and Q5. This will take several seconds. In the meantime, both power amplifiers have had sufficient time to stabilize, and the turn-on thumps have been eliminated.

While C6 is charging, before Q4 and Q5 have been turned on, the astable multivibrator is oscillating, causing LED1 to flash at about 2 Hz. This gives a visible indication that the protection circuit is working, and has not yet connected the speakers to the power amplifiers.

When the time delay has ended and Q4 and Q5 turn on, the collector of Q4 pulls the collector of Q3 low, stopping the oscillation of the multivibrator, and causing LED1 to stay bright continuously. The relay energizes simultaneously. By staying bright continuously, LED1 is giving a visual indication that the circuit is working, and that the speakers "are" connected to the power amplifiers. At this point, the protection circuit has no further effect within the amplification system unless a dc voltage occurs on one of the power amplifier outputs.

Under normal conditions, when no dc voltage is present on the outputs, the amplified audio ac signal from both power amplifiers is applied simultaneously to R4, R5, C1, and C2. Because the time constant of this RC circuit is relatively long, C1 and C2 cannot charge to either polarity. In effect, they charge to the average value of the ac waveshape, which is zero. This is like trying to measure an ac voltage with your DVM set to measure dc; pure ac will provide a zero reading.

If a significant dc voltage appears on the output of either power amplifier, C1 and C2 will charge to that voltage regardless of the polarity. This dc voltage is applied to the bridge rectifier (D4 through D7). Although it might seem strange to apply dc to a bridge rectifier, the effect it has in this circuit is to convert it to the correct polarity for forward-biasing Q1. When Q1 is forward biased, it pulls the base of the darlington pair (Q4 and Q5) low, de-energizing the relay, and disconnecting the speakers from the power amplifiers before any damage can result. At the same time, the astable multivibrator is enabled once more, causing LED1 to flash, which gives a visual indication that a malfunction has occurred. The circuit will remain in this condition as long as dc appears on either of the inputs. Upon removal of the dc voltage, it will automatically resume normal operation. Pretty neat!

I left out some of the component values in this schematic because they will be determined by the voltage amplitude of the ac with which you wish to power it. If, for example, you chose a 12-Vac transformer as the source of operational power, ZD1 would be a 12-volt, 1-watt zener. The calculation of R1's value was discussed in the last chapter. The relay coil voltage would also need to be rated for 12-volts. The circuit operation is simple and non-critical. Transistors Q1 through Q5 can be any general-purpose type. Diodes D1 through D9 can be any type of general-purpose, low-current rectifier diodes. Remember to use capacitors with a high enough voltage rating to accommodate the power source you decide to use.

One final comment about this circuit; notice how C1 and C2 are connected. When two electrolytic capacitors are connected in this fashion, the result is a nonpolarized electrolytic capacitor. This can be handy if you need a large value of capacitance for an ac application.

# Power control

THE TERM *THYRISTOR* REFERS TO A BROAD FAMILY OF semiconductor devices used primarily for power control. Thyristors are basically fast-acting electronic switches. Common members of the thyristor family include SCRs, UJTs, TRIACs, Diacs, and (strangely enough) neon tubes.

## Silicon-controlled rectifiers (SCRs)

SCRs are probably the most common of all thyristors. An *SCR* is a three-lead device resembling a transistor. The three leads are referred to as the "gate," "cathode," and "anode." An illustration of SCR construction, lead designation, and electrical symbol is given in Fig. 9-1.

An SCR will only allow current to flow in one direction. Like a diode, the cathode must be negative, in relation to the anode, for current flow to occur. However, forward current flow will not begin until a positive potential, relative to the cathode, is applied to the gate. Once current flow begins, the gate has no more control of the SCR action until it drops below its designated holding current.

One method of stopping the forward current flow, in a conducting SCR, is to reverse-bias it (cause the cathode to become positive relative to the anode). Another method is to allow the forward current flow to drop below the SCR's holding current. The *holding current* is a manufacturer's specification defining the minimum current required to hold the SCR in a conductive state.

If the forward current flow drops below the specified holding current, the SCR will drop out of conduction. When this happens (due to either a voltage polarity reversal, or a loss of minimum holding current), control is again returned to the gate, and the SCR will not conduct (even if forward-biased) until another positive potential (or pulse) is applied to the gate.

Like a transistor, the SCR is considered a *current device* because the gate current causes the SCR to begin to conduct (if forward-

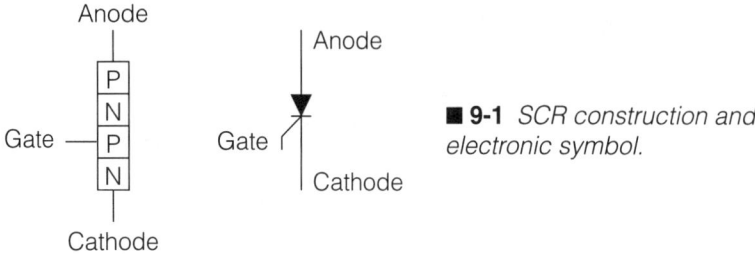

**9-1** *SCR construction and electronic symbol.*

biased). Also, the forward current flow is the variable that maintains conduction, once the SCR is turned on by the gate current.

Because SCRs can be turned off when they are reverse-biased, they are very commonly used in ac power applications. Because ac power reverses polarity periodically, an SCR used in an ac circuit will automatically be reverse-biased (causing it to turn off) during one-half of each cycle. During the other half of each cycle, it will be forward-biased, but it will not conduct unless a positive gate pulse is applied. By controlling the *coincidental point* at which the gate pulse is applied together with a forward-biasing half-cycle, the SCR can control the amount of given power to a load during the half-cycles it is forward-biased.

Consider the circuit shown in Fig. 9-2. As long as SW1 remains open, $R_L$ will not receive any power from the ac source because the SCR will not conduct during either half-cycle. If SW1 is closed (providing a continuous positive potential to the gate), the load will receive 1/2 of the available power from the ac source. In this condition, the SCR acts like a diode, and only conducts current during the half-cycles when it is forward-biased. (Resistor $R_G$ is placed in the gate circuit to keep the gate current from exceeding the specified maximum.)

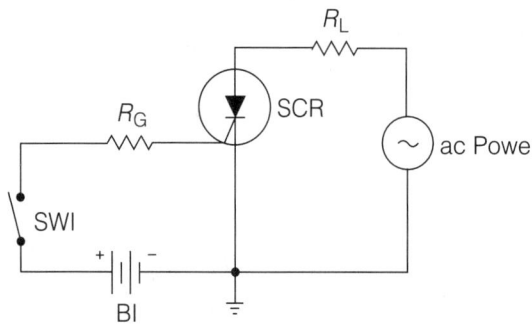

**9-2** *Demonstration of SCR operating principles.*

If it were possible to rapidly turn SW1 on and off, so that the SCR received a gate pulse at the "peak" of each forward-biasing half-cycle, the SCR would only conduct for "half" of the half-cycle. This condition would cause the load to receive 1/4 of the total power available from the ac source. By accurately varying the timing relationship between the gate pulses and the forward-biasing half-cycles, the SCR could be made to supply any percentage of power desired to the load, up to 50%. It cannot supply any greater than 50% power to the load, because it cannot conduct during the half-cycles when it is reverse-biased.

There are several important points to understand about the operation of this simple circuit. First, once the SCR has been turned on by closing S1, it cannot be turned off again during the remainder of the forward-biasing half-cycle. As long as the SCR is conducting a current flow higher than its minimum holding current, the gate circuit loses all control. Secondly, before the SCR receives a gate pulse and begins to conduct, there is virtually no power consumption in the circuit; it looks like an open switch. Once the SCR begins to conduct, virtually all of the power is delivered to the load.

An SCR wastes very little power when controlling power to a load, because it functions in either of two states: ON (looking like a closed switch), or OFF (looking like an open switch). A closed switch might have a high amplitude of current flow through it, but it poses no opposition (resistance). Therefore, the voltage drop across a closed switch is virtually zero. Because power dissipation is equal to current times voltage ($P = IE$), when the voltage is close to zero, so is power dissipation. In contrast, an open switch might drop a high voltage, but does not allow current flow. Again, it becomes irrelevant how high the voltage is, if there is no current flow; power dissipation is still zero.

To understand the importance of efficient power control, consider another method of varying power to a load. Figure 9-3 illustrates a circuit in which a rheostat is used to vary the power delivered to a 50-ohm load ($R_L$).

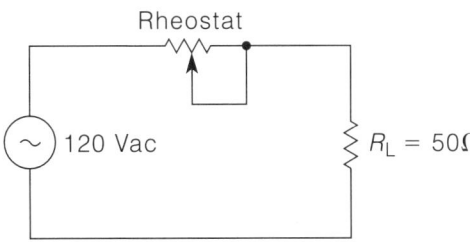

■ **9-3** *Rheostat power control circuit.*

The rheostat is adjustable from 0 to 50 ohms. When it is adjusted to 0 ohms, it will appear to be a short and the entire 120-Vac source will be dropped across $R_L$. The power dissipated by $R_L$ can be calculated as follows:

$$P = \frac{E^2}{R} = \frac{120^2}{50} = \frac{14400}{50} = 288 \text{ watts}$$

The power dissipated by the rheostat will be:

$$P = \frac{E^2}{R} = \frac{120^2}{0} = 0$$

If the rheostat is adjusted to present 50 ohms of resistance, the voltage dropped by the rheostat will be equal to the voltage dropped by $R_L$. Therefore, 60 Vac will be dropped by both. Because they are both equal in voltage drop and resistance, the power dissipation will also be equal in both. Therefore, under these circumstances, the power dissipation of either is:

$$P = \frac{E^2}{R} = \frac{60^2}{50} = \frac{3600}{50} = 72 \text{ watts}$$

Because the purpose of the circuit shown in Fig. 9-3 is to control the power delivered to the load ($R_L$), all of the power dissipated by the rheostat is wasted. In the previous example, the efficiency of the power control is 50%. At different settings of the rheostat, different efficiency levels occur; but it is obvious that this level of waste is unacceptable in high-power electrical applications.

A disadvantage of using a single SCR for power control, as illustrated in Fig. 9-2, is that it is not possible to obtain full 360-degree control of an ac waveform (only 180 degrees, that is, the forward-biasing half-cycle, can be controlled). To overcome this problem, two SCRs might be incorporated into a circuit for full-wave power control.

## The triac

Another member of the thyristor family, the triac can be used for full-wave power control. A *triac* has three leads designated as the "gate", "M1," and "M2." Triacs are triggered ON by either a positive or negative pulse to the gate lead, in reference to the M1 terminal. Triacs can also conduct current in either direction between the M1 and M2 terminals. Like an SCR, once a triac has been triggered, the gate looses all control until the current flow through the M1 and M2 terminals drops below the manufacturer's specified holding current. Triacs are considered current devices. The electrical symbol used for triacs is illustrated in Fig. 9-4.

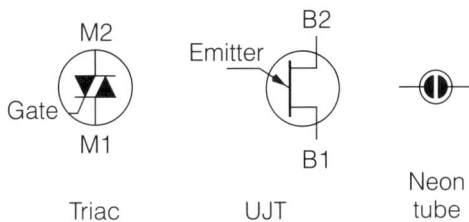

■ **9-4** *Additional thyristor symbols.*

The principle of efficient power control is essentially the same for the triac as it is for the SCR. Because a triac operates in only two modes (ON or OFF), full-wave power control can be obtained without appreciable power losses in the triac itself. The primary advantage of a triac is its capability of being triggered in either polarity, and controlling power throughout the entire ac cycle.

Triacs are commonly used for smaller power control applications (light dimmers, small dc motors and power supplies). Unfortunately, triacs have the disadvantage of being somewhat difficult to turn off (especially when used to control inductive loads). Because of this problem, SCRs (rather than triacs) are used almost exclusively in high-power applications.

## UJTs, diacs, and neon tubes

So far in this chapter, you have examined the theoretical possibility of controlling power with SCRs or triacs. If you could vary the gate *trigger pulse timing* relative to the ac cycle (this is called varying the *phase angle* of the trigger pulse), you would have an efficient electronic power control tool. Obviously, it would be impossible for a human to turn a switch on and off at a 60-Hz rate to provide trigger pulses for power control. UJTs, diacs, and neon tubes are commonly used to accomplish this function. (The neon tube is not actually a member of the thyristor family, but its function is identical to the diac. Some equipment still uses neon tubes for triggering, because they serve a dual purpose as "power on" indicators.)

Like the transistor, the *UJT (unijunction transistor)* is a three-lead device. The three leads are referred to as the "emitter", "B1", and "B2". The schematic symbol for UJTs is given in Fig. 9-4. Unlike the previously discussed SCR and triac, the UJT is a voltage device. When the voltage between the emitter and B1 leads reaches a certain value (a ratio of the applied voltage between the

B1 and B2 leads, and the manufactured characteristics of the UJT), the resistance between the emitter and B1 decreases to a very low value.

Contrastingly, if the voltage between the emitter and B1 decreases to a value below the established ratio, the resistance between the emitter and B1 increases to a high value. In other words, a UJT can be thought of as a voltage-breakdown device. It will avalanche into a highly conductive state (between the emitter and B1 leads) when a peak voltage level (referred to as $V_p$) is reached. It will continue to remain highly conductive until the voltage is reduced to a much lower level called the "valley voltage" ($V_v$).

*Firing circuits* are electronic circuits that vary the amplitude and phase of an ac trigger voltage applied to the gate lead of an SCR (or other thyristor). Using a combination of amplitude adjustment and phase-shifting techniques, the $V_p$ level (resulting in UJT voltage breakdown) can be made to occur precisely and repeatedly at any "phase angle relative to an ac power waveform applied to a triac or SCR." The conductive breakdown characteristics of the UJT are used as a means of providing the gate trigger pulses to "fire" the SCRs or triacs. Therefore, the SCRs or triacs can be repeatedly fired at any point on the ac waveform resulting in full-wave (0 to 100% duty cycle) power control.

Another voltage breakdown device is the diac. Because diacs are special-purpose diodes, their operational description and symbols have already been discussed in chapter 7. Diacs are simply the solid-state replacement for neon tubes. The schematic symbol for neon tubes is given in Fig. 9-4. Neon tubes and diacs are devices which will remain in a non-conductive state until the voltage across them exceeds a breakover voltage (or ionization voltage in the case of neon tubes). At breakover, they will become conductive and remain so until the voltage across them drops below a holding voltage (a much lower voltage than the breakover voltage), at which time they will become nonconductive again.

## Building a soldering iron controller

Soldering irons are available in many power ranges. The smallest sizes, around 15 watts, are recommended for very small and precise work, such as SMC (surface mount component) work. Medium sizes, about 30 watts, are recommended for most general electronic work, including PC board soldering. Larger sizes, from 60 watts and up, are for large soldering jobs, such as making solder connections to large buss bars or stud mount diodes.

If you run into situations where you need to do a variety of different types of soldering, there are several solutions. The obvious solution is to buy several different types of soldering irons. Another solution is to buy a soldering station, with an automatic tip temperature regulator (starting at about $100.00). A good "middle-of-the-road" solution is to buy a 60-watt soldering iron, and use it with the circuit illustrated in Fig. 9-5. In addition to being a useful and convenient tool, this circuit will help illustrate most of what has just been discussed concerning thyristors and power control.

Referring to Fig. 9-5, the incoming 120-Vac power is applied across the triac through the load (light bulb and soldering iron, in parallel). Assuming SW1 is turned on at the instantaneous point in time that the ac voltage is at zero, the triac is off (nonconductive). As the ac voltage begins to increase through a half-cycle (the polarity is irrelevant because the diac and triac are both bilateral in operation), all of the ac voltage is dropped across the triac because it looks like an open switch.

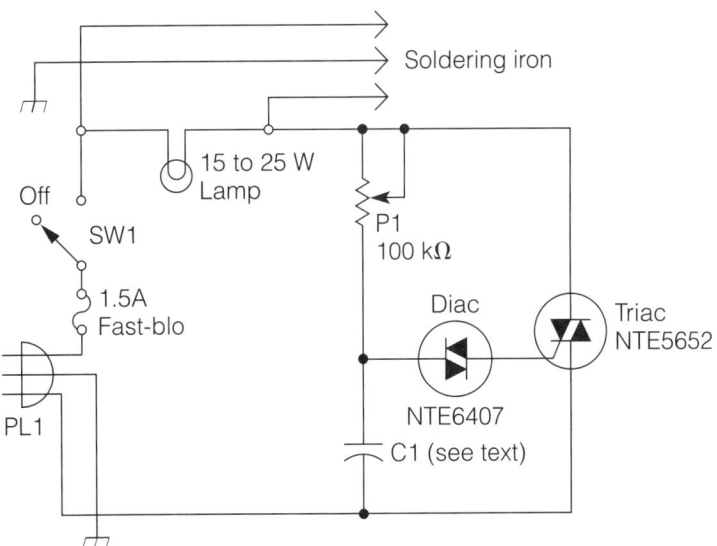

■ **9-5** *A soldering iron controller.*

Similarly, the same voltage is dropped across the firing circuit, or trigger circuit (P1, the diac, and C1), because it is in parallel with the triac. C1 will begin to charge at a rate relative to the setting of P1. As the ac half-cycle continues, C1 will eventually charge to the specified breakover voltage of the diac, causing the diac to

*Building a soldering iron controller*

avalanche, and a current pulse (trigger) to flow through the gate and M1 terminals of the triac.

This trigger pulse causes the triac to turn ON (much like a closed switch) resulting in the remainder of the ac half-cycle being applied to the load (lamp and soldering iron). When the ac power has completed the half-cycle and approaches zero voltage (prior to changing polarity), the current flow through the triac drops below the holding current and the triac returns to a nonconductive state. This entire process continues to repeat with each half-cycle of the incoming ac power.

There are several important points to understand about the operation of this circuit. The diac will reach its breakover voltage and trigger the triac at the same relative point during each half-cycle of the ac waveform. This relative point will depend on the charge rate of C1, which is controlled by the setting of P1. In effect, the setting of P1 controls the average power delivered to the load. P1 can control the majority of the ac half-cycle because C1 also introduces a "voltage-lagging" phase shift. Without the phase shift, control would be lost after the peak of the ac power cycle was reached.

Throughout the entire power control range of this circuit, the power wasted by the triac is negligible, compared to the power delivered to the load.

PL1 is a standard 120-Vac 3-prong plug. If you build this project in an aluminum project box, the ground prong (round prong) should be connected to the aluminum box (the chassis in this case). For safety sake, the 120-Vac hot lead should also be fuse protected. P1 is mounted to the front panel of the project box for easy access.

I used a flat, rectangular aluminum project box large enough to set the soldering iron holder on. I also connected the soldering iron internally to a phenolic solder strip with a strain relief to protect the cord. This, of course, is a matter of opinion. You might want to wire the circuit to a standard 120-Vac socket for use with a variety of soldering irons.

The triac, diac, and C1 can be assembled on a small universal perf board or wired to a phenolic solder strip.

The 15- to 25-watt lamp is a standard 120-Vac incandescent light bulb of any style or design you like (it might also be any wattage you desire, up to 60 watts). It is mounted on the outside front panel of the project box and serves several useful indicator functions. First, it indicates that the power is on and that the circuit is functioning. Secondly, with a little practice, the brightness of the bulb is a good indicator of about how much power you are applying to the soldering iron. For example, if you're using this circuit with a 60-watt soldering iron, and you adjust P1 until the bulb is about half as bright as normal, you're supplying about 30 watts of power to the tip. Thirdly, the light bulb makes a good reminder to turn off the soldering iron when you're finished working. (I can't count how many times I have come into my shop and found the soldering iron still turned on from the previous day.)

The best value for C1 will probably be about 0.1 $\mu F$. After building the circuit so that it can be tested using the light bulb as the load, try a few different values for C1, and choose the one giving the smoothest operation throughout the entire power range. C1 must be a nonpolarized capacitor rated for at least 200 volts.

In addition to controlling the power delivered to a soldering iron, this circuit is a basic light dimmer circuit. You might use it to control the power delivered to any "resistive" load up to about 150 watts. For controlling larger loads, you will need to use a larger triac and, depending on the triac, you might need to use different values for P1 and C1. For controlling large loads, it's also a good idea to place a varistor (MOV) across the incoming 120-Vac line (such as an NTE 2V115).

## Circuit potpourri

The theories and postulations are done. Now's the time for applications and fun!

### Watts an easier way?

They don't get any simpler than this; a very simple soldering iron power control in Fig. 9-6. This circuit inserts a diode (with SW1 open) in series with the soldering iron heating element. The diode will block one-half of the incoming ac voltage to the heating element, resulting in a power decrease.

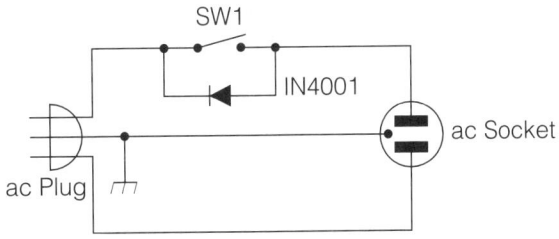

■ **9-6** *Simple soldering iron power control.*

Although it would seem logical that the soldering iron would operate at half-power with the diode in the power circuit (that is, a 60-watt iron would become a 30-watt iron), it is not quite that simple. The type of wire universally used in resistive heating elements is called "nichrome." *Nichrome*, like most resistive substances, has a positive temperature coefficient; as it becomes hotter, its resistance value goes up. When a resistive heating element designed for a 120-Vac application (such as in a soldering iron) experiences a decrease in applied power, its temperature goes down, resulting in a proportionate decrease in resistance. This decrease in resistance has the effect of causing the wire to dissipate more "power per volt" than it did at the rated voltage.

The circuit illustrated in Fig. 9-6, if used in conjunction with a 60-watt soldering iron, would decrease its operational power down to about 40 watts. In reality, this would handle the vast majority of soldering jobs you will encounter. If you used this circuit with another soldering iron rated at 30 watts, it would decrease it down to about 18 watts (just about right for SMC work). In other words, two soldering irons (a 60 watt, and a 30 watt), two general-purpose diodes, and two switches will provide you with the full gamut of soldering iron needs. One additional benefit; using a 50- or 60-watt soldering iron with the diode in the power circuit (and placing the wattage at about the optimum point for general-purpose electronic work) will cause the heating element to last about 40 times longer, and will thus increase tip life.

By the way, placing a general-purpose diode in series with incandescent (but *not* fluorescent!) light bulbs will cause the light bulbs to last about 40 times longer. The disadvantage resulting from this is the lower quantity and efficiency of the light produced.

### Isolation might be good for the soul

Figure 9-7 shows how to construct a simple isolation transformer for your electronics bench. Typically, isolation transformers are needed when testing "line-powered" equipment with "line-powered" test equipment.

For example, referring back to Fig. 9-5, suppose you wanted to observe the waveshape across the parallel circuit of the soldering iron and lamp with a line-powered oscilloscope. You connect the ground and scope probe across the lamp, turn on SW1, and "boom;" the fuse blows and the triac might be destroyed. What happened?

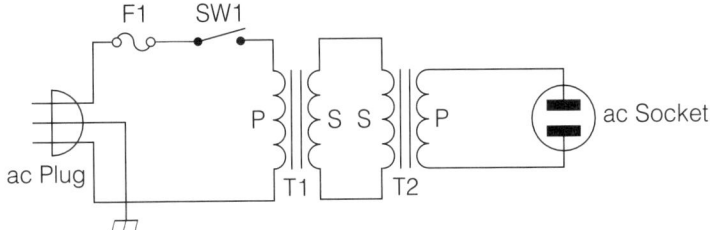

■ 9-7 *General purpose isolation circuit.*

For safety reasons, the ground (common) lead connections on most line-powered test equipment are connected to earth ground. Looking at Fig. 9-5, imagine the results of connecting earth ground to either side of the lamp/soldering iron connections. From a problematic viewpoint, there are several possibilities; two possibilities of how PL1 is wired (correct and incorrect), and many possibilities of how the scope can be connected. Out of these possibilities, three would result in improper operation, or destroyed components.

For example, if the iron controller's power cable was connected with the black wire going to the fuse (which is standard, safe, and correct), and the scope "common" probe was inadvertently touched to that part of the circuit (fuse, SW1, or that side of the lamp), then a direct short to ground would exist! If the probe connection was made to the cable side of the fuse, a catastrophic short would occur, hopefully blowing the fuse or breaker at the main box. If the connection was made beyond the fuse, this short could blow either the main fuse, the controller fuse, or both.

If the controller was isolated from the main power line, by using the isolation transformer, neither of the lines (black or white) going to the controller have any continuity to ground. Remember, in household wiring, the white wire is called the *neutral*, and it is kept at "ground" potential. The black and red wires are hot. Without the isolation provided by the transformer, the common scope probe, being connected to earth ground and neutral (through the house wiring) and being attached to the hot side of the controller, would present a dead short. For this reason, among others, isolation transformers are handy to have around.

In Fig. 9-7, T1 and T2 are any two "identical" power transformers, with appropriate ratings for the desired application. For example, assume T1 has a secondary rating of 25 volts at 2 amps. It is, therefore, a 50-VA (volt-amp) transformer (25 volts × 2 amps = 50 VA). To calculate the maximum current rating of the primary, simply divide the volt-amp rating by 120 volts (intended primary voltage). This comes out to a little over 410 milliamps. The primary and secondary of any individual power transformer will always have the same volt-amp rating.

If you decided to use two such transformers for the isolation circuit, the primary of T1 would be connected to the line (120 Vac). The secondary of T1 would be connected to the secondary of T2, whose primary would be the output. In other words, the function of the first transformer is to convert 120 Vac down to 25 Vac (step-down application), and the second transformer converts 25 Vac back up to 120 Vac (step-up operation). This "conversion from the original" and "conversion back to the original" process is the reason both transformers must be identical.

As explained earlier, 410 milliamps is the highest current load that can be drawn from T2's "intended primary" under ideal conditions. However, there are losses in power transformers and, in this case, the losses are doubled. Therefore, it is necessary to derate T2's primary maximum current value by 10%. Consequently, the maximum current output from this hypothetical isolation circuit will only be about 370 milliamps; a little too small for most general-purpose isolation requirements.

If you decide to build this type of isolation circuit for your bench, you'll probably want to use transformers with significantly higher VA ratings. Unfortunately, it soon becomes apparent that power transformers in the 400- to 500-VA range are expensive, but so are isolation transformers. However, some "odd" value industrial transformers with high VA ratings can be purchased very inexpensively from many surplus electronic dealers. Just make sure they are not ferroresonant transformers, or power transformers intended for use with 400-Hz ac power.

### Curiosity catcher

Figure 9-8 is an SCR latch circuit. When assembled as illustrated, depressing the "normally open" momentary switch (SW1) provides a positive gate pulse (relative to the cathode) and fires the SCR. Once conducting, the gate has no more control over the SCR, and it continues to conduct, powering the piezo buzzer, until the cathode/anode current flow is interrupted by depressing the "normally closed" momentary switch (SW2). Because the cathode/anode current flow drops below the holding current of the SCR (it actually drops to zero), control is returned to the gate, and the SCR will not conduct again until SW1 is depressed.

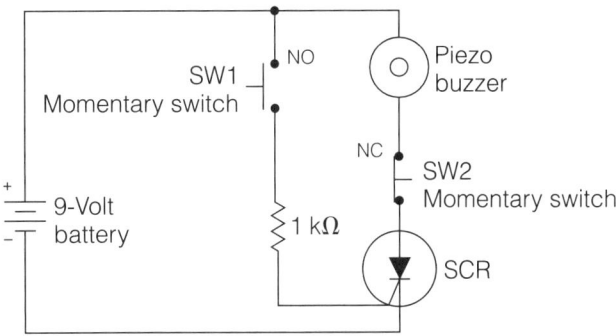

■ **9-8** *SCR latch circuit.*

A circuit of this type is called a *latch circuit*. (The electromechanical counterpart of this circuit is a latching relay.) Just about any commonly available SCR can be used. The piezo buzzer can be substituted for almost any type of load that you need to be latched. There are many, many applications for such a circuit, but I would like to describe one that I had a lot of fun with.

I mounted this circuit in a small black box, using a 9-volt transistor battery as the power source. The piezo buzzer is commonly available in any electronic parts store. The buzzer was mounted close to the front of the box, where "speaker holes" were drilled. SW2 was located "inside" the project box. I mounted it in a lower back corner, so that the only way it could be depressed was to use a straightened paper clip through an almost invisibly small hole drilled through the back of the box. SW1 was mounted to the front plate with a large, red, plastic knob. Underneath the SW1 push-button, I arranged stick-on letters to read, "DO NOT PUSH." You can probably guess the rest.

If the box is left in an obvious place, it will drive many people crazy, until they see what happens when the button is pressed. After their curiosity gets the better of them and they press the button, the piezo buzzer sounds off, and there isn't any apparent way to stop it.

After building this box, user discretion is advised! After playing this joke on a person lacking a really good sense of humor (while the buzzer was still sounding), I made the mistake of saying he could never figure out how to stop it. He promptly dropped it on the floor, and stomped it with a large, silver-tip boot. I stood corrected.

### See ya later, oscillator

The circuit illustrated in Fig. 9-9 is a *UJT oscillator*. UJTs make good oscillators for audio projects, and they have some distinct ad-

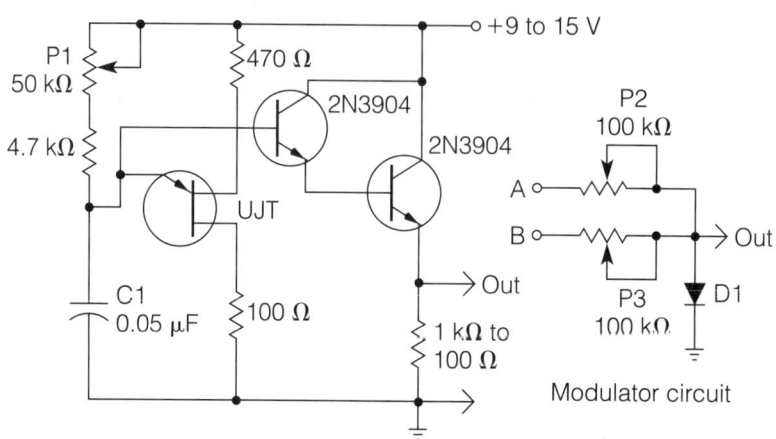

■ **9-9** *Buffered UJT audio oscillator.*

vantages over astable multivibrators. They are easier to construct, and only require the use of 1 UJT. Also, the output is in the form of a "sawtooth" wave, which provides more pleasing audio harmonics than square waves. (This circuit will be used as a foundation for later projects in this book.)

UJT oscillators do have a drawback. Their usable audio output, which is taken across the capacitor, has a very high impedance. Therefore, it is best to use an impedance matching circuit, such as the darlington pair common-collector amplifier shown in the illustration. (A JFET works very well for this application also. JFETs will be discussed in the next chapter.) Impedance matching circuits used for this type of application are often called *buffers*.

Using the values given in the illustration, a fairly wide range of the audio frequency spectrum can be produced by rotating P1 (frequency adjustment). The value of C1 can be increased for extremely low oscillations, and vice versa. The output of the darlington pair can drive a small speaker directly, but the sound quality will be greatly improved by coupling the output with an audio transformer or, better yet, using a separate audio amplifier and speaker.

If you would like to experiment with some really unique sounds, build two of these UJT oscillator circuits and connect their outputs into the modulator circuit, also illustrated in Fig. 9-9. One UJT oscillator output will connect to point A, and the other to point B. The output of the modulator circuit must be connected to the input of a separate amplifier and speaker. The diode can be any general-purpose rectifier diode.

An understanding of this modulator circuit requires an explanation of several new principles; the first involving diodes. When diodes are forward-biased with very low voltages and currents, they react in a highly nonlinear fashion. This area of diode conduction is called the *forward-biased knee* of the diode response.

A second new principle involves the process of "modulation." *Modulation* occurs when two different frequencies are mixed in a non-linear circuit. The effect of modulation causes some additional frequencies to be created. For example, if a 1-kHz and a 10-kHz frequency are applied to the inputs of a linear mixer, the output will be the original frequencies, only mixed together. If the same two frequencies are applied to the input of a nonlinear mixer circuit, the original frequencies will still appear at the output, but two additional frequencies, called *beat frequencies*, will also occur. The beat frequencies will be the sum and difference of the original frequencies. Using 1 kHz and 10 kHz as the originals, the beat frequencies will be 11 kHz (sum) and 9 kHz (difference).

When the two UJT oscillators are connected into the modulator circuit (with its output applied to the input of a power amplifier and speaker), P2 and P3 are adjusted to cause the outputs of the oscillators to fall into the "knee" area of D1's forward conduction response. You might have to add some additional series resistance between each oscillator and each modulator input. If this is the case, try increasing the resistance in 100-k$\Omega$ increments. When the two frequencies modulate, the sound produced will be instantly recognizable. This type of sound, or "tonality," has been in the background of most science fiction movies since the 1950s.

# Field-effect transistors

THE *FIELD-EFFECT TRANSISTOR (FET)* IS AN ACTIVE "voltage" device. Unlike bipolar transistors, FETs are not current amplifiers. Rather, they act much like vacuum tubes in basic operation. FETs are three-lead devices similar in appearance to bipolar transistors. The three leads are referred to as the *gate*, *source*, and *drain*. These three leads are somewhat analogous to the bipolar transistor's base, emitter, and collector leads respectively. There are two general types of FETs; junction field-effect transistors (JFETs) and insulated-gate metal oxide semiconductor field-effect transistors (MOSFET or IGFET).

FETs are manufactured as either N-channel or P-channel devices. *N-channel FETs* are used in applications requiring the drain to be positive relative to the source; the opposite is true of *P-channel FETs*. The schematic symbols for N-channel and P-channel JFETs and MOSFETs are shown in Fig. 10-1. Note that the arrow always points toward the channel (the interconnection between the source and drain) when symbolizing N-channel FETs, and away from it in P-channel symbologies.

All types of FETs have very high input impedances (1 M$\Omega$ to over 1,000,000 M$\Omega$). This is the primary advantage to using FETs in the majority of applications. The complete independence of FET operation from its input current is the reason they are classified as voltage devices. Because FETs do not need gate current to function, they do not have an appreciable loading effect on preceding stages or transducers. Also, Because their operation does not depend upon "junction recombination" of majority carriers (as do bipolar transistors), they are inherently low-noise devices.

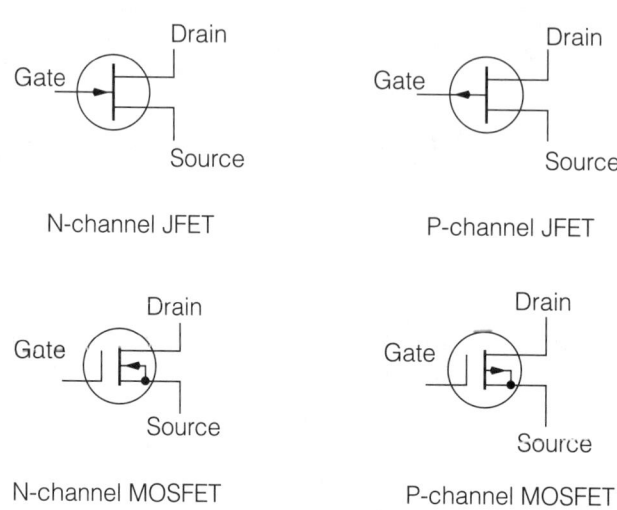

■ 10-1  *JFET and MOSFET symbols.*

## FET operational principles

The basic operational principles of FETs are actually much simpler than bipolar transistors. FETs control current flow through a semiconductor "channel" by means of an electrostatic field.

Referring to Fig. 10-2, examine the construction of a JFET. Notice that there are two junctions, with the P material being connected to the gate, and the N material making up the channel. Assume that the source lead is connected to a circuit common, and a positive potential is applied to the drain lead. Current will begin to flow from source to drain with little restriction. The N-channel semiconductor material, although not a good conductor, will conduct a substantial current.

Under these conditions, if a "negative" voltage is applied to the gate, the PN junctions between the gate and channel material will be "reverse-biased" (negative on the P-material). The reverse-biased condition will cause a depletion region extending outward from the gate/channel junctions. As you might recall, a depletion region becomes an insulator because of the lack of majority charge carriers. As the depletion region spreads out from the gate/channel junction deeper into the channel region, it begins to restrict some of the current flow between the source and drain. In effect, it reduces the conductive area of the channel, acting like a water valve closing on a stream of flowing water. This depletion region will increase outward in proportion to the increase in amplitude of the negative voltage applied to the gate.

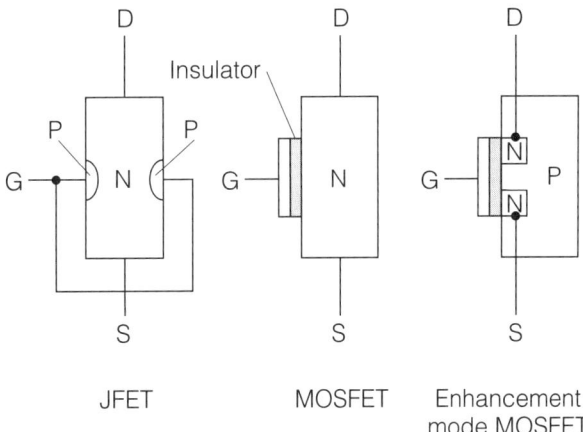

■ **10-2** *FET construction.*

If the negative gate voltage is increased to a high enough potential, a point will be reached when the depletion region entirely "pinches off" the current flow through the channel. At this point, the FET is said to be "pinched off" (this pinch-off region being analogous to cutoff in bipolar transistors), and all current flow through the channel stops. The basic principle involved is controlling the channel current with an electrostatic field. This field effect is the reason for the name field-effect transistor.

Continuing to refer to Fig. 10-2, notice the difference in construction between a MOSFET and JFET. Although a JFET's input impedance is very high (due to the reverse-biased gate junction), there can still be a small gate current (because of leakage current through the junction), which translates to a reduced input impedance. However, gate current through a MOSFET is totally restricted by an insulating layer between the gate and channel.

A MOSFET functions in the same basic way as a JFET. If a negative voltage is applied to the gate of an N-channel MOSFET, the negative electrostatic charge around the gate area repels the negative charge carriers in the N-channel material, forming a resultant depletion region. As the negative gate voltage varies, the depletion region varies proportionally. The variance in this depletion region controls the current flow through the channel and, once again, current flow is controlled by an electrostatic field.

A third type of FET, called an enhancement-mode MOSFET, utilizes an electrostatic field to "create a channel," rather than deplete a channel. Referring again to Fig. 10-2, notice the construction of an enhancement mode MOSFET. The normal N-channel is separated by a section of a P-material block, called the *substrate*. N-channel enhancement-mode MOSFETs, such as the one illustrated, require a positive voltage applied to the gate. The positive potential at the gate attracts "minority" carriers out of the P-material substrate, forming a layer of "N-material" around the gate area.

This has the effect of connecting the two sections of N-material (attached to the source and drain) together to form a continuous channel, and thus allows current to flow. As the positive gate potential increases, the size of the channel increases proportionally, which results in a proportional increase in conductivity. Once more, current flow is controlled by an electro–static field.

All of the operating principles discussed in this section have been applied to N-channel FETs. P-channel FET devices will operate identically, with the only difference being in the reversal of voltage polarities.

## FET parameters

As previously discussed, the primary gain parameter of a standard bipolar transistor is beta. Beta defines the ratio of the current flow through the base relative to the current flow through the collector. In reference to FETs, the primary gain parameter is called transconductance ($G_m$). The *transconductance* is a ratio defining the effect that a gate-to-source voltage ($V_{GS}$) change will have on the drain current ($I_D$). Transconductance is typically defined in terms of *micromhos* (the "mho" is the basic unit for expressing conductance). Typical transconductance values for common FETs range from 2,000 to 15,000 micromhos. The equation for calculating transconductance is:

$$G_m = \frac{\text{change in drain current}}{\text{change in gate-to-source voltage}}$$

For example, assume you were testing an unknown FET. A 1-volt change in the gate-to-source voltage caused a 10-milliamp change in the drain current. The calculation for its transconductance value would be:

$$G_m = \frac{I_D}{V_{GS}} = \frac{10 \text{ mA}}{1 \text{ volt}} = 0.01 \text{ mho} = 10{,}000 \text{ micromhos}$$

Referring to Fig. 10-3, assume the FET in this illustration has the same transconductance value as calculated in the previous example. A 1-volt change in the gate-to-source voltage (input) will cause a 10-milliamp change in the drain current. According to ohm's law, a 10-milliamp current change through the 1-k$\Omega$ drain resistor ($R_D$) will cause a 10-volt change across the drain resistor (10 milliamps × 1000 ohms = 10 volts). This 10-volt change will appear at the output. Therefore, because a 1-volt change at the gate resulted in a 10-volt change at the output, this circuit has a voltage gain ($A_e$) of 10.

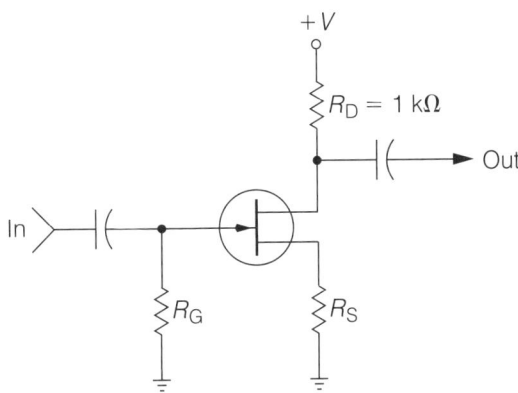

■ **10-3** *JFET common-source amplifier.*

In numerous ways, FET circuits can be compared with standard bipolar transistor circuits. The circuit shown in Fig. 10-3 is analogous to the common-emitter configuration, and it is appropriately called a *common-source configuration*. The output is inverted from the input, and it is capable of voltage gain. If the output were taken from the source, instead of the drain, it would then be a common drain configuration. The output would not be inverted, and the voltage gain would be approximately 1. Of course, the common drain FET amplifier is analogous to the common-collector amplifier in bipolar design.

## FET biasing considerations

Referring again to Fig. 10-3, note that the gate is effectively placed near the same potential as circuit common through resistor $R_G$. With no input applied, the gate voltage (relative to circuit common) is zero. However, this does not mean the gate-to-source voltage is zero. Assume the source resistor ($R_S$) is 100 ohms, and that the drain current, which is the same as the source current, is 15 milliamps. This 15-milliamp current flow through RS would cause it to drop 1.5 volts, placing the source lead of the FET at a positive 1.5-volt potential "above circuit common." If the source is 1.5 volts more positive than the gate; it could also be said that the gate is 1.5 volts more negative than the source. (Is an 8-oz. glass, with 4 ounces of water in it, half-full or half-empty?) Therefore, the gate-to-source voltage in this case is "negative" 1.5 volts. This also means that the gate has a $-1.5$-volt negative bias. If a signal voltage is applied to the input, causing the gate to become more negative, the FET will become less conductive (more resistive), and vice versa. A JFET exhibits maximum conductivity (minimum resistance), from the source to the drain, with no bias voltage applied to the gate.

MOSFETs are biased in similar ways to JFETs, except in the case of enhancement mode MOSFETs. As previously explained, enhancement-mode MOSFETs are biased with a gate voltage of the opposite polarity to their other FET counterparts. Some enhancement mode MOSFETs are designed to operate in either mode of operation.

In general, FETs provide a circuit designer with a higher degree of simplicity and flexibility, because of their lack of interstage loading considerations (a transistor stage with a high input impedance will not load down the output of a previous stage). This can also result in the need for fewer stages, and less complexity in many circuit designs.

## Static electricity: an unseen danger

The introduction of MOS (metal-oxide semiconductor) devices brought on a whole new era in the electronic world. Today, MOS technology has been incorporated into discrete and integrated components, allowing lower power consumption, improved circuit design and operation, higher component densities, and more sophisticated operation. Unfortunately, a major problem exists with all MOS devices. They are very susceptible to destruction by static electricity.

Inadvertent static electricity is usually caused by friction. Under the proper conditions, friction can force electrons to build up on non-conductive surfaces creating a charge. When a charged substance is brought in contact with a conductive substance of lesser charge, the charged substance will discharge to the other conductor until the potentials are equal.

Everyone is "jolted" by static electricity from time to time. Static electrical charges can be built up on the human body by changing clothes, walking over certain types of carpeting, sliding across a car seat, or even friction from moving air. The actual potential of typical static charges is surprising. A static charge of sufficient potential to provide a small "zap" upon touching a conductive object is probably in the range of 2 to 4 thousand volts!

Most MOS devices can be destroyed by static discharges as low as 50 volts. The static discharge punctures the oxide insulator (which is almost indescribably thin) and forms a tiny carbon arc path through it. This renders the MOS device useless.

The point is, whenever you work with any type of MOS device, **your body and tools must be free of static charges**. There are many good methods available to do this. The most common is a "grounding strap," made from conductive plastic, that might be worn around the wrist or ankle and attached to a grounded object. Soldering irons should have a grounded tip, and special "antistatic" desoldering tools are available. Conductive work mats are also advisable. MOS devices must be stored in specially manufactured small parts cabinets, anti-static bags, and conductive foam.

*Do not try to make your own grounding straps out of common wire or conductive cable of any type!*

This is very dangerous. It is like working on electrical equipment while standing in water. Specially designed grounding straps, for the removal of static charges, are made from conductive plastic exhibiting very high resistance. Consequently, static charges can be drained safely, without increasing an electrocution risk in the event of an accident.

The susceptibility to static charges has led many people to believe that MOS devices are somehow "fragile." There is some evidence to support this notion; but in actuality, the problem is usually the result of an inexperienced design engineer incorporating a MOS device into an application where it doesn't belong. If properly implemented, MOS devices are as reliable as any other type of semiconductor device. However, care should be exercised in handling PC boards containing MOS devices, because some designers might extend an open, unprotected MOS device lead to an edge connector where it is susceptible to static voltages once unplugged.

## Building a high-quality MOSFET audio amplifier

Many audiophiles today are adamant supporters of the virtues of MOSFETs used as output drivers in audio amplifiers. They claim that MOSFETs provide a softer, richer sound; one more reminiscent of vacuum tube amplifiers. Although I won't get involved in that dispute, I will say that MOSFETs provide a more "pleasing" type of distortion if overdriven.

There are good economic and functional reasons to use MOSFETs as output drivers, however. At lower power levels, power MOSFETs display the same negative temperature coefficient as bipolar transistors. But at higher power levels, they begin to take on the characteristic of devices with a positive temperature coefficient. Because of this highly desirable attribute, temperature compensation circuits are not required and there is no danger of thermal runaway. Also, power MOSFETs, being voltage devices, do not require the high current drive that must be provided for their bipolar counterparts. The result is a simpler, more temperature-stable amplifier circuit. The only disadvantage in using power MOSFETs (that I have discovered), is the lack of availability of high-power, complementary pairs.

Figure 10-4 is a schematic diagram of the modifications required to convert the professional quality audio amplifier discussed in chapter 8 into a high-quality MOSFET audio amplifier. The "only" type of MOSFETs I recommend for this project are the 2SK134 (N-channel) and 2SJ49 (P-channel). These are simply the best MOSFETs available at any price for this application.

One of my primary goals in the design of the professional-quality amplifier was versatility. I wanted a basic amplifier building block that would provide the builder with a circuit that could be easily modified for a variety of audio amplification needs. Whereas it becomes a little too risky for the amateur to start adding parallel bipolar output stages (due to thermal instability), this is not a problem with power MOSFETs. The circuit modification illustrated in Fig. 10-4 should allow you to increase the power output of the amplifier to about 200-watts rms into an 8-ohm load.

Obviously, this also means that you must use a higher-voltage, higher-current (dual-voltage) power supply (about +/− 55 to 60 Vdc at 6 amps). This will, of course, also require larger power supply filter capacitors at higher voltage ratings (at least 10,000 $\mu$F per voltage at 75 WVdc), and a much larger heatsink (or dual heatsinks). If less power output is desired, all of these items can be derated accordingly.

Referring to Fig. 10-4, notice Q18 and Q19 are labeled as optional. You will only need to include these MOSFETs if you desire a power output greater than about 100-watts rms (that is, if you use a dual power supply higher than about 50 volts for each polarity).

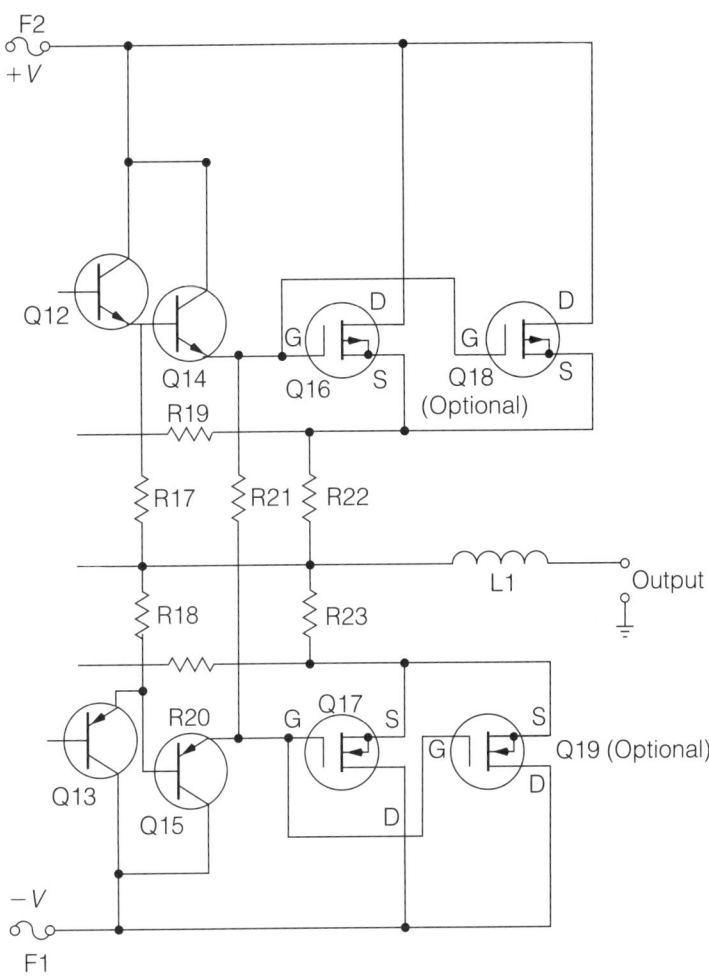

■ **10-4** *A MOSFET output stage incorporated into the professional-quality audio amplifier covered in chapter 8.*

The original bipolar output transistors are removed, and Q16 through Q19 are connected as illustrated in the Fig. 10-4 schematic (remember Q18 and Q19 are optional). Q16 and Q18 are the 2SK134 transistors, and Q17 and Q19 are the 2SJ49 transistors. All of these output MOSFETs must be installed on a heatsink using standard insulating and thermal conductivity techniques. Referring back to Fig. 8-9 in chapter 8, Q9 (which was originally used for thermal tracking) can be removed from the heatsink and installed directly on the PC board.

If you want to use the original power supply for about a 50-watt rms output, no further modifications are required. P3 will have to

*Building a high-quality MOSFET audio amplifier*

be re-adjusted, because MOSFETs require a little higher bias voltage than bipolar transistors. For an approximate 100-watt rms output, R22 and R23 should be changed to 0.22 ohm at 5 watts each. F1 and F2 need to be replaced with 3-A, fast-blow fuses. For even higher power outputs, Q18 and Q19 should be incorporated into the amplifier with R22 and R23 changed to 0.12 ohm at 5 watts each. F1 and F2 are increased to 5-A, fast-blow fuses.

After the amplifier operation is tested, and it is functioning well, Q14, Q15, and R21 can be removed from the circuit, if so desired. (MOSFETs don't need the high current drive required for bipolar outputs.) After this removal, a jumper must be installed, on the PC board, from the emitter of Q12 to the emitter of Q14. Another jumper is needed from the emitter of Q13 to the emitter of Q15. The jumpers are required to connect the Q12 and Q13 emitters to the MOSFET gates. Don't forget to adjust P3 for the correct bias setting after making this modification.

During the initial testing of high power amplifiers, the lab power supply (the first project given in this book) can be used to detect major faults or wiring errors, without posing the same risk to expensive components that a high-powered dc supply will. Although a high-power amplifier cannot operate at peak performance levels using lower currents and voltages, an initial test (with the current limiting feature of the lab power supply) can check the basic functional aspect of the amplifier; thereby, it can provide a greater assurance of success with the high-power dc supply.

## Circuit potpourri

Time for more practical fun.

### Sounds like fun

Figure 10-5 is my favorite project in this book. It can actually be "played" similar to a musical instrument to produce a variety of pleasing and unusual sounds. It is also capable of running in "automatic mode" for unattended fascination.

The heart of this light-controlled sound generator is the basic UJT oscillator illustrated in chapter 9, Fig. 9-9. Referring to this illustration, the 4.7-k$\Omega$ resistor in series with P1 is replaced with a photoresistor. Three such oscillators are needed for the Fig. 10-5 circuit. Each oscillator should have a different C1 value; the lowest C1 value chosen should be placed in oscillator #1 (to produce the highest audio frequency), the intermediate value in oscillator #2,

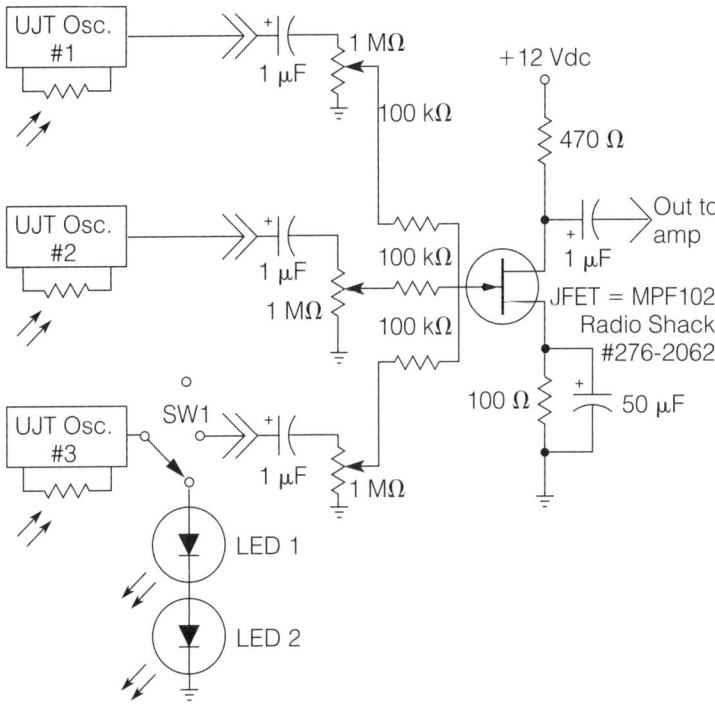

■ **10-5** *Light-controlled sound generator.*

and the highest capacity value (producing the lowest frequency) in oscillator #3.

The outputs of the three oscillators are capacitor-coupled to the input of a JFET audio mixer. The output of the mixer is connected to a "line level" input on any audio amplification system.

The P1 potentiometer in each oscillator is adjusted for a good reference frequency under ambient lighting conditions. By waving your hand over the photoresistor (causing a shadow), the frequency will decrease accordingly. With all three oscillators running, the waving of both hands over the three photoresistors can produce a wide variety of sounds. By experimenting with different P1 settings in each oscillator, the effects can be extraordinary.

Another feature, added for automatic operation, is two "high-brightness" type LEDs on the output of oscillator #3. Referring to Fig. 10-5, when SW1 is in the position to connect the oscillator #3 output to the LEDs, the LEDs will flash on and off at the oscillator's frequency. If these LEDs are placed in close proximity to the photoresistors used to control the frequency of oscillators #1 and

#2, their frequency shifts will occur at the oscillator #3 frequency. In addition, even subtle changes in ambient light will cause variances. The possibilities are infinite. Although not shown in Fig. 10-5, you will need to add some series resistance between the output of oscillator #3 and the LEDs to limit the current to an appropriate level (depending on the type of LEDs used).

This circuit works very well with the 12-watt amplifier discussed in chapter 8. Placed in an attractive enclosure, it is truly an impressive project.

The JFET mixer (Fig. 10-5) is a high-quality audio frequency mixer for any audio application. If additional inputs are needed, additional 1-M$\Omega$ potentiometers and 100 k$\Omega$-resistor combinations can be added.

### Emergency automobile flasher

Figure 10-6 illustrates how one HEXFET (a type of MOSFET) can be used to control a high-current automobile headlight for an emergency flasher. A UJT oscillator (chapter 9, Fig. 9-9) is modified for extremely low-frequency (ELF) operation (about 1 Hz), and its output is applied to the gate of the HEXFET as a switching voltage. The 1-k$\Omega$ resistor and 1000-$\mu$F capacitor are used to decouple the oscillator from the power circuit.

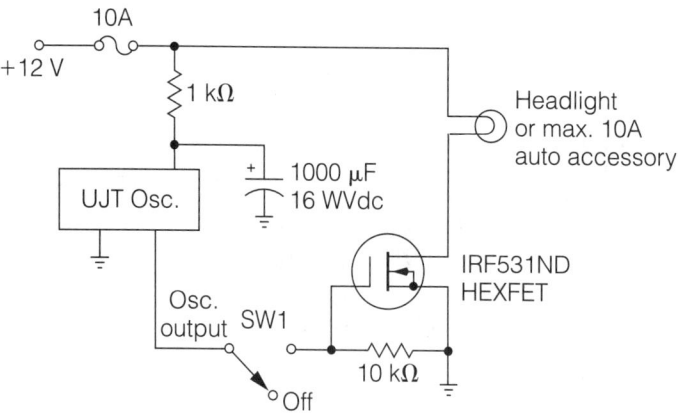

■ **10-6** *Automobile headlight flasher.*

Any high-current automotive accessory (up to 10 amps) can be controlled with this circuit—even inductive loads, such as winch motors.

## Home-made ac

The circuit illustrated in Fig. 10-7 is used to convert a low-voltage dc power source (usually 12 volts from an automobile battery) to a higher-voltage ac source. Circuits of this type are called *inverters*. The most common application for this type of circuit is the operation of line-powered (120 Vac) equipment from a car battery. There are, of course, many other applications.

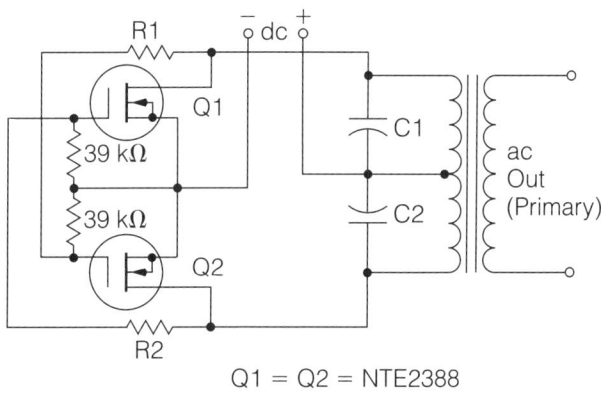

Q1 = Q2 = NTE2388

■ **10-7** *Power MOSFET inverter.*

For example, if you wanted to use this circuit for the previously mentioned application, the 12-volt dc source from the car battery would be applied to the dc terminals shown in Fig. 10-7 (observing the correct polarity, and fuse protecting the +12-volt line from the battery). A standard 12.6-volt ct secondary/ 120-volt primary power transformer is used. The VA rating of the transformer will depend on the load of the line-powered equipment intended for use with this circuit. If the line-powered device required 120 Vac at 1 amp (for example), a minimum size of 120 VA is needed (I recommend using at least a 10 to 20% higher VA rating to compensate for certain losses). With the components specified, a 200-VA transformer is the largest transformer that can be used.

The combination of C1, C2, and the transformer secondary make up a resonant circuit (resonance will be discussed in a later chapter). Used in conjunction with the active components (Q1 and Q2), this circuit becomes a free-running oscillator, with the frequency being determined primarily by the value of C1 and C2. The transformer will operate the most efficiently at about a 60-Hz frequency, so the value of C1 and C2 should be chosen to "tune" the oscillator as closely to that frequency as possible.

C1 and C2 should be equal in capacitance value. Start with 0.01 $\mu$F for C1 and C2, and use a resistance value of 100 k$\Omega$ for R1 and R2. These values should bring you close to 60 Hz. If the frequency is too high, decrease the values of the capacitors slightly and vice versa for a lower-frequency condition.

# Batteries

BECAUSE MANY HOBBY AND COMMERCIAL ELECTRONIC products receive their operational power from batteries, I thought it suitable to include this chapter. Before getting into details, you must begin by learning some basics.

The term *battery* actually refers to an electrochemical dc power source containing multiple cells. Nominal voltage levels for individual cells can vary, depending on the type, from 1.25 to 2 volts per cell. The 1.5-volt electrical devices most people refer to as "flashlight batteries" are, technically speaking, cells.

*Primary batteries*, or "dry cells", are typically thought of as nonrechargeable, although this description is not always true. The electrolyte used in primary batteries is not always "dry" either.

*Secondary batteries* are rechargeable, and usually do contain a liquid or paste-type electrolyte.

The *amp-hour (Ah) rating* is a term used to define the amount of power that a battery, or cell, can deliver.

## Battery types

All batteries use a chemical reaction to produce electrical current. Battery types are usually defined by the types of chemicals or materials used in their construction.

There are seven main types of commercially available batteries: zinc, alkaline, rechargeable alkaline, nickel-cadmium, lead-acid, "gelled" electrolyte, and lithium.

*Zinc batteries* are the most common type of "flashlight" battery. They are available in regular and heavy-duty types, with neither being very exemplary in regards to performance levels. They are not recommended for the majority of electronic projects.

Regular *alkaline batteries* last 300% to 800% longer than zinc batteries, depending on their application.

*Nickel-cadmium batteries,* or "nicads," are very popular secondary batteries. These are the type used in the majority of commercially available rechargeable products. A newer type of high-capacity nicad is also being used extensively today. It boasts a much longer service life, and a faster recharge rate (typically 5 to 6 hours). A bothersome peculiarity of nicads is their tendency to develop a "memory." Because of this characteristic, it is recommended that they be fully discharged before recharging.

*Lead-acid batteries* are most commonly used as automobile batteries. They are also available in smaller sizes that are sealed (except for a blow-hole to allow gases to escape). "Motorcycle-type" lead-acid batteries are available in a wide range of amp-hour ratings, and are a good choice for heavy-duty projects.

*Gelled electrolyte batteries* fall into the same category as lead-acid types. They are most often used in "uninterruptable power supplies" (abbreviated as "UPS" supplies; power supplies intended to supply 120-Vac power in the event of a power failure), burglar alarms, and emergency lights.

*Lithium batteries* are designed to supply a small amount of power for a long period of time. Lithium batteries are most often used to power memory back-up systems in computers, and in wrist watches. They are very expensive.

The most recent entry into the common marketplace is called the *rechargeable alkaline battery.* It was developed and patented by Rayovac Corp. under the trade name of "Renewal." These batteries offer two to three times the energy storage capacity of nicad batteries, higher terminal voltage (1.5 volts vs. 1.2 volts), retention of full charge for up to five years in storage, lower operating temperatures, and they are "environmentally friendly." Rechargeable alkaline batteries can be used through well over 25 full charge/discharge cycles with very little degradation in performance, so that can add up to quite a cost savings over buying "throw-away" batteries.

## Battery ratings

As stated previously, the power delivering ability of a battery is rated in "amp-hours." However, this does not mean that a 5-Ah battery will deliver a full 5 amps for one hour, and then suddenly quit. The battery manufacturer will calculate this rating over a longer period of time and then "back-calculate" to find the rating. For example, a 5-Ah battery should be capable of providing 500 milliamps for 10 hours. This calculates out to 5 amps for 1 hour, 1 amp for 5 hours, or 500 milliamps for 10 hours.

Trying to operate a battery at its maximum current rating for 1 hour is destructive to the battery. A 5-Ah battery would be a good choice for a project requiring one amp of current, possibly even two amps. Primary battery life and secondary service life (the total number of charge/discharge cycles it can withstand before failure) can be extended (by factors ranging from 100% to 400%) by avoiding excessive current drain.

### Battery care

In addition to excessive-duty operation, the most destructive variable to batteries is heat. Batteries should always be stored in a cool place, even a refrigerator (not a freezer!). Batteries being used in vehicles that are left outside in the sunlight should be removed and brought indoors, if possible.

Secondary batteries should be maintained in a charged state. Lead acid and gelled electrolyte batteries need to be recharged about once every 3 to 6 months if not used. Nicads, on the other hand, have a self-discharge rate of about 1% per day! For optimum performance, they need to be left on a trickle charge continuously when not in use.

### A few words of caution

Some types of batteries contain very potent acids, caustic substances, or highly poisonous materials. These materials include mercury, lead, hydrochloric acid, and other substances so toxic that the EPA classifies them as *toxic waste*. If you want to experiment with chemistry, buy a Gilbert Chemistry Set, but don't attempt to mess around with this stuff.

**Don't try to recharge a battery that is not designed for recharging**. It could explode. And if the explosion doesn't do enough damage in itself, toxic waste can be sprayed in the eyes and mouth. Even rechargeable batteries can explode if recharged too fast, or if overcharged.

Lead-acid batteries can produce enormous current flows. An accidental direct short, with a lead-acid battery as the power source, can literally blow up in your face; this could result in eye damage, or fires. **Large batteries should be fuse-protected right at the battery terminals**.

### Recharging batteries

The recharging of secondary batteries is not a complex process, but there are a few rules to follow. Nicad batteries should be completely discharged before trying to recharge them. They should not be discharged too quickly, however. One good way to discharge them properly is to connect them to a small incandescent lamp rated for the same voltage as the battery. When the lamp goes out, they're discharged. Of course, other types of resistive loads will perform this function as well as a light bulb, but they won't provide a visual indication of when the discharge has been completed. As stated previously, it is destructive to discharge any battery too quickly. Nicads, however, are the only secondary battery type in which mandatory discharge becomes a concern. Other types of batteries can be recharged without having to be fully discharged.

Recharging procedures are generally the same for all types of secondary batteries. The primary rule to keep in mind, in regard to charging rates, is to not try to recharge them too fast. In some cases, trying to recharge a secondary battery too fast may cause it to explode. A good rule-of-thumb to follow is: do not allow the charging current to exceed one-tenth the value of the amp-hour rating. For example, a 5-Ah battery should not receive a charging current any higher than 500 milliamps (1/10 of 5 amps).

In many cases, the recharge voltage applied to a secondary battery is higher than the rated battery voltage. For example, a typical automobile battery is rated at 12 volts (six cells at 2 volts each). The charging voltage applied to an auto battery is 13.8 volts. There are practical reasons for doing this with a car battery, but it is seldom advisable to follow this practice with batteries in other applications. In the majority of situations, the use of higher voltages ends up being a waste of power, and a potential risk toward overheating the battery.

## Building a general-purpose battery charger

By now, you may have come to the conclusion that all you need to properly recharge a battery is a variable dc power supply and the properly sized series resistor to limit the current. You're right! For example, if you wanted to recharge a 5-Ah, 12-volt battery, you will want to limit the charging current to 500 milliamps.

When the "dead" battery is first connected to the power supply, assume it to "look" like a short (it will come close to that if it is totally discharged). That means you need a resistor to drop 12 volts at 500 milliamps. Using ohm's law, that comes out to 24 ohms. Because 24 ohms is not a standard resistor value, a 27-ohm resistor will do nicely. In the beginning of the charge cycle, this resistor will be required to dissipate almost six watts of power, so use one with a 10-watt rating. Alternatively, you could use two 56-ohm, 5-watt resistors in parallel; or, you could use three 8-ohm, 2-watt (or higher, 5-watt being preferable) resistors in series. Set the power supply to 12 volts, put the 27-ohm resistor in series with the battery, and the battery should recharge properly.

There is a small problem with this simple resistor-power supply method. As the battery begins to charge, the voltage across it will increase. This causes a subsequent decrease of voltage across the resistor and the charging current decreases. When the battery gets close to being totally recharged, the charging current drops to a very low value. The result is that the recharging process takes more time. This might or might not be a problem, depending on your needs.

If a more rapid recharge is desired, the circuit illustrated in Fig. 11-1 will provide a constant charge current, regardless of the battery voltage. This charge current will be maintained until the battery is fully charged, then it will automatically drop to a safe level. This circuit can be used with any type of variable power supply (as long as the power supply can be adjusted to a voltage slightly higher than the battery voltage), and it can charge any type of battery up to about a 15-Ah rating.

Referring to Fig. 11-1, the variable power supply is connected to the positive and negative input terminals of the charging circuit, and adjusted to be about 1 volt higher than the battery voltage. (The "lab quality power supply" project, discussed in earlier chapters, will work well with this circuit.) The rotary switch (RS1) is set to the desired charging current position:

- [ ] Position #1 = 15 Ah batteries (or larger)
- [ ] Position #2 = 7.5 Ah batteries
- [ ] Position #3 = 2 Ah batteries
- [ ] Position #4 = 1 Ah batteries
- [ ] Position #5 = 0.2 Ah batteries (or smaller)

RS1 is set to the position rated at, or below, the actual Ah rating of the battery. For example, a battery rated at 5 Ah should be recharged in the #3 position, not in the #2 position!

Figure 11-1 is a simple current-limiting circuit. It is identical in function to the current limit circuit used in the "lab power supply" project discussed earlier in this book. However, some component values have been changed to provide different current limit values. D1 is to protect the circuit in the event the battery terminals are connected in the wrong polarity.

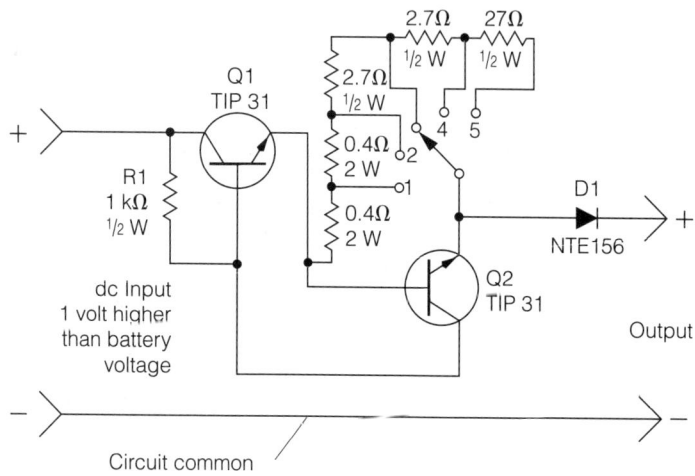

■ **11-1** *Battery charger circuit.*

If you would like to modify this circuit to provide different charging currents for special needs, this can be easily accomplished. Simply divide 600 millivolts by the desired charging current in milliamps. The answer, in ohms, will be the total resistance needed between Q2's base-emitter leads to provide the desired current.

# Integrated circuits

THE PROCESS OF MINIATURIZING MULTIDEVICE CIRCUITS (transistors, resistors, diodes, etc.) is called *integration*. An *integrated circuit (IC)* is a chip that contains (or can perform the function of) many discrete (non-integrated) devices. In appearance, most common types of ICs are small, rectangular packages with 8 to 16 pins (or legs) extending from the package. This physical configuration is called a *dual in-line package (DIP)*. Other common package styles include round metal casings with multiple leads extending from them, and larger rectangular packages with up to 40 pins.

Most ICs are manufactured as general purpose building blocks within a specific functional area. For example, the common 741 operational amplifier is designed specifically for amplification type functions, but its external component design can modify its performance for literally thousands of different applications. In contrast, many ICs are manufactured for very specific applications, especially in the consumer electronics field. For example, an IC specified for use as a rotational speed control for VCR heads can be used for little else.

It is easy to integrate semiconductor components (or components that can be made from semiconductor materials). *VLSI chips* (very large-scale integration) can contain as many as 250,000 transistors. Resistors can be made very accurately from semiconductor material. However, large-value "reactive" components, such as inductors and capacitors, cannot be reduced in size. Also, semiconductor components designed to dissipate large quantities of power cannot be integrated very successfully. For these reasons, many ICs are nothing more that the total low-power semiconductor part of a larger circuit. The reactive and high-power components must be added for a functioning circuit. "Switching regulator control" ICs are a good example of this kind of design.

Because of the vast number of integrated circuits available, it is absolutely necessary to have a good selection of "manufacturers' data books" in your electronics library. You will have to depend on these data books to obtain pin-out diagrams (illustrations of the functional aspects of each connection lead on an IC), application information, and functional specifications.

ICs can be divided into two main families; "digital" and "linear" (or "analog"). Digital ICs will be discussed in the next chapter, so in this chapter, we will concentrate primarily on linear devices.

Linear ICs can be further subdivided down into several classifications; "operational amplifiers," "audio amplifiers," "voltage regulators," and "special purpose devices."

For very common applications, a manufacturer may build a complete circuit, using ICs and discrete components, and encapsulate the complete circuit into a block. These devices are called *hybrid modules*. Common hybrid modules include high-power audio amplifiers, power supply regulators, motor controls, and certain types of high-power, high-voltage devices.

## Operational amplifiers

*Operational amplifiers* are basic amplification building blocks. They consist of multiple high-gain differential amplifiers (exhibiting high common-mode rejection) without any kind of feedback loop. Depending on the configuration of the external components, they can be used for voltage amplifiers, transconductance amplifiers (voltage-to-current converters), transimpedance amplifiers (current-to-voltage converters), differentiators, integrators, comparators, oscillators, and regulators.

Virtually all operational amplifiers have two inputs marked with a positive (+) sign and negative (−) sign. These are the noninverting inputs and inverting inputs, respectively. Don't confuse these with the power supply connections. Many op amps have frequency compensation inputs. Normally, a capacitor or resistor/capacitor combination is connected between these pins for controlling high frequency characteristics. Offset null inputs are provided on many op amps to bias the output at a desired dc quiescent level.

Op amps are not specified in regard to frequency response, because it will vary depending on the way the op amp is used in a circuit. For example, an op amp used as a voltage amplifier (with the external components setting its gain at 10) will have a much broader frequency response than if the external components set its gain at 1000. Therefore, the frequency response characteristic of op amps is defined by the term *slew rate*. *Slew rate* defines the speed (given in microseconds or nanoseconds) in an op amp's output change in accordance with an instantaneous change on the inputs. The higher the slew rate, the higher the maximum usable frequency response.

Most op amps are designed to be used with dual-polarity power supplies. The power supply voltages should be equal, but opposite in polarity. Op amps designed for dual-polarity power supplies can work with single polarity supplies for certain applications. Also, some op amps are designed exclusively for single-polarity supplies.

The "perfect" op amp would have infinite gain, zero output impedance, infinite input impedance, instantaneous slew rate, and be totally immune to common mode noise on the power supply inputs and signal voltage inputs. If a perfect op amp existed, the circuit designer could use external components to design any feasible type of op amp circuit without any consideration of op amp parameters or limitations. Of course, perfection is not possible, but modern op amps, especially "FET input" op amps, come close to perfection parameters. Therefore, the equations used for designing amplifiers, filters, oscillators, and other circuits, seldom include any op amp variables.

Figure 12-1 illustrates some examples of common op amp circuit configurations. These illustrations do not give pin numbers or

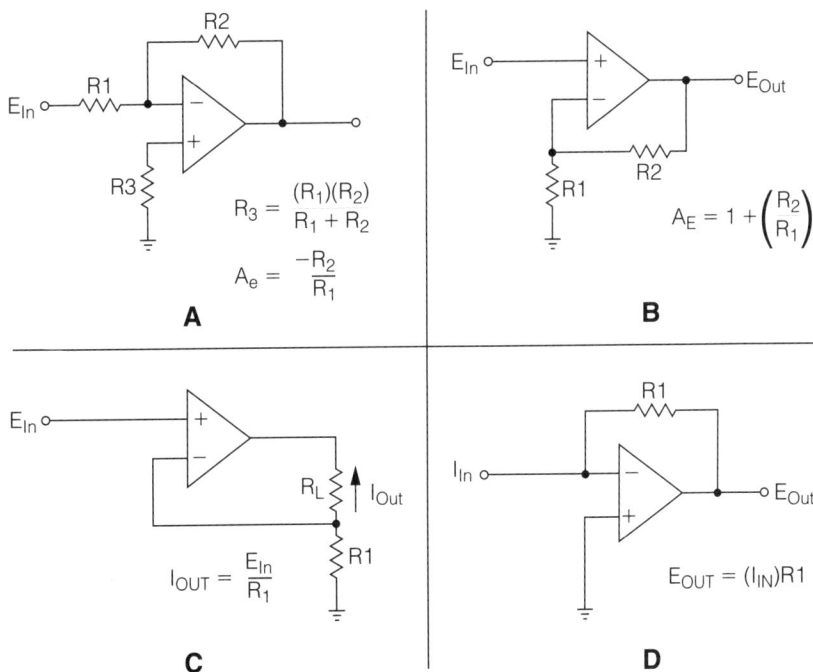

■ **12-1** *Common op-amp circuit configurations.*

power supply connections because they are general in nature. Virtually any general-purpose op amp will work in these circuits, and many high-quality, low-noise op amps will work better. For these circuits, and almost any type of op amp circuit, I highly recommend the low-noise NE5532 (NTE 778A) dual op amp ICs, although the "industry standard" 741 type will probably perform satisfactorily in most cases. For critical audio circuits (preamplifiers, mixers, tone controls, etc.), I recommend the TL-074 op amps, which boast a high slew rate and 0.005% THD (total harmonic distortion).

Referring again to Figure 12-1, examine some details of each circuit illustrated. Circuit A is an inverting voltage amplifier. The equation for calculating voltage gain shows that it is solely dependent upon the ratio of R1 and R2. For example, if R1 is 10 Kohm and R2 is 100 Kohm, the voltage gain is 10. The negative sign (−) is placed in front of the R1/R2 ratio expression to show that the output will be inverted. R2 in this circuit is the negative feedback resistor. Note that it is connected from the inverting input to the output. The output portion being "fed back" to the input will be inverted, or opposing in nature. Consequently, it is negative feedback. R3 is used for circuit stabilization, and its calculation is included in the illustration.

Circuit B is a noninverting voltage amplifier and, as the gain equation shows, its gain is also dependent upon the ratio of R1 and R2.

Circuit C is a transconductance amplifier, or a voltage-to-current converter. The current output, which is the current flow through $R_L$, will not be dependent upon the resistance value of $R_L$. Rather, it will be dependent upon the ratio of the input voltage to the value of R1.

Circuit D is a transimpedance amplifier, which functions in an opposite manner to circuit C. Circuit D converts an input current into a proportional voltage. The associated equation for this function defines the component relationship.

All of the circuits illustrated in Fig. 12-1 must be designed with their power supply limitations in mind. For example, circuit A cannot be designed so that a desired input and voltage gain will drive the output in excess of the power supply voltages used. In the same way, the manufacturer's specified maximum power dissipation cannot be exceeded without destroying the chip. In multiple op amp circuit designs, it may also be necessary to decouple the individual op amp circuits (usually with about a 100-$\mu$ electrolytic capacitor and a 0.1-$\mu$F nonpolarized capacitor).

## IC or hybrid audio amplifiers

Many currently available audio amplifiers are offered as "totally integrated" or hybrid. To the electronic "user," the only major difference is in reference to size; hybrid circuits are typically larger, and might require a larger heatsink.

In general, audio amplifier ICs might be thought of as "simple-to-use power op amps," because that is essentially what they are. Many types are designed for single polarity power supplies (which make them excellent for battery-powered applications), and they frequently have internal gain and frequency compensation networks. These attributes, together with others, add up to a quick and simple, general-purpose audio amplification system for low to medium-power applications. Heatsinks are typically required for outputs in excess of 1 watt.

Although great progress has been made in the integration techniques of audio amplifiers, the performance levels of many medium to high-power systems leaves much to be desired. For this reason, most audiophiles still prefer discrete audio power amplifiers in the majority of cases.

## IC voltage regulators

This is an area where ICs have virtually "taken over." Typical IC voltage regulators provide almost unexcelled voltage regulation, overvoltage protection, current limiting, and automatic thermal shut-down (shuts off when it gets too hot); and, many are adjustable. In addition, they are commonly available and very inexpensive. Its no wonder they are popular.

In addition to all of their other desirable attributes, voltage regulator ICs are very easy to use and implement for almost any application. Most have only three connection terminals; one is the "raw" dc input, another is the output, and the third simply connects to circuit common, or to a single potentiometer for voltage adjustment. You can't get much simpler than that!

The small size and low cost of regulator ICs has made it practical to use them as on-board regulators for each individual printed circuit board within an entire system. This has reduced the need (and cost) of large, high-current regulated power supplies for larger electronic systems. And, because this puts them in close proximity to each PC board within a system, the regulation and noise immunity is usually better than can be achieved with a large central power supply.

Regulator ICs commonly require heatsinking for maximum output. Most are available in either TO-220 or TO-3 packages. This causes them to look like power transistors, but it provides them the flexibility to be mounted on the many styles of transistor heatsinks already available.

## Special-purpose ICs

There are many more special-purpose integrated circuits than there are pages in this book. The determining factor as to whether or not any particular circuit will be available in IC form is purely economic. It is expensive for a manufacturer to do the research and development required to produce a new type of IC. Consequently, manufacturers do extensive marketing research (to ensure the demand, in great quantities, for a special-purpose IC) before they can justify the cost of the initial investment. If they do not believe that they can regain their R & D costs, the IC is not manufactured.

Skim through several IC data books to acquire a basic knowledge of "what's out there" regarding special purpose ICs. Before using a special-purpose IC, closely examine its specifications. There are many examples where discrete versions of circuits will perform better than their IC counterparts.

## Improvement of lab-quality power supply

The lab-quality power supply, as discussed previously in chapters 3 through 6 of this book, has a minor shortcoming. If you have used this supply for many of the prior projects, you might have noticed that the output voltage will drop slightly when the supply is heavily loaded. Because this power supply was offered as the first project in this book, I felt it was prudent to keep the parts count and complexity to a minimum. Although many commercial power supplies utilize this same basic design for their unregulated supplies, it might not measure up to some of your future requirements in more critical circuits. Therefore, this section will detail two ways of improving the voltage regulation performance; it's up to you to decide on either method. You might also decide to leave this supply as it is, and build the quad power supply detailed in the next section instead.

First, refer back to Fig. 6-9 in chapter six. To improve the voltage regulation of this circuit, it is necessary to increase the gain of the

Q3 and Q4 stages. One method of increasing gain would be to replace Q3 and Q4 with darlington pairs. Another method, providing even tighter control, is illustrated in Fig. 12-2.

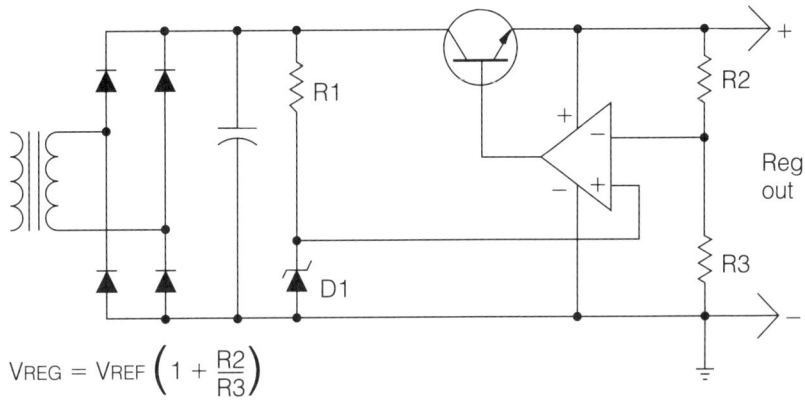

$$V_{REG} = V_{REF}\left(1 + \frac{R2}{R3}\right)$$

■ **12-2** *Improvement of lab-quality power supply.*

Although this circuit only shows the positive regulator, you would simply duplicate the basic theory and design in the negative regulator. (The current-limiting circuit of Fig. 6-9 is not shown in Fig. 12-2 for the sake of clarity). Q3 of Fig. 6-9 would be replaced by the general-purpose op amp shown in Fig. 12-2. The inverting input of the op amp would connect to the wiper of the voltage control potentiometer, P1 (Fig. 6-9), and the noninverting input would connect to the anode of D1 (the voltage reference diodes in Fig. 6-9). The output of the op amp connects to the base of Q1 (the series-pass transistor, Fig. 6-9) and the op-amp power supply connections are made to the positive and circuit common points as illustrated.

The circuit, shown in Fig. 12-2, is included as a basic reference for all of your future power supply needs. To use this design, start by designing a simple zener regulated power supply (the transformer, bridge rectifier, filter capacitor, R1, and D1). Be sure that the zener voltage of D1 is somewhat less than the anticipated voltage drop across R3. The equation for calculating the value of R2 and R3 is provided in the illustration. Of course, Q1 is chosen based on the current and power dissipation requirements of the proposed loads to be applied to the circuit.

The best method of improving the lab power supply is illustrated in Fig. 12-3. The full regulator circuit design has been included so that the circuit modifications might be more easily located. These alter-

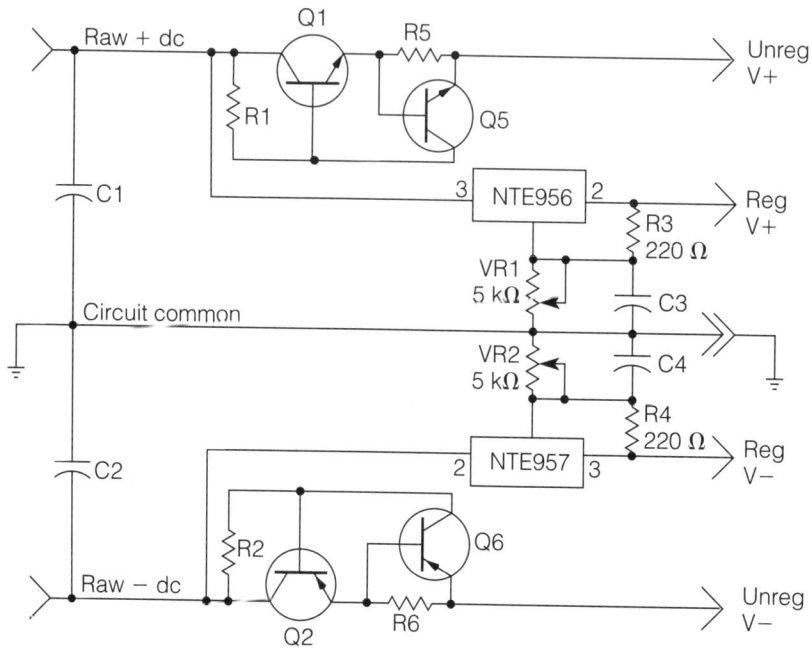

■ **12-3** *Improvement of original lab power supply illustrated in Fig. 6-7.*

ations should be self-explanatory by comparing Fig. 6-9 with Fig. 12-3. Note that you will need to mount two additional binding posts for the positive and negative regulated outputs. Keep the original two posts intact, connected to the unregulated current limit circuit, for testing audio power amplifiers (and other projects requiring higher-voltage, dual-polarity supplies). Also note that P1 and P2 (Fig. 6-9) must be replaced with 5-kΩ potentiometers.

## Building a quad power supply

Figure 12-4 illustrates a quad-output lab power supply. This is an extremely versatile lab power supply. It provides the most frequently used voltages in dual polarity. Each output is current limited at 1.5 amps, and all of the regulator ICs are internally protected from overvoltage and overtemperature conditions. The regulator ICs are the "fixed voltage" type, meaning that they are not voltage adjustable, but they provide extremely good voltage regulation. All of the power supply components have been chosen so that all four outputs can be loaded to maximum capacity simultaneously.

*Integrated circuits*

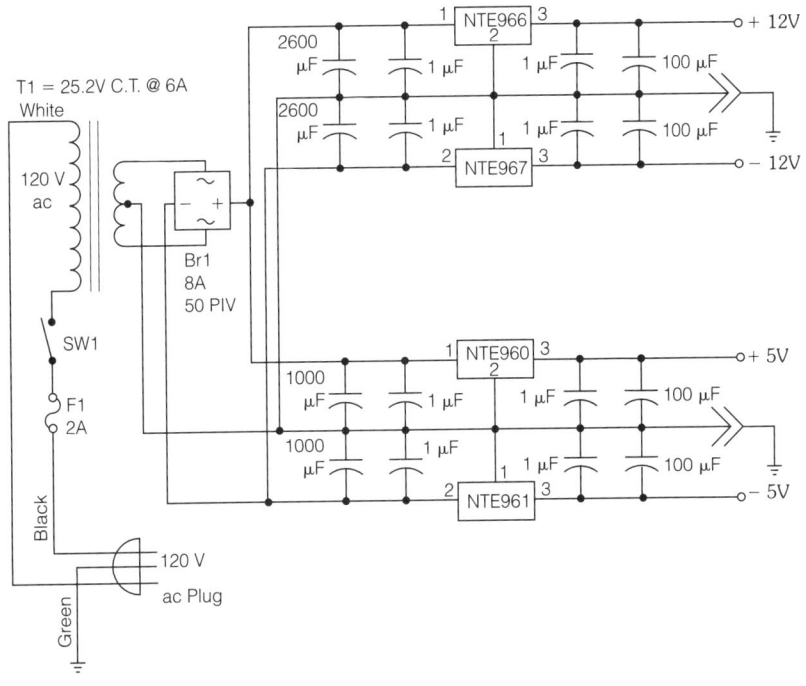

■ **12-4** *A quad-output lab power supply.*

All four regulator ICs will require some form of heatsinking. Mounting the ICs to a metal enclosure will probably be sufficient, but the combination of the chassis and the "sinks" is recommended. After constructing the power supply, load down all four outputs, and let it run for a while. If the enclosure (or the ICs) start to become too hot, you will need to include some additional heatsinking between the enclosure and IC mounting tabs.

## Circuit potpourri

### Noise hangs in the balance

Most high-quality microphones, those designed for professional entertainment use, are low-impedance devices and have balanced XLR-type connector outputs. Although these "mikes" can be impedance matched for a high-impedance input with a single transistor stage, this will defeat the whole purpose of having a balanced line.

The circuit illustrated in Fig. 12-5 will match the impedance correctly, and utilize the high common-mode rejection characteristic

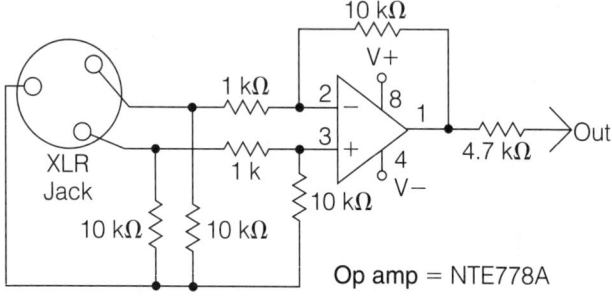

■ **12-5** *A balanced input circuit for low-impedance microphones.*

of operational amplifiers to remove the unwanted "hum" picked up by long microphone wires.

### Hum reducer

Furthermore, on the subject of hum (stray 60-Hz noise, inductively coupled to an audio circuit), the circuit illustrated in Fig. 12-6 can do a very effective job of removing almost all hum content, even after it has already been mixed with other audio material. However, it will also reduce some low bass frequencies (from about 40 to 100 Hz).

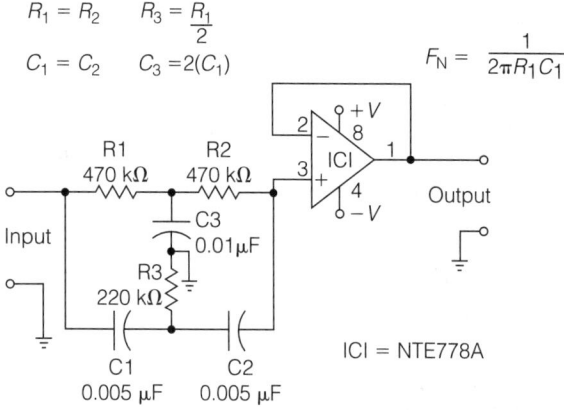

■ **12-6** *A 60-Hz notch filter.*

Figure 12-6 is called a *twin-tee notch filter*. The component values chosen will cause the output to drastically attenuate 60-Hz frequencies, but allowing most other frequencies to pass unaltered. Attenuation at 60 Hz should be on the order of about 12 dB.

Equations are provided in the illustration to design notch filters for other frequencies, also. Notice the equation for finding the notch frequency ($F_n$). The little symbol with the two legs and a wavy top is called *pi*. Pi is a mathematical standard, and it is approximately equal to 3.1416.

## Let the band pass

Figure 12-7 is an example of a bandpass filter. A *bandpass filter* only passes a narrow band of frequencies, and severely attenuates all others. This circuit illustrates how a JFET can be used as a voltage-controlled resistor; to "tune" the bandpass frequencies of the filter by means of a control voltage (about 0 to 2 Vdc). The usable frequency range is from sub-audio to about 3 kHz. This circuit is commonly used to produce "wah-wah" effects for electric guitars and other instruments.

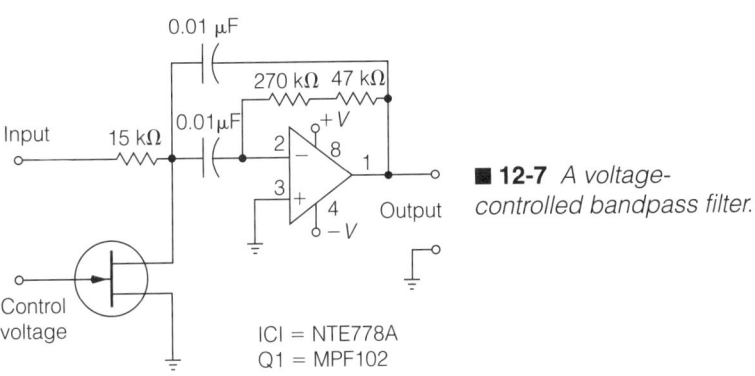

■ **12-7** *A voltage-controlled bandpass filter.*

When several of these circuits are placed in parallel with component values chosen to provide different ranges of bandpass, the result is a *parametric filter*. If a manual control of the bandpass is desired, Q1 might be replaced with a 5-kΩ resistor.

## High-versatility filters

Figures 12-8 and 12-9 are both examples of equal component, *Sallen-Key active filter* circuits. Equal component filters are easier to build, because other filter designs require exact multiples or divisions of the values of key components which might not be standard in value. Figure 12-8 is a low-pass filter, and Fig. 12-9 is a high-pass design. Both designs provide about 12 dB-per-octave of signal attenuation for the unwanted frequencies (called *rolloff*).

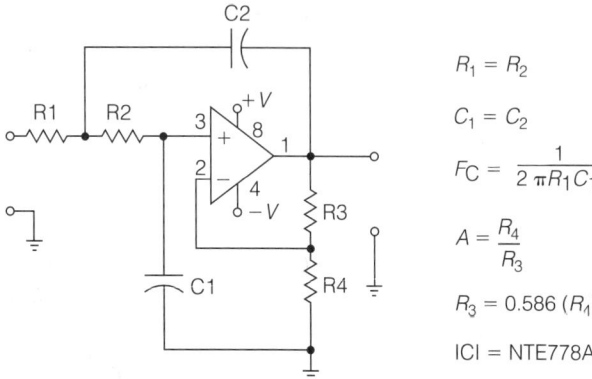

■ **12-8** Low-pass filter.

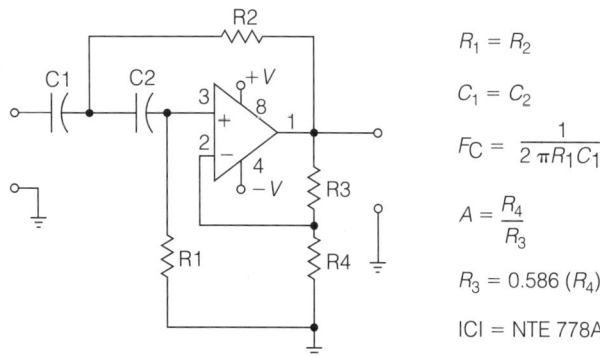

■ **12-9** High-pass filter.

To design these filter circuits for your own applications, start with: R3 = 12 kΩ and R4 = 22 kΩ. Using R1 and R2 values of 4.7 kΩ; and C1 and C2 values of 0.005 μF; $F_c$ should be a little over 6 kHz. From this point, you can adjust component values, as needed, by following the equations provided in the illustrations.

These circuits can be cascaded (in series) for sharper cut-off responses. For example, if two Fig. 12-8 circuits were built, each having a cut-off frequency of 6 kHz, and they were placed in series, the final output would have a 24-dB/octave rolloff. The Fig. 12-8 and Fig. 12-9 circuits can also be placed in series to provide precise bandpass responses with both upper and lower cutoff frequencies. For example, assume you need a filter circuit to reject all frequencies except those occurring between 6 kHz and 8 kHz. You would design the low-pass filter for a cut-off frequency of 8 kHz, and the high-pass filter for a cut-off of 6 kHz. By placing the two filter circuits in series, the total response would be that of a bandpass filter with a "pass-band" of 6 kHz to 8 kHz.

## How was your trip?

Figure 12-10 illustrates an adjustable trip-point relay driver. This is an example of how an operational amplifier can be used as a "comparator." A comparator circuit "compares" two voltages. When one exceeds the other, a signal is provided.

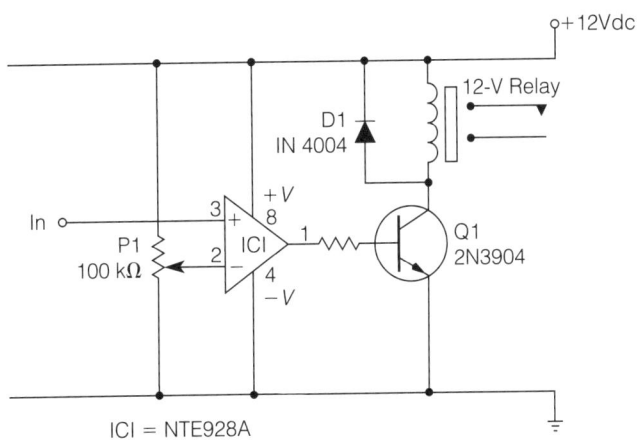

■ **12-10** *Adjustable trip-point relay driver.*

The op amp compares an adjustable reference voltage, provided by P1, to an input voltage. When the input voltage exceeds the reference voltage, the op amp output goes high, turning on Q1, which energizes the relay. The relay, of course, can be used to turn-on, or provide power to, just about any kind of device. For precise, repeatable performance, the positive rail should be regulated by an LM7812 (NTE 966) positive voltage regulator.

Circuits of this nature are used primarily for control applications. For example, different variables such as temperature, humidity, or light could be utilized to control other circuits, appliances, or machinery. For a security light controller, a photoresistor and fixed resistor, in series as a voltage divider network, could be connected between the positive rail and ground. The input to the op amp could be connected to the junction between the two. If this circuit were to be placed outdoors, the resistance of the photoresistor would increase as the sun started to set. When the voltage drop across the photoresistor became high enough to exceed the reference voltage applied to the op amp, the relay would be energized turning on an outside light.

When two op amps are configured as comparators, one referenced to a low voltage and the other to a higher voltage (called *high and low trip points*), the resulting circuit is called a *window comparator*. Window comparators are used to control a variable within a set of operating parameters.

### Watt an amplifier

As a final entry into this section of circuit potpourri, examine the simple 1-watt IC audio amplifier of Fig. 12-11. P1 is the volume control. The rest of the circuit is self-explanatory.

This circuit works very well in battery-powered applications, and makes an excellent little amplifier for many of the sound-effect circuits covered previously in this book.

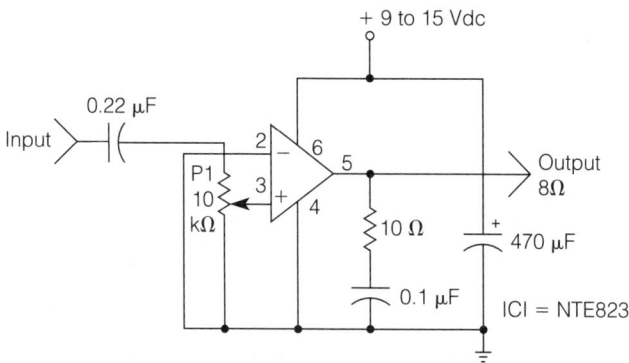

■ **12-11** *1-watt audio amplifier.*

# Digital electronics

# 13

DIGITAL ELECTRONICS IS A LARGE BRANCH OF THE electrical/electronic field relating to those electronic functions used in performing logical functions. It encompasses every type of logical system from simple combinational logic control in a coffee pot to the largest computer systems. The concept of digital control is not new. Even before the first computer appeared, or solid-state electronics came into being, large banks of relays were performing logical control functions in industrial facilities.

The earliest types of computers were called *analog computers*. They were enormous machines made from thousands of vacuum tubes. Computations were performed using voltage levels based on the decimal (base 10) numbering system. Although this method seems natural because "you" think in terms of 10's (being that you were created with 10 fingers), the old analog computers soon gave way to the more modern binary computers (base 2). There are very good reasons for this change, which will be explained as this chapter continues.

Although many people seem to have trouble comprehending different numbering systems, it's really quite simple. The key is in understanding the mechanics behind the decimal system, and then applying those principles to any other numbering system.

The term *decimal* means "base 10." If the number 1543 is broken down into decimal column weights, it comes out to 3 units (or ones), 4 tens, 5 hundreds, and 1 thousand. Notice how each succeeding weight is actually the base number (10) times the "weight" of the preceding column. In other words, $1 \times 10 = \mathbf{10}$, $10 \times 10 = \mathbf{100}$, $10 \times 100 = \mathbf{1000}$, etc.

The *binary numbering system* works exactly the same way, except that it is based on 2 instead of 10. For example, the first weight (or least significant digit) is the units, or ones, column. The weight of the second column is $2 \times 1$, or **2**. The next column is $2 \times 2$, or **4**. The next column is $2 \times 4$, or **8**. Instead of the column weights being ones, tens, hundreds, thousands, ten-thousands, etc.; the binary column weights will be ones, twos, fours, eights, sixteens, etc.

In the decimal system, there are ten possible numbers which can be placed in any weight column (0, 1, 2, 3, 4, 5, 6, 7, 8, 9). In the binary system, there are only two possible numbers for any one weight column (a 0 or a 1; a "yes" or a "no;" an ON or an OFF). If a binary number such as 0111 is broken down into weights, it means 1 one, 1 two, 1 four, and 0 eights. By adding the weights together, the binary number can be converted to decimal. In the previous example, $1 + 2 + 4 = 7$. Therefore, 0111 is the binary equivalent to decimal 7.

The following example demonstrates how it is possible to count up to 9 using the binary numbering system:

| 8 4 2 1 | Column Values | 8 4 2 1 |
|---|---|---|
| 0 0 0 0 = 0 | | 0 1 0 1 = 5 |
| 0 0 0 1 = 1 | | 0 1 1 0 = 6 |
| 0 0 1 0 = 2 | | 0 1 1 1 = 7 |
| 0 0 1 1 = 3 | | 1 0 0 0 = 8 |
| 0 1 0 0 = 4 | | 1 0 0 1 = 9 |

The binary numbering system is used in digital electronics because the binary digits 1 and 0 can be represented by an electronic device being either ON or OFF. For example, a relay can be energized or de-energized; a transistor can be saturated or cutoff; etc. The advantage to a simple on/off status is that the "absolute value or voltage level is not important." In other words, it is totally irrelevant whether a transistor in cutoff has 4.5 volts, or 5.5 volts, on its collector. The only important data from a binary point of view is that it is OFF.

There are other numbering systems used extensively in digital electronics besides the binary system. The two most common ones are the octal system (base 8) and the hexadecimal system (base 16 system). These different numbering systems come in handy when interfacing with humans, but at the actual component level, everything is performed in binary.

## Logic gates

Various forms of logical building blocks are available in integrated circuit form. These logical building blocks are called "gates," with each gate having a distinct function. *Logic gates* can be further combined into more complex digital building blocks to perform a variety of counting, memory, and timing functions.

Digital ICs are grouped into families, with each family possessing certain desirable traits that make them more or less suited to a variety of applications. One logic family might not be compatible with another family, so it is typical for the designer to use only one family type for each application. The most commonly used *logic families* in the present market are CMOS (complementary metal oxide silicon) and TTL (transistor-transistor logic).

Logic gates respond to "high" or "low" voltage levels. The specific voltage level for a "high" or "low" condition will vary from one logic family to another. For example, a logical one (high) in TTL logic is about 5 volts; in contrast, a possible 12-volt level might be used for CMOS logic. However, the functional operation and symbolic representation is universal throughout all of the families. Figure 13-1 lists some of the more common logic devices, and their associated symbols.

■ **13-1** *Common logic symbols.*

Logic gates, and other logic devices, are functionally defined by using truth tables. Figure 13-2 illustrates a variety of truth tables for some common logic gates. Compare the AND gate illustration in Fig. 13-1 with its corresponding truth table in Fig. 13-2. Because the AND gate has two input leads, there are a total of four possible

| IN | Out |
|----|-----|
| AB | X |
| 00 | 0 |
| 01 | 0 |
| 10 | 0 |
| 11 | 1 |

Truth table for 2-input AND gate

| IN | Out |
|-----|-----|
| ABC | X |
| 000 | 0 |
| 001 | 0 |
| 010 | 0 |
| 011 | 0 |
| 100 | 0 |
| 101 | 0 |
| 110 | 0 |
| 111 | 1 |

Truth table for 3-input AND gate

| IN | Out |
|----|-----|
| AB | X |
| 00 | 0 |
| 01 | 1 |
| 10 | 1 |
| 11 | 1 |

Truth table for OR gate

| IN | Out |
|----|-----|
| AB | X |
| 00 | 1 |
| 01 | 0 |
| 10 | 0 |
| 11 | 0 |

Truth table for NOR gate

| IN | Out |
|----|-----|
| AB | X |
| 00 | 1 |
| 01 | 1 |
| 10 | 1 |
| 11 | 0 |

Truth table for NAND gate

| IN | Out |
|----|-----|
| AB | X |
| 00 | 0 |
| 01 | 1 |
| 10 | 1 |
| 11 | 0 |

Truth table for Exclusive OR gate

■ **13-2** *Truth tables for common logic gates.*

logic conditions that could occur on the inputs. Notice the truth table lists the four possible input conditions; together with each of their resultant outputs for each condition. As shown by the truth table, the only time that the output goes "high" is when the A input "and" the B input are high.

Logic gates can have more than two inputs. Figure 13-2 illustrates the truth table for a 3-input AND gate. Common logic gates are available with up to eight inputs.

Referring again to Fig. 13-1, note the OR gate and its associated truth table in Fig. 13-2. As the name implies, its output goes high whenever a high appears on the A input, "or" on the B input, (or both).

In digital terminology, a "not" function means a logical condition is inverted, or reversed. A NAND gate (short for Not AND) is an AND gate, with the output inverted. Notice the outputs in the truth tables for the AND gate, and the NAND gate, are simply inverted. This same principle holds true for the OR and NOR gates.

It is common for the output of one logic gate to provide inputs for several other logic gates. The maximum number of inputs that can be driven by a particular logic gate is specified as its *fanout*. Typical logic gates have fanouts ranging from 5 to 20. If it becomes necessary to drive a greater number of inputs than the fanout of a particular gate, a buffer is used to increase the fanout capability. The symbol for a buffer is illustrated in Fig. 13-1.

The need often arises to invert a logic signal. The symbol for an inverter is shown in Fig. 13-1. An inverter is sometimes called a "not" gate. Note that it has a small circle on its output just like the "Not AND" (NAND) and "Not OR" (NOR) gates. Anytime a small circle appears on an input or output of a logic device, it is symbolizing the inversion of the logic signals (or "data"). Also notice the horizontal line above the A output of the inverter. It is called a *"not" symbol*. Whenever a horizontal line is placed above a logic expression, it means that it is inverted.

Another common type of logic gate is the *exclusive OR gate*. Refer to its symbol and the associated truth table in Fig. 13-1 and 13-2. As the truth table indicates, its output only goes high when its inputs are different from each other. The *exclusive NOR gate* provides the same logic function with an inverted output.

## Combining logic gates

The most basic type of digital system is designed to provide a logical output based on a set of input conditions. This is referred to as *conditional logic*. Figure 13-3 illustrates a hypothetical copy machine using a simple conditional logic system to monitor a copier's operation, and to provide operator feedback or shutdown.

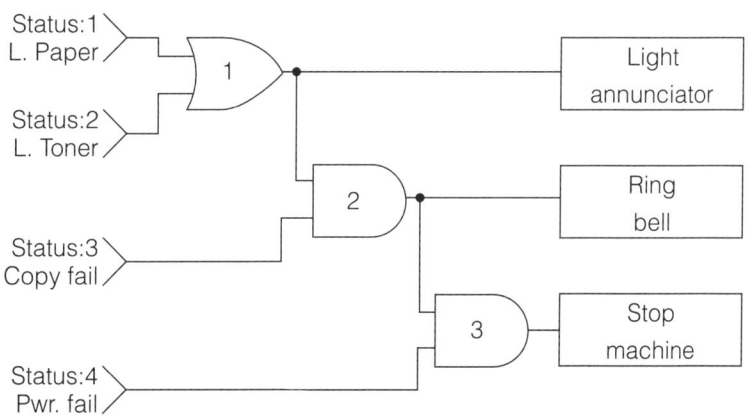

■ **13-3** *Basic example of conditional logic.*

The four status signals are designed to go high if a problem arises. The status #1 and status #2 signals will go high if a "low paper" or "low toner" situation occurs. Neither of these are major problems, so it is desirable to simply light an annunciator panel lamp, which usually says something similar to "check machine." The #1 OR gate output will go high and turn on the annunciator, if either (or both) of these problems arise.

However, if the status #3 goes high (copy failure), it means that the machine is completely out of paper or toner, which is a little more serious. Therefore, the #2 AND gate is used to indicate (in digital language), "I have run out of paper OR toner AND I can no longer make a copy. Wake up!!" So it rings a bell, and the annunciator is left on.

Meanwhile, AND gate #3 is watching for a serious problem. If a "low paper" OR "low toner" condition happens AND there is a "copy failure" AND a "power failure" all at the same time, it shuts down the copy machine. There aren't any copy machines out in the real world with status controls as basic as the one illustrated in Fig. 13-3, but a good example has been provided of how some of these logic gates can be combined.

The most commonly used types of conditional logic systems have been integrated. Integrated circuits such as "BCD to decimal decoders," "BCD to 7-segment decoders," "data selectors," and "data routers" are just a few examples. Almost any original design will incorporate some unique conditional logic.

## Multivibrators

*Multivibrators* are used extensively in digital electronics to provide clock signals (oscillators), count and store data, and control timing sequences. They can be divided down into three major groups, or types; astable multivibrators (called *clocks* or *oscillators*), bistable multivibrators (flip flops), and monostable multivibrators (one-shots).

You should already be familiar with *astable multivibrators* from our previous discussions regarding their use in sound circuits. The term *astable* means "not stable;" they cannot come to rest in either a high or low state. In other words, they *oscillate*. Because their outputs are in the form of a square wave, they are naturally suited to digital systems. IC forms of astable multivibrators are designed to operate at very high speeds.

Three common examples of *bistable multivibrators* are illustrated in Fig. 13-1. As the name suggests, bistable multivibrators have two stable states; "set" and "reset." They are usually called "flip-flops," abbreviated F-F.

Referring to Figure 13-1, notice that each F-F has a "Q" and "NOT Q" output. The NOT Q output is always logically opposite of the Q output. When a F-F is "reset," the Q output will be a logic 0, thus meaning that the NOT Q is a logical 1. If a F-F is "set," the logical states of the Q and NOT Q will be reversed.

The first F-F illustrated in Fig. 13-1 is a "set-reset F-F," or *RS flip-flop*. If a logical 1 is applied to the R (reset) input, the Q output goes to logical 0. Similarly, if a logical 1 is applied to the S (set) input, the Q output will go to a logical 1. RS flip-flops have limited use in most digital systems. Their importance lies in their ability to latch, or remember, a logical status (if the logic levels to the RS inputs are not altered).

The second type of F-F illustrated is the *JK flip-flop*. The JK inputs are "clocked inputs." This means that the logical levels applied to the JK inputs have no effect without a coincidental pulse applied to the clock input. For example, if the K input is 1, and the J input is 0, the F-F will reset as soon as a clock pulse is applied to the clock input. If the K input changes to 0, and the J input goes to 1, the F-F will set as soon as (but not before) another pulse is received at the clock input.

If both the J and K inputs are held at logical 1, a JK flip-flop becomes a *toggle F-F*. The output, or Q status, of a toggle F-F will change state every time the correct *transitional change* occurs at the clock input. Toggle F-Fs are designed to change state, or toggle, on either the "leading edge" or "trailing edge" of input clock pulses. For example, if a toggle F-F is specified to toggle on the trailing edge of the input clock pulses, a transitional change of the clock from a 0 to a 1 state will have no effect on the F-F. However, when the clock pulse changes from a 1 to a 0 (trailing edge), the F-F will toggle. Therefore, if a steady stream of clock pulses is applied to the clock input of a toggle F-F, the Q output of the F-F will be one "half" of the frequency of the input clock.

Referring to Fig. 13-4, three "trailing edge" toggle F-Fs are connected together to form a *counter*. The lower part of the illustration shows a series of eight clock pulses that are applied to the clock input of the first toggle F-F. The right hand table lists the output status of the Q outputs after the "trailing edge" of each pulse. Remember, the "leading edge" is when the clock goes from

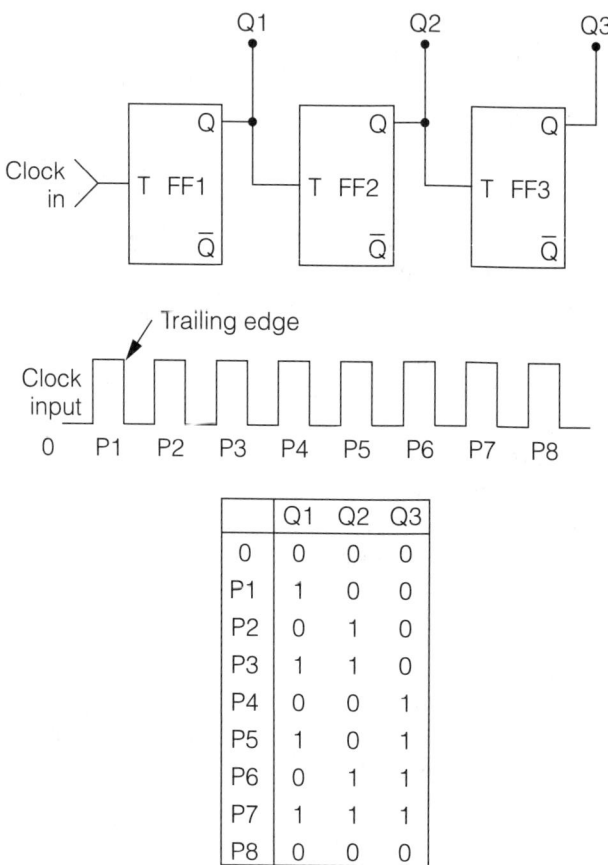

■ 13-4 *Basic modulo eight counter.*

a 0 (low) to 1 (high). In contrast, the "trailing edge" occurs when the clock goes from a 1 to a 0.

Continuing to refer to Fig. 13-4, note that a trailing edge occurs eight times out of the eight pulses applied to the clock input. This means F-F$_1$ will toggle (or change state) eight times, as shown in the "Q$_1$" column of the chart. F-F$_1$ has to toggle eight times to produce four complete output pulses (a complete pulse requires two transitions; 0 to 1 and then 1 to 0). The result is a Q1 output of "four" pulses. These four pulses are applied to the clock input of F-F$_2$, which also divides its clock input by two, resulting in a Q2 output of two pulses. Similarly, these two pulses become the clock input of F-F$_3$, and the resultant output of Q3 is one pulse. The overall response of all three F-Fs, as shown in the accompanying chart, is a "binary upcount," with Q1 being the least significant digit and Q3 the most significant digit.

If continuous clock pulses are applied to the counter in Fig. 13-4, it will continue to cycle as shown in the chart. One full pulse will occur at the Q3 output for every eight pulses applied to the input. A counter of this configuration is known as a *modulo eight counter*. There are two primary uses for such a circuit. The first use, quite obviously, is to count and accumulate totals (by adding additional F-F stages, virtually any size number can be counted, or accumulated). The second use is to "divide" a high frequency down to a lower frequency. For example, a 16-kHz "square wave" frequency applied to the input of the Fig. 13-4 counter would be divided down to 2 kHz on the Q3 output. Logic gates can be incorporated with similar counter circuits to provide any *modulo divider* desired.

It is very common to use four toggle F-F stages with some associated logic gates to reset all four F-Fs when the counter "attempts" to increment up to a 10 count. This results in a 0 through 9 count, which is compatible with the decimal numbering system. The Q outputs of such a counter is called *binary coded decimal*, or BCD for short. There are many integrated circuits available to decode BCD outputs into "humanly recognizable" forms. One common IC of this type is called a *seven-segment decoder/driver*. It decodes BCD outputs into the outputs needed to display decimal numbers on seven-segment LEDs. Many types of counters and decoders are available in IC form for a wide variety of applications.

If you would like to begin experimenting with some counter circuits, Fig. 13-5 illustrates a simple circuit for counting and displaying the count in BCD. The NTE 4029B is a good example of a versatile counter in IC form. By applying logic levels to various pins, it can be made to count up or down, in BCD or binary. It can also be pre-programmed with a beginning number, reset, and set. Additional 4029B chips can be connected together (to the "carry in" or "carry out" pins) for a count as high as desired. The circuit illustrated in Fig. 13-5 will begin with a BCD 9, and count down with every pulse received.

The circuit illustrated in Fig. 13-6 adds a seven-segment decoder/driver IC (NTE 4511) to display the count in decimal numbers. Any type of "common cathode" seven-segment LED display can be used.

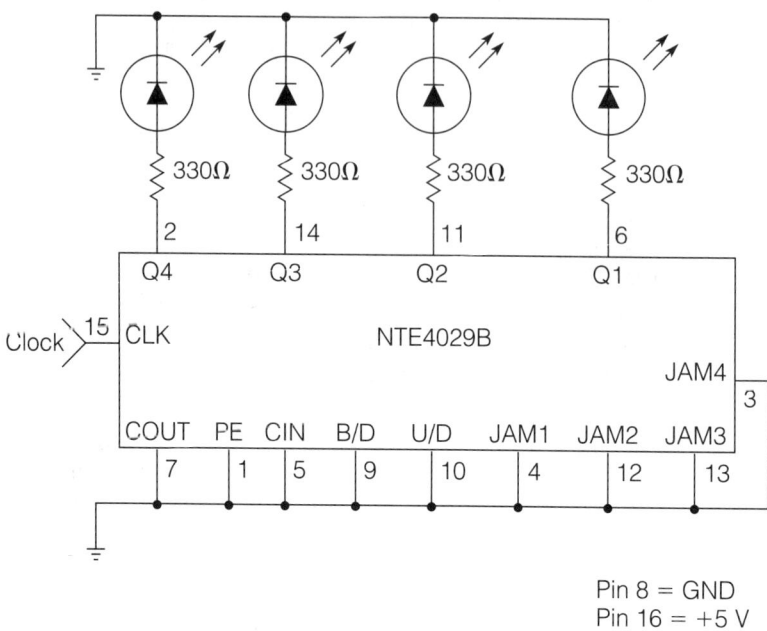

■ **13-5** *A digital decade counter with LED display.*

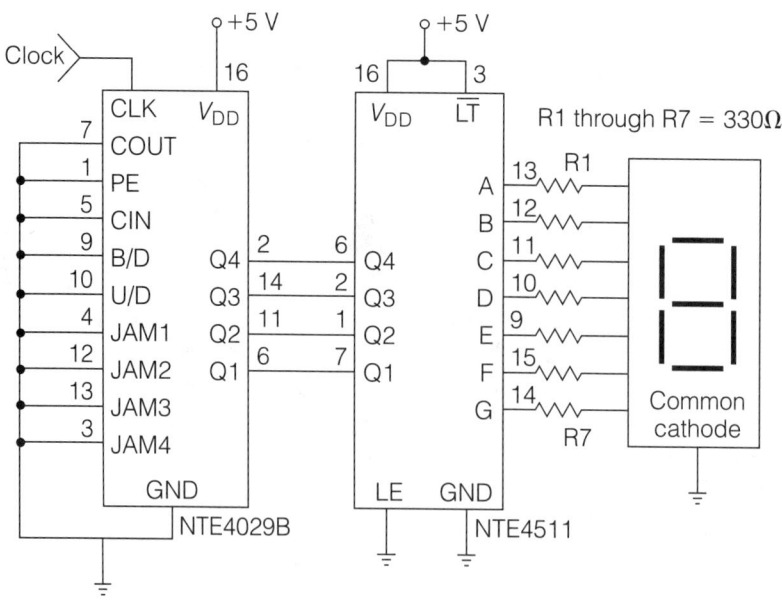

■ **13-6** *Decade counter with 7-segment decoder and display.*

The circuits illustrated in Fig. 13-5 and 13-6 can be used as basic building blocks for a multitude of counting/dividing applications. A few cautions should be observed, however. First, these are CMOS devices, meaning that they are very susceptible to damage by static charges. Secondly, these circuits, like most other digital counting circuits, need a "good" quality clock signal for proper operation.

## Digital clocks

*Digital clocks* can be in the form of continuous oscillations or gated pulses from other sources. They are always in the form of square waves, but they do not necessarily have to be symmetrical. In other words, the off time can be much longer than the on time, or vice versa.

For "solid" and repeatable operation of digital circuits, the clock pulses must have short transitional time periods. That is, the change from 0 to 1 and 1 to 0 must be very quick (this is called *rise time* and *fall time* respectively).

It is often desirable to use a mechanical switch to provide counting or control pulses to digital circuits. All mechanical switches, however, have an inherent characteristic called *bounce*. When switch contacts come together to make contact, some mechanical vibration will result in the physical movement. For most applications, this vibration is unnoticed, but because digital circuits are designed to operate at extremely high speeds, they will respond to switch bounce. Therefore, if a mechanical switch is used in conjunction with digital circuits, it must be made "bounceless" for reliable operation.

Some examples of bounceless switches and digital clocks will be given in the circuit potpourri section of this chapter.

## Shift registers

A *shift register* is actually a modified form of digital counter. It consists of flip-flops, like a counter, but its purpose is to temporarily hold numbers for processing, display, or rerouting.

Shift register action is defined by how a number is put into a shift register, and how it is taken out. The possible combinations are "serial-in, serial-out;" "parallel-in, parallel-out;" "serial-in, parallel-out;" and "parallel-in, serial-out." A shift register can be designed so that the complete digital number is applied to the data inputs simultaneously and after one clock pulse, the number is loaded in. This is *parallel-in operation*. In contrast, if a number is loaded into a shift register one bit (short for "binary digit") at a time (requiring one clock pulse per bit loaded), it is classified as *serial-in operation*. The same principle applies in removing a number from a shift register.

Shift registers are utilized within a digital system, in the same way a human uses a "scratch pad," to hold various sub-totals and interim calculations until they are needed for the final solution to a sequential problem. In this sense, they can be referred to as *latches* or *memory*.

## Digital memory devices

The term *memory* is usually given to a large number of digital devices arranged specifically for the purpose of retaining large quantities of digital information, or "data." It is technically accurate, however, to classify a single flip-flop as a 1-bit memory.

Digital memory is classified into two major categories; "volatile" and "nonvolatile." *Volatile memory* will lose all accumulated data when the power is removed. For example, the most common type of volatile memory consists of thousands of flip-flops with their inputs and outputs arranged in a grid pattern. Data can be addressed (located) and read into, or read out of, the memory. However, when the circuit power is turned off and then back on, the thousands of flip-flops will "power- up" in random fashion and all previously stored data is lost. This, of course, is a major disadvantage.

There are many types of *nonvolatile memory devices* available on today's market. In integrated circuit form, nonvolatile memory is called *ROM (read-only memory)*. IC ROMs operate on a variety of principles. Some retain data by converting it into electrostatic charges (which are retained after power is lost), and other types are permanently "programmed" by the physical destruction (burning in) of many diodes integrated into the chip (EPROM).

In general, permanently programmed IC memory devices are called *ROM*, and IC volatile memory devices are referred to as *RAM (random-access memory)*. Some types of RAM memory chips can be connected to a battery that supplies operating power for memory retention. This is called a *battery back-up* memory system. Long-life lithium batteries are typically used for this purpose.

## Summary

Many logic devices are classified by their operational use, instead of by their internal design. In the same way that a transistor can be used for an amplifier, oscillator, buffer, or regulator; a flip-flop can be used as part of a counter, shift register, or memory.

The field of digital electronics has literally revolutionized the world. Although I would not classify it as difficult from a conceptual point of view, it is an extremely broad field that cannot be fully covered in any one book, and most certainly not a single chapter.

If you wish to learn more about digital electronics, experiment with some of the circuits included in this chapter, and spend considerable time researching available product information from the manufacturer's data books. If you anticipate a long-term hobby interest in digital electronic circuits, it will be necessary to acquire a good dual-trace oscilloscope (at least 40 MHz) for observance of the high-frequency circuit operations.

## Circuit potpourri

### How is your pulse?

Figure 13-7 incorporates the very popular NE555 timer IC to form a *digital pulser*, or clock. The output is a "clean" square wave, and it can be adjusted to a very slow rate; thus, the operation of a circuit under test can be visually checked. If a flashing LED is desired to give visual indication of the output, the anode of any common type LED can be connected to the output in series with a 330-ohm resistor to ground. P1 and C1 are the primary frequency determining components. If higher output frequencies are required, their values can be reduced, and vice versa.

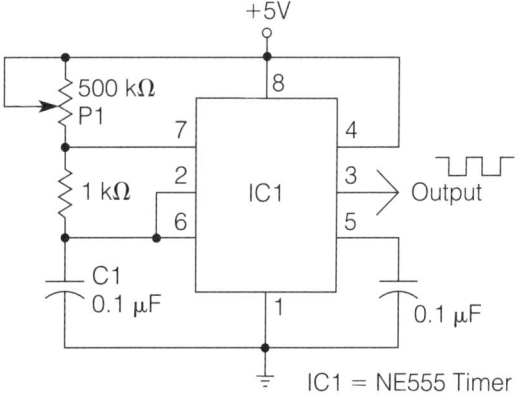

■ **13-7** *Digital pulser circuit.*

## Improved digital pulser

Figure 13-8 illustrates an improvement over the simpler pulser circuit detailed in Fig. 13-7. The oscillator section is basically the same, and can be modified for different frequencies as described for the circuit in Fig. 13-7.

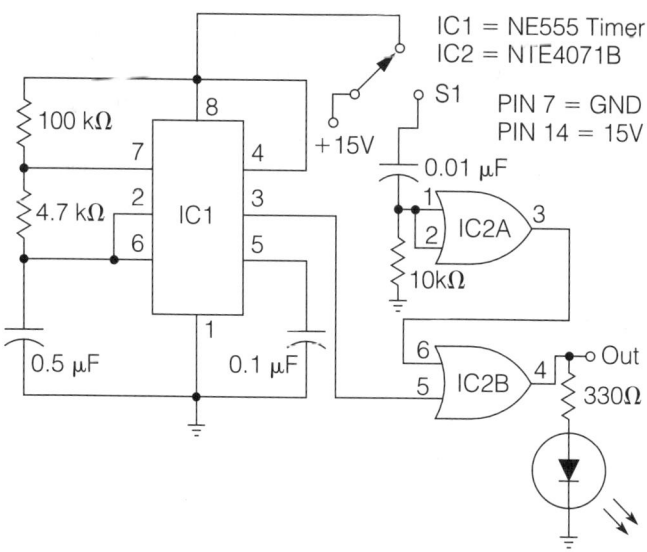

■ **13-8** *Improved logic pulser.*

S1 should be a momentary push-button switch with the "normally closed" contacts wired, as shown in the illustration. In this position, a constant train of logic pulses will be at the output (pin 4) of IC2. When S1 is depressed, it forces the output to a continuous high level. This circuit is very handy for checking the functional status of logic gates.

For convenient use, the circuit can be constructed on a narrow piece of universal breadboard and inserted in a round, plastic toothbrush holder, or in a similar hand-held package. The output is applied to a "probe tip" that can be purchased at an electronics parts store, or fabricated from a small nail. The power leads are typically channeled through the rear end of the holder.

IC2 is a CMOS device, and normal static electricity precautions must be observed during construction. Only two of the four gates within this chip are used in the circuit. The remaining "inputs" (not outputs) of the unused gates should be grounded.

## Digital logic probe

Figure 13-9 illustrates a logic probe circuit. This circuit differs from the logic pulser in several ways. It does not "inject" any pulses; it only reads the logic status of devices under test.

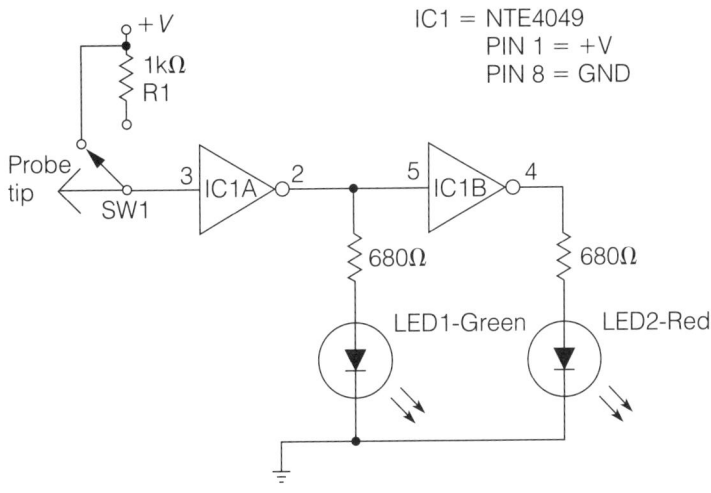

**13-9** *CMOS logic probe.*

When the probe tip is placed on a test point with a logical high level, the red LED will light; a low level will light the green LED. If the test point is toggling, both LEDs will light.

Logic probes typically receive their operational power from the circuit under test. The +V and GND leads should have small alligator clips on their ends so they can be attached to the power supply lines of the circuit under test. To protect this circuit against accidental polarity reversal, a common 1N4001 diode can be put in series with the +V lead (cathode connected to R1 and anode connected to +V). IC1 will function properly on any dc voltage from +5 to +15 volts.

SW1 is a SPDT toggle switch. In the position illustrated, the logic probe is used to test CMOS circuits. Switch SW1 to the other position for testing TTL circuits.

This circuit should be constructed and enclosed in the same manner as that described for the circuit of Fig. 13-8. The NTE 4049 is a CMOS chip; so watch the static electricity, and connect all unused "inputs" to ground.

## Bounceless is better

Figure 13-10 is a bounceless switch. As explained earlier in the text, any type of mechanical switch will produce very high frequency erratic pulses while "settling" from the vibrational shock of changing state. This is commonly known as *switch bounce*. High-speed digital circuits will respond to switch bounce, causing erroneous operation. The circuit illustrated in Fig. 13-10 can be used to manually trigger digital circuits. SW1 can be any type of SPDT switch. The NTE 4049 is a CMOS device; so take the necessary precautions against static electricity during construction, and connect the unused inputs to ground.

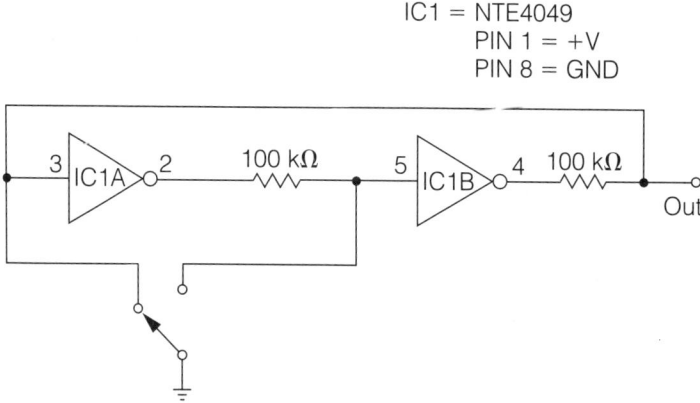

■ **13-10** *CMOS bounceless switch.*

## High-stability crystal timebase

All of the types of oscillators that have been discussed thus far have one inherent problem; their operational frequencies will vary with temperature changes. For digital circuits requiring high-stability oscillators, a *crystal-based oscillator* is usually employed.

Crystals are "piezo-electric" elements with vibrational characteristics that are highly immune to temperature changes. Figure 13-11 is a crystal timebase oscillator. Depending on the crystal frequency chosen, start with a 10-M$\Omega$ resistor for R1, and increase its value as needed.

The NTE 4011B is a CMOS device, so take precautions in assembly, as detailed previously.

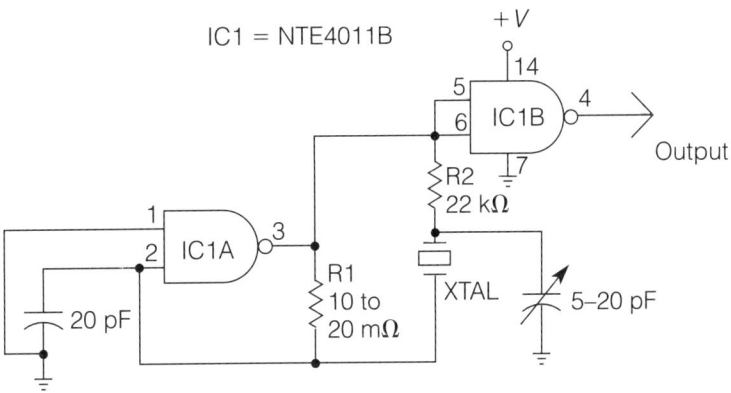

■ **13-11** *High-stability crystal timebase.*

Crystal oscillators operate at very high frequencies. To lower the output frequency, an appropriate "modulo" flip-flop divider circuit is required.

## PWM motor control

Small dc motors are used extensively in the electrical/electronic fields, but they have one common shortcoming. As the applied voltage to a dc motor is reduced to reduce the speed (which is necessary for many applications), the motor loses its torque. At low voltages, dc motor operation becomes erratic, or it might fail to start at desired speeds. The circuit in Fig. 13-12 solves that problem.

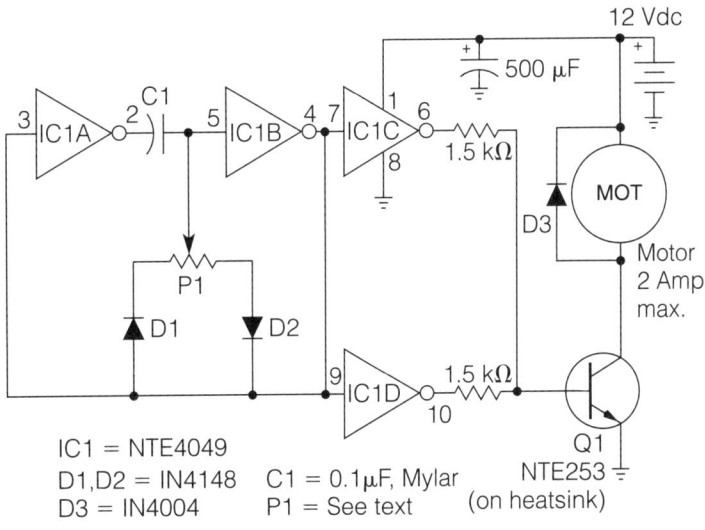

■ **13-12** *PWM small motor speed control.*

*Circuit potpourri*

The principle behind the operation of this type of motor control is called *pulse width modulation (PWM)*. The "amplitude" of the pulses applied to the motor always stays the same, but this circuit varies the *duty cycle* (on-time verses off-time) of the pulses. The motor cannot respond to the rapid voltage changes, but it "integrates" the duty cycle into an equivalent "power" value. The speed of the motor will vary in accordance with any changes in duty cycle, but the torque characteristics are much improved.

Q1 is a darlington transistor providing the additional current gain needed for this circuit. IC1C and IC1D are paralleled to provide more current drive for Q1. Any type of dc motor, drawing up to 2 amps, can be speed controlled. Depending on certain variables, you might have to experiment with different values of P1 for best results. Start with a 10-k$\Omega$ potentiometer as a baseline.

The NTE 4049 is a CMOS device, so use appropriate precautions during construction.

# Computers

IN TODAY'S WORLD, THE MOMENT YOU HEAR THE WORD "digital," you tend to think of a computer. Computer advocates have sensationalized the minds and imaginations of many people with terms like "electronic brains," "artificial intelligence," and "cyberspace." Science fiction writers have had a field day with computers starting as far back as the 1950s. So, in (not virtual) reality, what is a computer?

A *computer* is a machine capable of performing analytical calculations, storing data, and controlling or monitoring redundant operations. Computers are very useful tools, because they can be "programmed" to do one thing unerringly at very high speeds. However, to even try to compare the largest computers, with the lowest forms of human intelligence, is ludicrous. At this very moment, as you are reading this text, your brain is receiving over 100 million bits of information per second. These "bits" of brain information, however, cannot be compared to the bits, or "binary digits," of computer information. A bit of computer data is simply a 1 or a 0, but a bit of human brain information is complex electrochemical data, which has an almost infinite variation.

## How a computer works

As Fig. 14-1 illustrates, a computer can be subdivided into four basic sections:

- ☐ Input
- ☐ Output
- ☐ CPU (Central Processing Unit)
- ☐ Memory

For any logical system to be called a computer, it must meet five essential criteria:

- ☐ It must have input capability.
- ☐ It must have data storage capability.

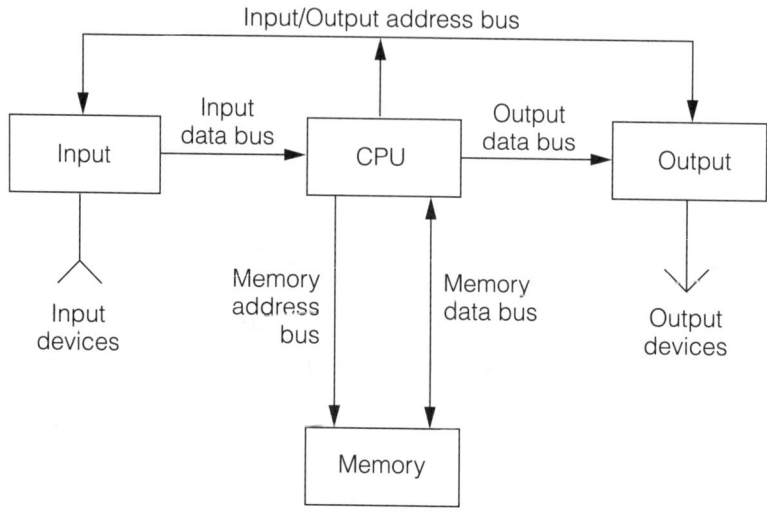

■ 14-1 *Block diagram illustrating the basic sections of a computer.*

☐ It must be capable of performing analytical calculations.
☐ It must be capable of making logical decisions.
☐ It must have output capability.

Referring to Fig. 14-1, the *input section* accepts information from a selected input device, and converts it into digital information, which can be understood by the CPU. The CPU controls the timing and data selection points involved with accepting inputs and providing outputs by means of the *input/output bus*. The CPU also performs all of the arithmetical calculations and memory storage/retrieval operations. The *memory address bus* defines a specific area in the memory to be worked upon, and the *memory data bus* either stores or retrieves data from that specific location. The *output section* accepts the digital information from the CPU, converts the information into a usable form, and routes it to the appropriate output device.

The analytical part of a computer is called the *CPU (central processing unit)*. In the not-too-distant past, CPUs were relatively large printed circuit boards containing many integrated circuits. In modern computers, CPU printed circuit boards have been replaced with a single VLSI (very large-scale integration) chip called a *microprocessor*.

A microprocessor can be broken down into two main parts; the ALU (arithmetic logic unit) and the ROM (read-only memory).

The ALU controls the logical steps and orders for performing arithmetical functions. It interacts with the ROM to obtain instructions for performing redundant operations. The ROM also contains instructions pertaining to start-up, and to power-loss conditions.

The digital information sent to the microprocessor can also be broken down into two main types; "data" and "instructions." These two types of digital information work with the two main parts of the microprocessor to perform all analytical operations.

Digital information is received at the microprocessor in the form of digital words, called *bytes*. In modern computers, a byte consists of 16 bits of data. A *bit (binary digit)* is simply a logical level; it can have only two states, either high or low. All 16 bits, comprising the byte, are applied to the microprocessor simultaneously. If the byte of information is an "instruction" word, it will be interpreted by the instruction code residing in the ROM section of the microprocessor. Instruction words tell the microprocessor what mathematical operations to perform on the data also being received.

Digital information is transferred throughout a computer system by means of "buses" (see Fig. 14-1). You might think of a *bus* as a communication pathway. Because bytes of information usually contain 16 bits, modern computer buses are either 16 or 32 electrical lines, with each line carrying one bit. This results in the bus being capable of transferring complete bytes of instruction and data information simultaneously.

Modern microprocessors contain numerous *registers*, or temporary memory cells. Mathematical operations are performed by shifting data words back and forth through the registers. Other registers serve a variety of purposes, such as storing instructions and keeping track of the current place in the program.

Basically, a computer is only capable of adding, subtracting, and accumulating data. Because it is capable of performing these simple operations at amazingly high speeds, complex mathematical calculations can be broken down into simple steps which the computer can then calculate. For example, a computer actually multiplies by *redundant addition* of the same number. In solving a multiplication problem such as 5 × 10, the digital form of the number 10 is placed in an accumulator register. The computer continues to accumulate (or add) 10s in the same register until five 10s have been added together. Of course, adding five 10s together, or multiplying 10 by 5, provides the same answer. Division is performed by *redundant subtraction* in the same manner. By shifting and manipulating digital information in the various registers, which you might think of as "scratch-pad" memory, a computer can perform virtually any mathematical calculation. The important point to recognize is that a computer does not perform calculations in the same manner as a human being does. In reality, its operation is much more similar to an ancient calculating instrument called an "abacus."

With computers, timing is crucial. In order to maintain precise timing of all functions, a computer must have an *internal clock*, or oscillator, to provide timing pulses to all of the sections simultaneously. In this way, exact synchronization can be maintained. Because the frequency of this clock controls the ultimate speed of the computer, it is desirable to set this frequency as high as possible, while still maintaining reliable operation. Modern computers have speeds as high as 100 MHz.

The smallest single operation performed by a computer is the *machine cycle*. It consists of two stages; the "fetch" cycle and the "execute" cycle. During the *fetch cycle*, the processor fetches an instruction from memory. Then, during the *execute cycle*, the computer performs some action based upon the content of that instruction. The processor knows which instruction to go to next from the address stored in the "program counter." The *program counter* always contains the address of the next instruction. When computer programs (called *software*) are written, they are arranged in a sequential order. The program counter keeps track of the next instruction to be acted upon by simply incrementing (by 1) for each machine cycle. So, in essence, a computer follows a set of sequential instructions (called a *program*) by counting from beginning to end. When the end of the program is reached, it starts over again. When the computer is continually repeating the sequential steps in the program, the program is said to be "running."

### Hardware, software, and firmware

The physical pieces making up a computer system are called "hardware." Common examples of computer *hardware* include keyboards, video monitors, floppy disc drives, compact disc (CD) drives, hard disk drives, modems, power supplies, printed circuit boards, and electronic components. In other words, hardware is almost everything you can physically touch.

In contrast, "software" is all of the "application oriented" programs used with or contained within a computer system. *Software programs* can be changed, manipulated, or customized at will. Depending on how the software programs are stored, they can be lost when operational power is removed.

Internal computer programs are also stored in the form of "firmware." *Firmware programs*, usually referred to as ROM (read-only memory), are general-purpose programs that cannot be modified during normal computer operation. Neither can they be lost when the operational power is removed. Examples of firmware include power-up programs, instruction code sets, and other computer routines which are common to all main, functional computer applications.

## Memory and data storage

There are many different mediums for storing digital information. In this section, a few of the more common storage methods will be discussed.

Most RAM memories are "volatile," meaning the information is lost when operational power is removed. Basically, RAM memories consist of many thousands, or millions, of flip-flops. Each individual flip-flop can "remember" one bit (i.e., latch into a high or low status). Programs are pulled out of a "nonvolatile" memory system (such as a hard disk drive, a compact disc drive, or a floppy disc) and loaded into the RAM memory, from which the computer actually runs the program.

Technically speaking, hard and floppy discs are randomly accessible, and they are *nonvolatile* (meaning that the digital information is not lost at power down). However, colloquially speaking, they are not classified as RAM memory.

ROM can take on many forms. Some IC ROM chips can only be programmed one time. These types of chips are simply called ROMs. Other types of ROM chips can be physically removed from a computer mainframe and re-programmed, using special ROM programmers. They are called *PROMs (programmable read-only memory), EAROMs (electrically alterable read-only memory), EPROMs (erasable programmable read-only memory)*, and *UVROMs* (ROM chips that can be erased by exposure to strong ultra-violet light and re-programmed). The newest type of ROM memory is called *CD-ROM*. CD-ROMs can store tremendous quantities of digital information for their size, and they are randomly accessible.

Older computer systems used magnetic tape for operational data storage. Although some modern computer systems use magnetic tape for "backup" storage (permanent storage of important programs or data, in case of accidental destruction of the original), magnetic tape is no longer used as an operational storage medium because the data is not randomly accessible.

### Input/output devices

Common examples of I/O (input/output) devices are keyboards, video monitors, printers, and modems (modulator-demodulators). All *I/O devices* are connected to the computer by means of ports. A *port* is just another name for a plug or a connector.

There are two main types of I/O ports used to connect the computer to the outside world; parallel and serial. *Parallel ports* have many parallel lines to enable data to be sent (or received) at a rate of 16 (or 8) bits at a time. *Serial ports* send (or receive) data on only one line; one bit at a time. Generally speaking, printers and data acquisition systems typically use parallel ports, but keyboards and modems use serial ports.

Computers communicate with I/O ports by means of *addressing*. Each I/O port has a specific code, or address, that is applicable to it alone. When an I/O port "sees" its specific address on the I/O bus, it will activate for communication with the microprocessor. The microprocessor will "look" at the various I/O ports as it is instructed to do so by the running program. The functions of the I/O ports are defined by firmware or software.

### Programming computers

Generally speaking, computers can be programmed at three different levels. If you program the microprocessor directly using binary number commands, the process is called *machine-language programming*, or "MLP" for short. This method is difficult and cumbersome, because the 16-bit digital commands are not easily recognizable to a human.

The next programming level above MLP is called *assembly-language programming*. When using assembly language, each binary number command is replaced with an easier to understand "mnemonic" (pronounced "nih-monic"). For example, instead of programming with commands like 0011010010011011, which is MLP, programming is performed with mnemonics like ADD A,B, which instructs the microprocessor to add the contents of the A and the B registers.

For long, complex programs, various high-level languages are used. Common examples are "Pascal," "Fortran," "APT," and many others. High-level program languages are easy to use because they are very much like human language. However, after the program is written, it must be converted into MLP by a special computer program called a *compiler*.

## Computer processing of analog signals

When a computer needs to examine (or to output) a continuously variable signal (referred to as *analog* or *linear*), the analog signal must be converted into digital "words" before the microprocessor can understand it. Special *data acquisition I/O modules* used for this purpose are called *D/A converters (digital to analog converters)* or *A/D converters (analog to digital converters)*. A/D converters change analog input levels into equivalent digital numbers, understandable to the microprocessor. In contrast, D/A converters change digital words, from the microprocessor, into equivalent voltage or current outputs.

The process of converting analog levels to digital words, and digital words to analog levels, is called *digitizing*. To digitize an analog level, the A/D converter actually divides the level into thousands of incremental "steps," or pieces. The incremental level of the individual steps, which ultimately defines the accuracy of the conversion, is called the *resolution*. D/A converters simply reverse the process.

# 15

# More about capacitors and inductors

AS STATED PREVIOUSLY, CAPACITORS AND INDUCTORS ARE reactive components. This means that they "react," or oppose, changes in electrical variables. For example, a capacitor opposes, or reacts, to a change in voltage. Inductors, on the other hand, "react" to a change in current flow. This reactive effect has a profound relationship to frequency.

## Inductive reactance

As the frequency of the applied voltage to an inductor is increased, the inductor's opposition to ac current flow increases. This is because the amount of energy capable of being stored in the inductor's electromagnetic field (its inductance value) remains constant, but the time period of the applied ac voltage decreases. As the ac time period decreases, less energy is required from the inductor's electromagnetic field to oppose voltage alternations. For example, it would take 10 times the energy to oppose 100 volts for 10 seconds than it would to oppose the same voltage for one second. The same principle applies with an increase in the frequency (decrease in time period) of the applied ac. Another way of stating this basic principle would be to say that *the reactance of an inductor increases with an increase in frequency.*

This frequency-dependent opposition to ac current flow through an inductor is called *inductive reactance*. Inductive reactance $(X_L)$, just as with impedance $(Z)$ and resistance $(R)$, is measured in ohms. The equation for calculating inductive reactance is:

$$X_L = 2\pi f L$$

This equation states that inductive reactance (in ohms) is equal to 6.28 (the approximate value of $2\pi$) times frequency times the inductance value (in henries). For example, the inductive reactance of a 1-henry coil with 60-Hz ac applied to it would be:

$$X_L = (6.28)(60)(1) = 376.8 \text{ ohms}$$

If the frequency of the applied ac voltage was increased to 100 Hz, the inductive reactance of the same inductor would be:

$$X_L = (6.28)(100)(1) = 628 \text{ ohms}$$

Notice that the inductive reactance increases as the frequency of the applied ac increases.

## Capacitive reactance

Capacitors, like inductors, have a frequency-dependent opposition to ac current flow called *capacitive reactance* ($X_C$). When an ac voltage is applied to a capacitor, it will charge and discharge in an effort to maintain a constant voltage.

As the frequency of the applied voltage to a capacitor is increased, the capacitor's opposition to ac current flow decreases. This is because the amount of energy capable of being stored in the capacitor's electrostatic field (capacitance value) remains constant, but the time period of the applied ac voltage decreases. As the ac time period decreases, it becomes easier for the capacitor to fully absorb the charge of each half-cycle. In other words, from a relative point of view, it would require ten times the capacity for a capacitor to charge for 10 milliseconds than it would for 1 millisecond. Therefore, capacitive reactance decreases as the frequency of an applied voltage increases.

The equation for calculating capacitive reactance ($X_C$) is:

$$X_C = \frac{1}{2\pi f C}$$

This equation states that capacitive reactance (in ohms) is equal to the reciprocal of 6.28 ($2\pi$), times the frequency, times the capacitance value. For example, at 60 Hz, the capacitive reactance of a 10-$\mu$F capacitor would be:

$$X_C = \frac{1}{(6.28)(60)(0.00001)} = \frac{1}{0.003768} = 265 \text{ ohms}$$

*More about capacitors and inductors*

(Note: $0.00001\ F = 10\ \mu F$) If the frequency of the applied ac voltage was increased to 100 Hz, the capacitive reactance of the same capacitor would be:

$$X_C = \frac{1}{(6.28)(100)(0.00001)} = \frac{1}{0.00628} = 159\ \text{ohms}$$

Notice that the capacitive reactance decreases as the frequency increases.

After examining the previous examples of inductive and capacitive reactance, it should become apparent that: if an electrical circuit contained a combination of resistors, capacitors, and inductors, the overall "impedance" of the circuit would be frequency dependent also.

Calculating the impedance of a resistive/reactive circuit is not as easy as simply adding reactance and resistance values together. True impedance requires compensation for the phase difference between voltage and current. The higher level geometric equations used for calculating impedance are included in Appendix A; but, depending on your personal interests and goals, you might never need to use them.

## Capacitive and inductive comparison

This section is devoted to studying the comparative nature of inductors and capacitors. The particular qualities of each component can be more fully appreciated in this way.

| **Inductance** | **Capacitance** |
|---|---|
| Voltage leads the current. | Voltage lags the current. |
| Tries to maintain a constant current. | Tries to maintain a constant voltage. |
| $Time\ constant = \dfrac{L}{R}$ | $Time\ constant = RC$ |
| As frequency increases, reactance increases. | As frequency increases, reactance decreases. |
| Ac power dissipation is zero in a purely inductive circuit. | Ac power dissipation is zero in a purely capacitive circuit. |
| Exhibits minimum reactance at dc. | Exhibits maximum reactance at dc. |
| Stores energy in an electromagnetic field. | Stores energy in an electrostatic field. |

## Combining inductors and capacitors

Figure 15-1 illustrates the effects of combining inductors in series and parallel configurations. Notice that the method used for calculating total inductance is the same method used for calculating total resistance. In a *series* arrangement, the individual inductance values are simply added together. In a *parallel* configuration, the individual reciprocal values are calculated, then added together, then the reciprocal of the total is solved.

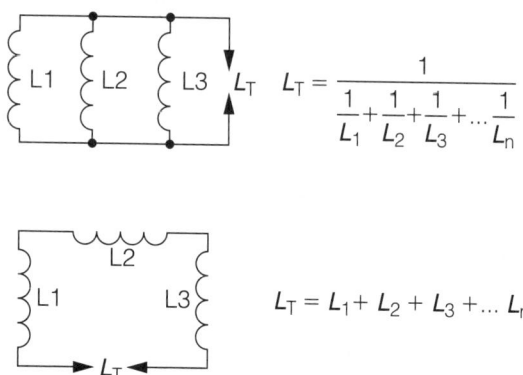

■ **15-1** *The effects of connecting inductors in series and parallel.*

Figure 15-2 illustrates how the opposite is true for capacitors. Capacitance values are summed in parallel; but in series, reciprocals must be used to calculate total capacitance. The most common purpose for placing capacitors in series is to increase the overall voltage rating. For example, two capacitors with a voltage rating of 100 volts will have a 200-volt rating when placed in series. However, as the equation in Fig. 15-2 indicates, their capacitance value will be halved.

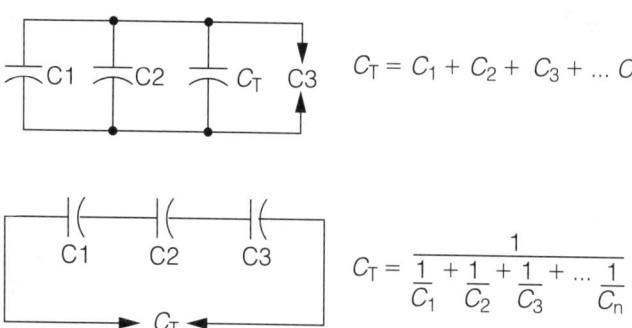

■ **15-2** *The effects of connecting capacitors in series and parallel.*

## Reflected impedance

If 120-Vac, 60-Hz power is applied to the primary of a power transformer with its secondary open, a very small current will flow through the primary winding. Only a small primary current will flow because the inductive reactance of the primary winding is typically very high. However, if a load is then placed on the secondary winding, the load impedance of the secondary will be "reflected" back to the primary winding. This effect is referred to as *reflected impedance*.

Reflected impedance is the effect that causes the primary current to automatically increase when the secondary current increases. Without it, a transformer could not function.

Reflected impedance is calculated by multiplying the square of the turns ratio, by the secondary load. For example, assume you have a transformer with a 10:1 turns ratio. Mathematically, a 10-to-1 ratio is the same as dividing 10 by 1, or simply 10. Therefore, the square of the turns ratio is 10 times 10, or 100. If a 5-ohm load is placed on the secondary of this transformer, the reflected impedance seen at the primary will be $5 \times 100$, or 500 ohms. This 500-ohm reflected impedance, seen at the primary, will be much less than the inductive reactance presented by the primary with the secondary open. Consequently, the primary current flow will automatically increase to compensate for the increased current flow in the secondary.

In the previous example, a "step-down" transformer was discussed. Reflected impedance works the same way in a "step-up" transformer. For example, assume that you have a step-up transformer with a turns ratio of 1:10. Mathematically, this is the same as 1 divided by 10, or 0.1. The square of 0.1 is 0.01. If a 5-ohm load is placed on the secondary of this hypothetical transformer, the reflected impedance seen at the primary will be 0.01 times 5, or 0.05 ohms.

## Resonance

You have learned that capacitors and inductors are reactive devices. As the applied frequency changes, the reactance values of inductors and capacitors will change. However, their reaction to frequency changes are exactly opposite. This leads to some interesting effects when inductors and capacitors are combined in electrical circuits.

Consider the *series-resonant circuit* illustrated in Fig. 15-3. The source is a variable-frequency ac source allowing the user to vary

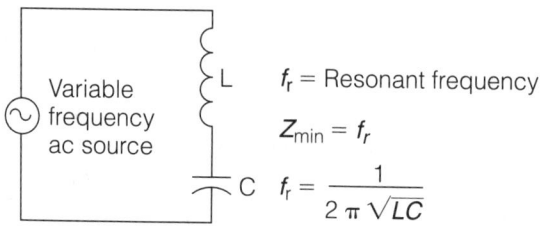

■ **15-3** *Series resonant circuit.*

the applied frequency throughout a wide range. This variable-frequency ac is applied to a series circuit containing some value of inductance and capacitance. Regardless of the ac frequency applied, both the inductor and capacitor will pose some value of reactance.

If the applied ac source is of relatively low frequency, the inductor will present little opposition, but the capacitive reactance will be high. On the other hand, if the ac frequency is increased to a high value, the capacitor will now present less opposition, but the inductive reactance will increase significantly. In either of these two cases, very little circuit current will flow because one reactive device, or the other, will pose a high opposition.

As the frequency of the variable ac source is adjusted throughout its entire range, a specific frequency will be reached that will cause the inductive reactance to equal the capacitive reactance. At this specific point in the frequency spectrum, the circuit current will be the highest because it will not be blocked by either reactive component. From a mathematical viewpoint, the inverse characteristics of the inductive reactance and the capacitive reactance cancel one another, causing only the resistive element to remain in the circuit. This is called the *point of resonance*. Resonance occurs when capacitive reactance is equal to inductive reactance.

Figure 15-4 illustrates the effect upon circuit current as the frequency of the applied ac is varied in the circuit shown in Fig. 15-3. Notice how the circuit current peaks at a specific frequency, but is substantially reduced at other frequencies above or below this point. The specific frequency causing the reactance values to equal each other is called the *resonant frequency*. Figure 15-3 shows the equation for calculating the resonant frequency when the inductance and capacitance values are known.

When a capacitor and an inductor are placed in series, the circuit current will always be highest at the point of resonance. Conversely, the circuit impedance at resonance is at its lowest point.

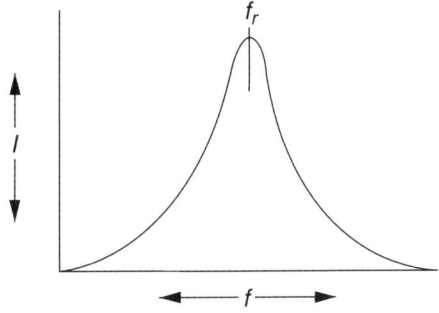

**15-4** *Current versus frequency response of Fig. 15-3.*

Consider the parallel resonant circuit shown in Fig. 15-5. As can be seen in this circuit, the inductor and capacitor are now in parallel. Once again, the source is a variable-frequency ac source. Assume the variable-frequency ac source is adjusted to a low frequency. The circuit current will be high because the inductor will present very little inductive reactance to the low frequency. This will allow the circuit current to freely pass through the inductive leg of the parallel circuit. If the ac source is adjusted to a high frequency, there will still be a high circuit current because the capacitor will present very little capacitive reactance to the high frequency, allowing the circuit current to freely pass through this parallel leg. In either of these two cases, the circuit current will be high because one reactive leg or the other will have a low reactance.

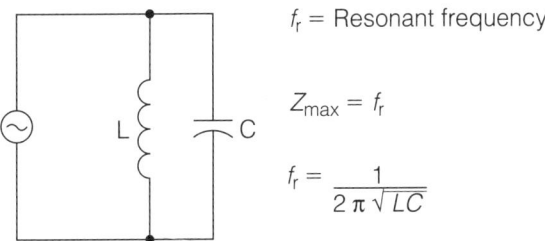

$f_r$ = Resonant frequency

$Z_{max} = f_r$

$f_r = \dfrac{1}{2\pi\sqrt{LC}}$

**15-5** *Parallel resonant circuit.*

As the frequency of the ac source is varied over a wide range, a specific frequency will be reached causing the inductive reactance to equal the capacitive reactance. At this frequency, the overall circuit current will be at its minimum. This is the resonant frequency for the parallel circuit. Any frequency above or below this point will cause the circuit current to increase dramatically.

Figure 15-6 illustrates the effect on circuit current as the frequency is varied over a wide range. Note that the circuit current is at its lowest point at the resonant frequency, but it climbs on either side of this frequency. In a parallel resonant circuit, the circuit impedance reaches its highest point at the resonant frequency, thus causing the circuit current to be at its lowest value. The equation for calculating the resonant frequency is shown in Fig. 15-5. Notice that it is the same equation as that used for a series resonant circuit.

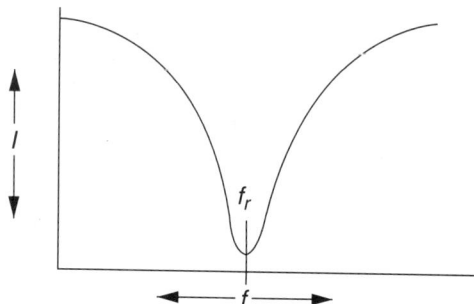

**15-6** *Current versus frequency response of Fig. 15-5.*

If a desired resonant frequency is known and you wish to calculate the needed inductance and capacitance values, the following *resonance equations* can be used.

$$L = \frac{1}{4\pi^2 f_r^2 C} \quad C = \frac{1}{4\pi^2 f_r^2 L}$$

where: $C$ is the capacitance in farads

$\pi$ is 3.14

$f_r$ is the resonant frequency in hertz

$L$ is the inductance value in henries

## Passive filters

Previously, operational amplifiers were discussed as the primary "active" element in the design of active filters. In the vast majority of cases, active filters are preferred because of their small size, low cost (inductors can be large, expensive, and difficult to find), and ease of design and construction. However, active filters do have one drawback; they require operational power to function. You might run into some situations where a passive filter (with reasonably high Q) would be preferable. A passive filter will always attenuate the applied signal to some extent, but they do not require a battery or dc power supply for operation.

There are four basic types of filters; "high-pass," "low-pass," "band-pass," and "bandstop" (often called *notch* or *band-reject*). The *LC filters* (inductor-capacitor filters), illustrated in Figs. 15-7 through 15-10, are the circuit diagrams for the four basic passive filter types. They are called *constant-K passive filters*. Because only reactive components are used (in contrast to RC- and LR-type filters), signal attenuation is held to a minimum, with relatively high Q values preserved. (A reminder: "Q" is the ratio of reactance value to resistance value).

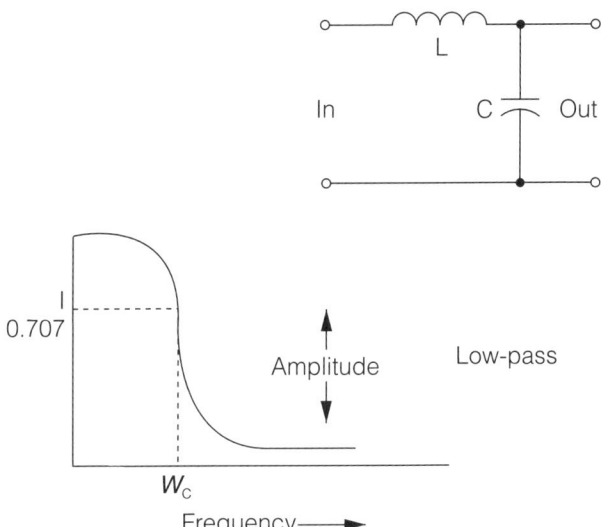

■ **15-7** *Low-pass filter and frequency response.*

Because of certain mathematical complexities in this type of filter design, the following equations will require the designer to mathematically substitute certain capacitor and inductor values until the best combination is achieved. In most cases, the determining factor will be the availability of inductor values.

For designing low-pass and high-pass filters, as illustrated in Figs. 15-7 and 15-8, the following equations are used:

$$C = \frac{1}{W_C R} \quad L = \frac{R}{W_C}$$

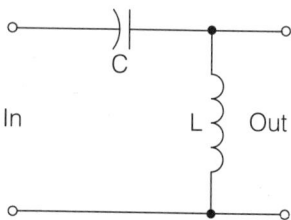

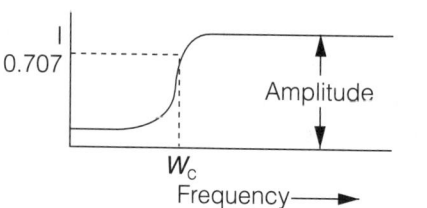

**15-8** *High-pass filter and frequency response.*

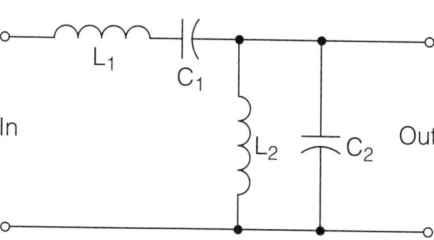

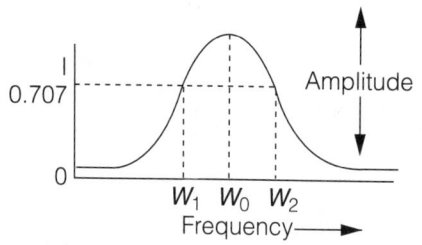

**15-9** *Bandpass filter and frequency response.*

where: $C$ = capacitance in farads

$L$ = inductance in henries

$R$ = *nominal terminating resistance* = $\sqrt{\dfrac{L}{C}}$

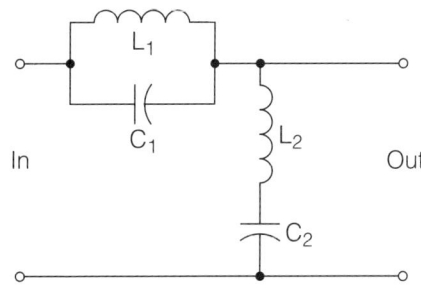

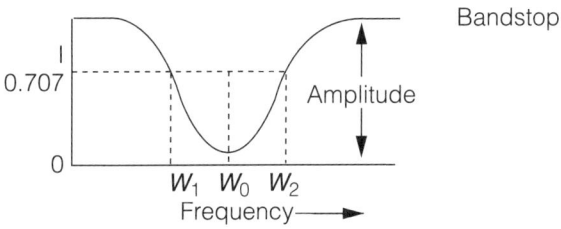

■ **15-10** *Bandstop (notch) filter and frequency response.*

$W_C$ = cutoff frequency × 6.28

For designing *band-pass filters*, as illustrated in Fig. 15-9, the following equations are used:

$$C_1 = \frac{W_2 - W_1}{R(W_0^2)} \qquad L_1 = \frac{R}{W_2 - W_1}$$

$$C_2 = \frac{1}{R(W_2 - W_1)} \qquad L_2 = R(W_2 - W_1)W_0^2$$

where: $C_1$ = Series capacitance in farads

$C_2$ = Shunt capacitance in farads

$L_1$ = Series inductance in henries

$L_2$ = Shunt inductance in henries

$$R = \text{Nominal terminating resistance} = \sqrt{\frac{L_1}{C_2}} = \sqrt{\frac{L_2}{C_1}}$$

$W_0$ = Midband frequency = 6.28($R$)

$W_1$ = Lower cutoff frequency × 6.28

$W_2$ = Upper cutoff frequency × 6.28

*Passive filters*

For designing *band-stop filters*, as illustrated in Fig. 15-10, the following equations are used:

$$C_2 = (W_2 - W_1)W_1 W_2 R \quad L_2 = \frac{R}{W_2 - W_1}$$

where: $W_O$ = midband frequency

$$= \left(\sqrt{W_1 W_2} = \frac{1}{\sqrt{L_1 C_1}} = \frac{1}{\sqrt{L_2 C_2}}\right) \times 6.28$$

All other expressions are the same as band-pass.

The mathematics involved in passive LC filter design are not as difficult as they appear at first glance. Remember, the absence of a mathematical sign between two variables means that they are multiplied together. For example, $RW$ means $R \times W$. Equality expressions contained within parenthesis mean that the expressions must be equal. For example, in calculating the midband frequency for the band-stop filter, all three expressions within the parenthesis must be equal; any "one" of the three equal expressions must be multiplied by 6.28.

## Circuit potpourri

Having fun with filters and oscillators.

### Proportional, integral, and differential action

In the days of computers' infancy, "analog" computers performed mathematical functions using thousands of dual vacuum-tube operational amplifiers. The "action" of these op-amp circuits was defined in mathematical terms which are still commonly used today. Figure 15-11 illustrates a modern IC op-amp, configured as a *proportional amplifier*. If you recognize this circuit as a simple non-inverting amplifier, you're correct. The term *proportional* is simply a mathematical definition of a "linear" function.

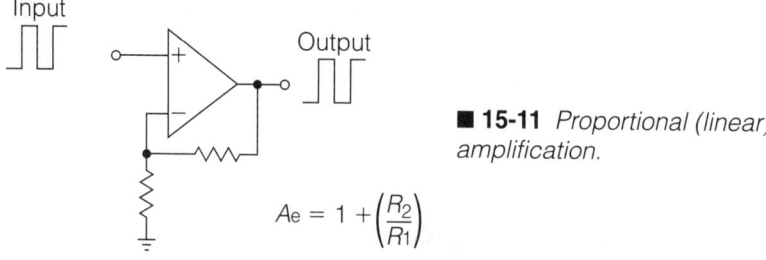

**15-11** *Proportional (linear) amplification.*

Figure 15-12 illustrates an *integrator circuit*. Integrators produce an output that is proportional to the amplitude of the input multiplied by its duration. In mathematical terms, it is an "integral extractor." Although the mathematical function is probably of little use, the practical aspects of this circuit are very useful. First, it is a type of low-pass filter, and it works very well for many low-pass applications. But its most common use is for converting square-wave signals into "linearized" triangular waves, for audio synthesizers and tone generators. Triangular waves have a pleasing, mellow tone when used for audio applications and are easily converted to sinusoidal waves for other applications. Also, triangular waves are predominantly composed of "even harmonics," which makes them useful for function generator applications.

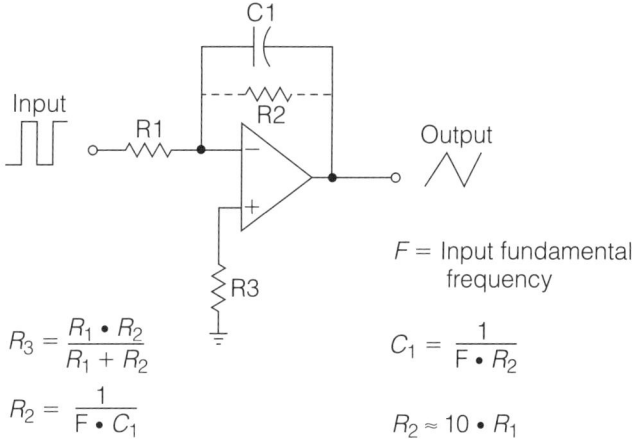

$$R_3 = \frac{R_1 \cdot R_2}{R_1 + R_2}$$

$$R_2 = \frac{1}{F \cdot C_1}$$

$F$ = Input fundamental frequency

$$C_1 = \frac{1}{F \cdot R_2}$$

$$R_2 \approx 10 \cdot R_1$$

■ **15-12** *Integrator.*

Note that in Fig. 15-12, the symbol $F$ is defined as the "fundamental frequency." The term *fundamental frequency* defines the frequency of a waveshape based on how often it repeats a full 360-degree cycle; exactly the same as you have learned for sine-wave ac voltages. However, every type of ac waveshape other than sinusoidal waveshapes contain "harmonics." *Harmonics* are multiples of the fundamental frequency added into the overall waveshape.

For example, theoretically speaking, a *square wave* contains all possible harmonics, however, the predominant harmonics are "odd" multiples of the fundamental frequency. By applying a square wave input to an integrator, the odd harmonics are removed; leaving the "even" harmonics in the form of a *triangular waveshape*. By removing the even harmonics, a triangular waveshape becomes a *sinusoidal waveshape*, which is "pure," or without harmonic content. Waveshapes containing harmonics are referred to as *complex ac waveshapes*. Therefore, to be "technically" accurate, the frequency of a complex ac waveshape should be defined in terms of its fundamental frequency.

A simple RC network will provide a waveshape similar to a triangular waveshape from a square-wave input, but it will not be a high-quality triangular wave because a capacitor does not charge in a linear fashion. In contrast, the circuit shown in Fig. 15-12 will provide a highly linear triangular wave output from a square wave input. You can use this circuit as a low-pass filter, a square-to-triangle wave converter, or to "mellow-out" the tone from many of the audio oscillator circuits illustrated in this book. R2 might (or might not) be needed, depending on the application and the operational amplifier used.

Figure 15-13 illustrates a *differentiator circuit*. Mathematically speaking, differentiators produce an output that is a derivative of the input. For most practical applications, you can think of a differentiator as a type of high-pass filter. Notice that the output of the differentiator in Fig. 15-13 with a square wave input. The high-frequency "shifts" of the square wave are extracted resulting in a series of dual-polarity pulses. Differentiators are useful in many types of pulse/timing circuits as high-pass filters, and they will also convert a triangle wave back into a square wave.

$F$ = Input fundamental frequency

$$C_1 = \frac{1}{F \cdot R_2}$$

$$R_1 = \frac{1}{F \cdot C_1}$$

■ **15-13** *Differentiator.*

## Square waves galore

IC operational amplifiers are easy to configure into *square-wave oscillators*. Figures 15-14 and 15-15 illustrate two common meth-

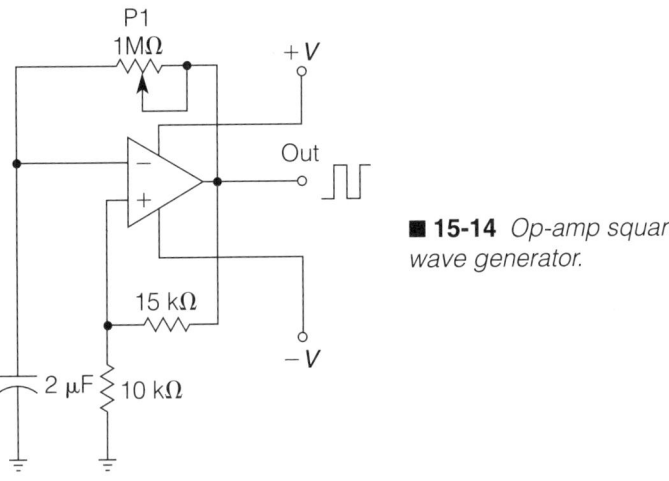

**15-14** *Op-amp square-wave generator.*

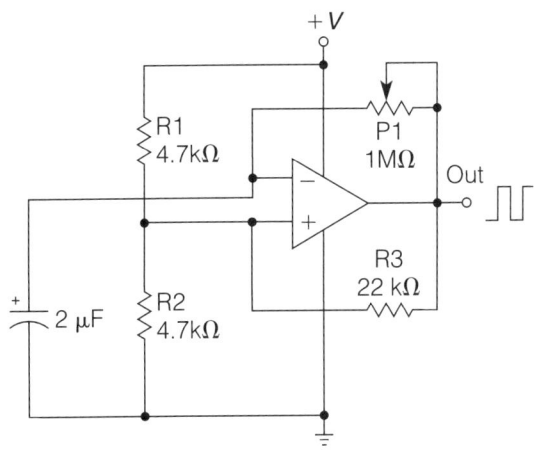

**15-15** *Single-supply square-wave generator.*

ods of accomplishing this end. Figure 15-14 has fewer components, but it requires a dual-polarity power supply. Figure 15-15 operates from a single supply. With the components illustrated, both will oscillate at about 2 to 3 Hz. By replacing the 2-$\mu$F capacitor with a 0.1-$\mu$F "cap," the frequency will increase to about 200 Hz; a 0.001-$\mu$F "cap" will set the frequency to about 10 kHz. The two feedback resistors and the capacitor constitute the frequency determining network components.

Depending on the op-amp used, these circuits might not produce square waves with a sufficiently fast rise and fall time for use in digital circuits. If you want to use these circuits for digital applications, apply the output of the oscillator to the input of an inverter gate. Then, use the output of the inverter to interface with other digital circuits.

### Basic tone generator

Figure 15-16 illustrates a very versatile, general-purpose *tone generator*. Virtually any kind of transistor, resistor, or capacitor combination will work. P1 controls the frequency.

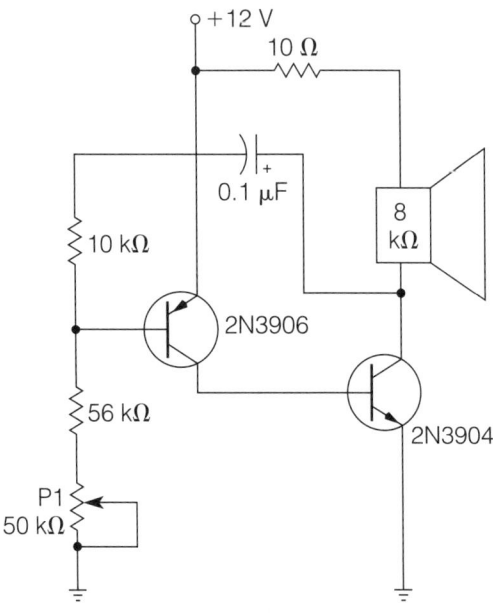

■ **15-16** *General-purpose tone generator.*

### Where's the fire?

Figure 15-17 is a variation of the circuit shown in Fig. 15-16. S1 should be a momentary push-button switch. When S1 is initially depressed, C1 begins to charge through R1. This applies an increasing bias to Q1, thus causing the pitch of the oscillations to increase. When S1 is released, C1 begins to slowly discharge through R2 and the Q1 base, causing the pitch to decrease. The overall effect is that of a siren.

To make the operation automatic, S1 can be removed, and the top of R1 can be connected to one of the low-frequency, op-amp

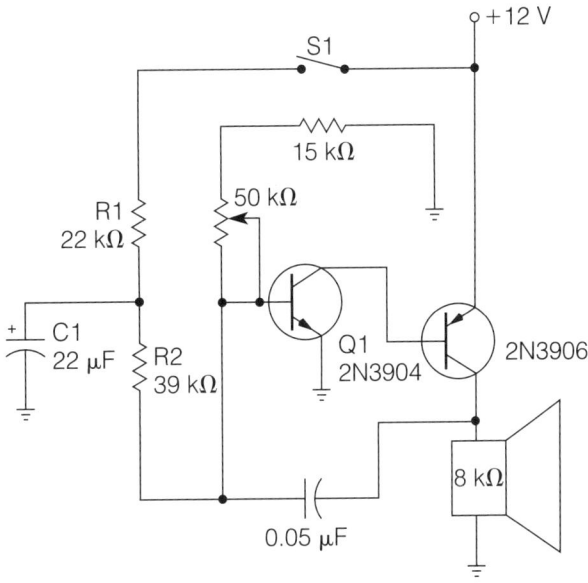

■ **15-17** *Electronic siren.*

square-wave generators illustrated in Figs. 15-14 and 15-15. The frequency of the square-wave generators might need to be reduced for optimum operation.

### Building a lab function generator

Figure 15-18 illustrates a very versatile square and triangle wave *function generator* for lab use or many other applications. It requires only two dual op amps, and it can be assembled on a small universal grid board. The square and triangle waves are independently amplitude-adjustable with low output impedances. The circuit is very forgiving of power supply variations and component tolerances. It is also easily modified for specific applications.

IC1A forms a comparator that is continually switched by the output of the IC1B integrator. Because the wiper of P1 provides the reference voltage, P1 acts as a symmetry control. The frequency of oscillation is controlled by P3, R1, and C1. IC2B serves as a buffer/attenuator amplifier with an adjustable dc offset controlled by P4 and IC2A.

When building the circuit, don't forget to connect the +12-V and −12-V power supply leads to their respective power input pins on the op amps (this is not shown on the schematic).

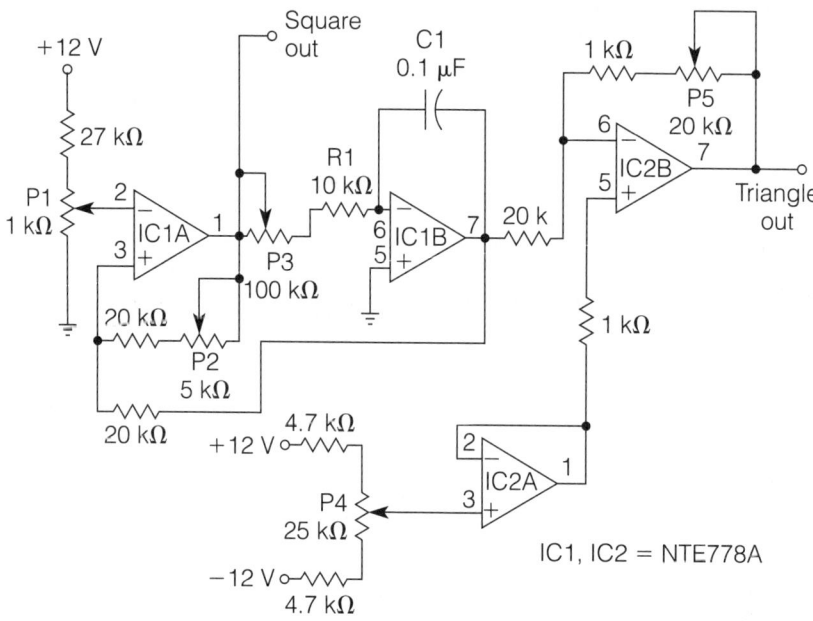

**15-18** *Square- and triangle-wave function generator.*

After building the circuit and applying power, adjust P1 (trim pot) for about 0.1 Vdc at pin 2 of IC1A. P2 controls the amplitude of the square wave output. P3 controls the frequency, and P5 controls the output level of the triangle wave. P4 (trim pot) should be adjusted for a 0-Vdc output at pin 7 of IC2B.

### Building a percussion synthesizer

The circuit illustrated in Fig. 15-19 is an extremely versatile circuit that can be used for a variety of applications. In its basic form, it is a *twin-tee filter/oscillator*, with adjustable Q and frequency. If the Q is adjusted high enough, it breaks into oscillation; and thus it becomes a low-distortion *sine-wave oscillator*. If the Q is adjusted to "far below" the oscillation point, it is a *high-Q filter*. If the Q is adjusted to "just below" the point of oscillation, it is a *dampened waveform generator*, or percussion synthesizer.

To use the circuit as an oscillator: D1, R2, R1, and SW1 are removed. P2, the Q-adjustment potentiometer, is adjusted to the point of oscillation, and the frequency is adjusted with P1 and P3.

To use the same circuit as a high-Q filter: D1, R1, and SW1 are removed, and R2 is connected directly to pin 2 of IC1. The input signal is applied to the other end of R2. P2 is adjusted until all

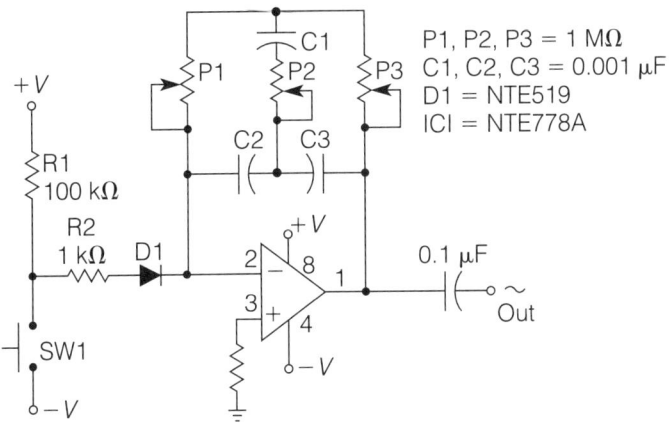

■ **15-19** *Twin-tee resonant frequency sine-wave oscillator (percussion synthesizer).*

"ringing" (dampened waveforms) disappears when an input signal is applied. P1 and P3 are then used to "tune" the filter.

This circuit also forms the heart of a *percussion synthesizer* (electronic drums) in the dampened mode. There are so many possible variations to this basic circuit, I can not possibly list all of them, but I trust your imagination will take over before you can try out the examples given in this context.

With the component values given, the circuit of Fig. 15-19 is the most basic form of the percussion synthesizer. The dual power supply voltages are not critical. They can be any level that is nominally used with operational amplifiers. SW1 should be a momentary, pushbutton switch. IC1 can be almost any type of internally frequency-compensated op amp (such as the common 741, NTE778A, and many others).

After building the circuit, connect the output to any type of audio amplifier. Set P1 and P3 in their (approximate) middle position, and adjust P2 until a constant tone is heard. Then, reverse the adjustment very slightly until the tone just dies out. Now, by pressing SW1, you should hear an electronic drum sound. By playing with the P1 and P3 values, the pitch of the drum should change, and even bell sounds can be produced.

You might also want to experiment with different triggering methods, depending on the type of op amp used. For example, try connecting the bottom side of SW1 to circuit common (instead of $-V$); or by disconnecting R1 from $+V$; or use a double-pole switch, and switch both. Use the method that provides the most "solid" initiating pulse.

To make a set of electronic bongos, refer to Fig. 15-20. One bongo is made by starting with two pieces of single-sided PC board material. They can be about the size illustrated in the diagram, or they might be cut into round pieces of suitable size, if so desired. Place one PC board on top of the other with the foil sides together (facing each other). Temporarily tape the boards together, and drill 5 small holes completely through both boards. Separate the boards, lay them foil side up, and form 5 "blobs" of solder over each hole on both of the boards (the holes are only needed to line up the solder contact points). Silver solder might be used to reduce the problem of oxidation on the surface of these switch "points." After soldering, lay the boards on a table, blob-side up; lay a wide file on top of the blobs on one of the boards, and lightly file the tops until they are all *planar* (no one blob higher than another); then, repeat this process on the other board.

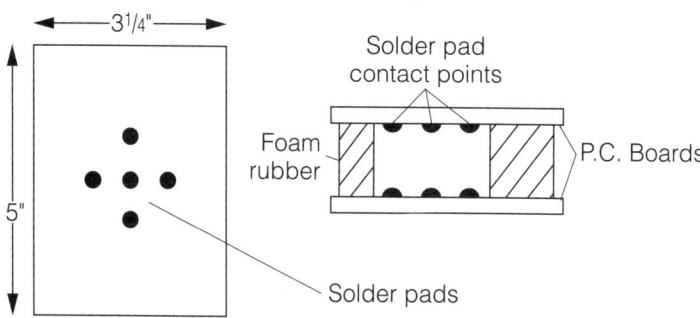

■ **15-20** *Physical construction of a drum pad.*

The final steps are to solder a piece of insulated hook-up wire to the foil side of each board to be able to substitute this assembly for SW1 in Fig. 15-19. A more professional-looking cable might be affected by substituting microphone cable, or audio coax, for the hook-up wires. Using some thin pieces of foam rubber as spacers, position one board above the other (with the contact points facing each other) as illustrated. The contact points should not touch each other, but they should be close enough so that a light pressure, or "tap," will cause them to make contact (the separation distance is exaggerated in the illustration for the sake of clarity).

This assembly can be held together by gluing both sides of the foam rubber to the PC boards (an epoxy "putty" does a good job).

If a more rugged assembly is desired, holes can be drilled in the corners of the PC boards and small bolts with locking nuts can be used to hold the boards together. This "large momentary switch" has a response similar to commercially available drum machine "pads." Obviously, two such assemblies (and two percussion synthesizer circuits) are needed for a "set" of bongos. One synthesizer circuit is tuned slightly higher than the other for a realistic bongo sound.

Figure 15-21 shows some other variations that can be incorporated into the basic percussion synthesizer. By removing R1 and SW1 (Fig. 15-19), a low-frequency, square-wave oscillator can be used to automatically operate the synthesizer. The oscillator output should be connected to the unconnected side of R2. Either one of the circuits illustrated in Figs. 15-14 and 15-15 will work very well for this application. Adjusting the frequency of the square-wave oscillator will set the "beat." A toggle switch could be installed to switch from "manual" (using SW1, or the drum pad assembly) to "automatic" (oscillator control).

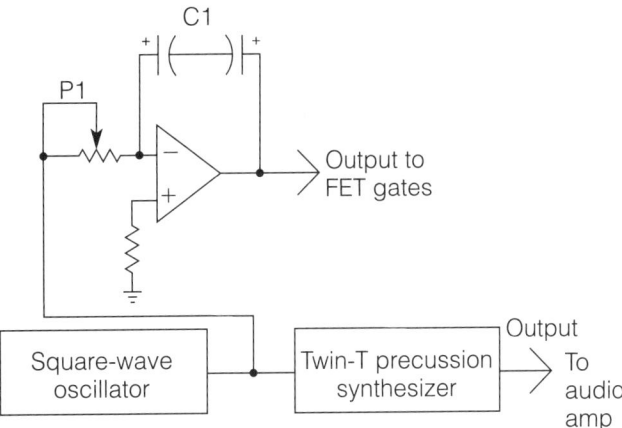

**15-21** *Block diagram of automatic operation, automatic tuning percussion synthesizer.*

A very impressive modification can be added to automatically tune the synthesizer while it is operating automatically. The effect is similar to that of a kettle drum; as the sound is decaying, it is also changing pitch. Referring to Fig. 15-19, P1 and P3 are replaced with two JFETs. The source and drain leads are connected to the same points as the two ends of the variable resistors, leaving the gate leads open. The orientation of the source and drain is not critical in this circuit. Referring back to Fig. 15-21, the output of the square-wave oscillator, which is automatically operating the synthesizer, is also applied to the input of an integrator. The triangle-wave output of the integrator is applied to the gates of both JFETs (be sure that the voltage polarity is correct for the type of JFET used). The two JFETs operate as voltage-controlled resistors, changing the pitch in perfect synchronization with the beat. Synchronization is maintained because both the beat, and the pitch shift, are being controlled by the same oscillator.

Continuing to refer to Fig. 15-21, you should experiment with a few different values of P1 and/or C1 to explore the great variety of complex sounds that can be generated from this circuit. A good starting point is to use a 100-k$\Omega$ potentiometer for P1, and two back-to-back 10-$\mu$F capacitors for C1 (as previously discussed, placing two electrolytic capacitors back-to-back creates a nonpolarized electrolytic capacitor).

There are some additional variations that could be incorporated into the Fig. 15-21 circuit for some really spectacular effects, depending on how far you want to go with it. The output of the integrator could be applied to an inverting amplifier to change the direction of the pitch shift. Another possibility is to increase the frequency of the square wave oscillator, and to use its output as the input to a digital counter. Two, three, four (or more) $Q$ outputs from the digital counter could be used to initiate drum sounds from multiple synthesizer circuits. These multiple synthesizer circuits would not beat at the same rate, but they would beat "in time" because each would be beating at some "harmonic" division of the oscillator frequency.

I leave the remaining infinite number of possibilities to you, and to your imagination and ingenuity.

# 16

# Radio and television

RADIOS AND TELEVISIONS ARE ELECTRONIC SYSTEMS. Although it is a little out of context with the goals of this book to become involved with systems discussions, I believe that a well-rounded foundational knowledge of the electrical/electronic fields should include the basic operational theory behind radio and television. I also believe you will find this chapter interesting because radio and television have always been the backbone of the modern electronics industry.

The majority of the building blocks incorporated into radio and TV systems have already been discussed. This chapter will focus on the three main areas of theory needed to "fill the gaps" so that a final systems discussion can be more easily understood. These three topics are: the modulation of signal intelligence, the transmission and reception of RF signals, and the cathode-ray tube.

## Modulation

*Modulation* is the use of one electrical signal to "control" a primary variable of another. For example, if an audio signal voltage is used to control the "amplitude" of a carrier signal, the result is *amplitude modulation*.

It is important that you do not confuse "mixing" with "modulation." *Mixing* occurs when two (or more) signals are simply combined in a linear network. Modulation, however, requires one signal to "control" a variable of another; variables such as the amplitude of an RF signal (amplitude modulation, or AM), the frequency (frequency modulation, or FM), the pulse width (pulse width modulation, or PWM), the phase (phase modulation, or PM) or the pulse code (pulse-code modulation, or PCM). Unfortunately, the electronics industry has traditionally retained many circuit names that are incorrect in this regard. For instance, when you examine the actual circuit operation of many circuits labeled as a "mixer/oscillator," you will discover that it is really a "modulator/oscillator."

Strictly speaking, when two signals are mixed, they combine without the creation of any additional frequencies. When two signals are modulated, they are said to "beat" with each other creating additional frequencies called "beat frequencies." If the two modulated signals are sinusoidal, the beat frequencies will be the sum and the difference of the original frequencies.

AM radio broadcast transmissions contain two signals of primary importance to the user: the "carrier signal," and the "audio signal," or the "program signal." The *carrier frequency* is the frequency to which the radio receiver is tuned for station selection. For example, the AM radio band (also referred to as the *medium wave broadcast band*, or simply the *broadcast band*) is legally designated from 535 kHz to 1605 kHz. If your favorite local radio station broadcasts on 830 kHz, this means that the carrier frequency being used for transmission is 830 kHz. The audio signal, or program, is riding on this carrier frequency.

Figure 16-1 illustrates an amplitude-modulated waveshape as it would appear when picked up by a radio antenna. Notice that the carrier frequency is much higher than the program signal riding on it. In actuality, there is not a literal program signal "on top of" the carrier. When the AM signal was broadcast, the program signal modulated the amplitude, or the level, of the carrier; this process formed an "envelope" of carrier amplitude, having the same "shape" as the program signal.

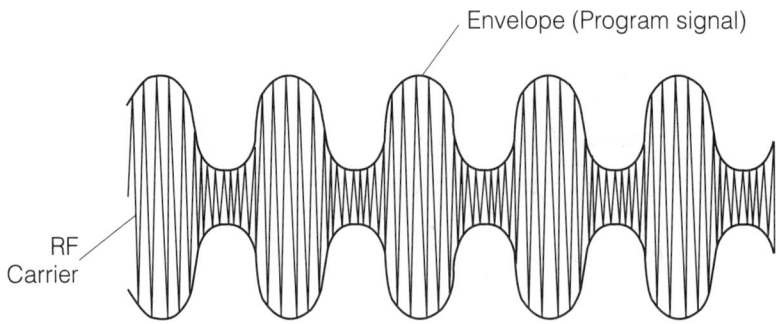

■ **16-1** *AM-modulated waveform.*

The "beat" frequencies, contained within the AM waveform of Fig. 16-1, will be the sum and the difference of the carrier and its "program" signal. For example, if the program signal was a constant 5-kHz tone, with a carrier frequency of 600 kHz, the beat frequencies would be 595 kHz (*difference frequency*) and 605 kHz (*sum frequency*). In a typical AM broadcast, the program signal will contain vocal and music information, making up a very wide "range" of frequencies. The highest frequency, of this range of frequencies, will determine the maximum separation of the beat frequencies from the carrier. For example, if the highest frequency in the program signal was limited to 1 kHz, then the beat frequencies would be 599 kHz and 601 kHz. However, as you can see from the earlier example, when the highest frequency is limited to 5 kHz, the width (or distance from the carrier frequency) increases. The range of beat frequencies above (and below) the carrier frequency are called *sidebands*. The width of the sidebands is closely monitored at AM broadcast stations; because, if they become too wide, they can interfere with adjacent stations.

*Frequency modulation (FM)* radio signals also have a carrier signal and a program signal. However, the program signal does not ride "on" the carrier frequency; it is contained "within" frequency variances modulated into the carrier signal. Because the program signal is not dependent upon carrier amplitudes (as is AM transmissions), FM radio is largely immune to many forms of interference. Figure 16-2 illustrates an exaggeration of an FM-modulated waveform.

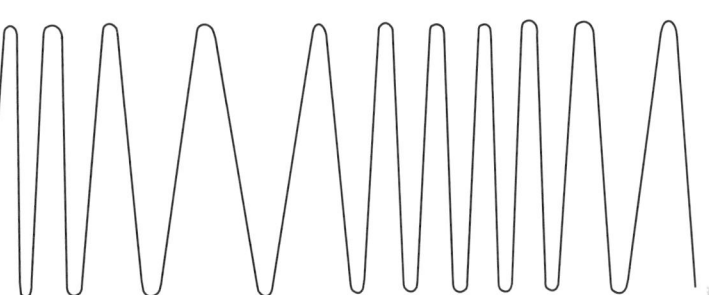

■ **16-2** *FM-modulated waveform.*

## Transmission and reception of RF signals

When ac current flows through a conductor, a moving electrical field is developed around the conductor. This field, consisting of both electromagnetic energy (current-related) and electrostatic energy (voltage-related) is radiated perpendicular to the wire, similarly to visible light radiating outward from a light source. If the frequency of the ac current is within the radio frequency spectrum (approximately 10 kHz to 30,000 MHz), the radiated energy is called *RF energy*. The most efficient transmission of RF energy occurs when the ac conductor is resonant at the frequency of the ac current. Special conductors designed to be resonant at specific RF frequencies are called *antennas* (either transmitting or receiving).

Think of the antenna as a transducer. A *transducer* is a bi-directional device that will take an electronic signal and convert it into another form of signal energy. For instance, piezoelectric elements can be used as speakers, translating the electronic signal into an audio emission. Conversely, they can also receive an audio signal, and then convert that signal into it's electronic equivalent (a microphone). An antenna can receive an electronic signal, and then convert that signal into an electromagnetic wave that will radiate through space. It can also receive an electromagnetic signal, and then convert that signal into a series of electronic pulses. This conversion of energies can then be utilized, in a receiver, to extract the signal intelligence that had originally been impressed, or modulated, into it at the transmitter.

RF energy travels at the speed of light (186,280 miles, or 299,800,000 meters, per second). Every frequency of RF energy has a specific "wavelength" in meters, calculated by dividing 300,000 by the frequency in kHz. The most efficient transmission (or reception) of RF energy will take place when the antenna length is the same as the RF energy wavelength. At this length, the antenna becomes *resonant* to the RF energy. Slightly less efficient omni-directional transmission/reception will also occur at even-harmonic, sub-divisions of the wavelength, such as at 1/2-wave or 1/4-wave lengths. Conversely, somewhat greater efficiencies are found in even-harmonic lengths that are greater than a wavelength. These variations of the efficiencies in an antenna occur because of the changes in the directional properties of the antenna at the various lengths. Because the wavelengths of the lower RF frequencies can be very long, 1/2-wave or 1/4-wave antennas are usually much more practical.

Now that some of the major definitions are out of the way, consider how an actual amplitude-modulated RF transmission takes place. To begin with, the RF transmitter must have an *oscillator* to produce the desired carrier frequency. The resonant frequency of the transmission antenna will be chosen to match the carrier frequency produced by this master oscillator. The signal from the oscillator proceeds through the *intermediate power amplifier (IPA)*, sometimes referred to as the *driver*. The carrier frequency from the IPA is then applied to a *final power amplifier* circuit, for boosting to an efficient level for transmission.

The audio signal, having been amplified through the *speech amplifier* sections, then drives the high-powered modulator circuit. This modulator section might be connected into the plate supply (for tubes units), or into the collector/or/drain supply (for solid-state amplifiers). By varying the plate current, or the drain/collector current, the modulator impresses (or modulates) the program signal onto the carrier signal. This composite signal is then passed on to a final amplifier *tank circuit* for impedance matching, and then conveyed to the antenna. This system is often referred to as being a *high-level modulation*.

If the modulating signal is applied to the final amplifier through its cathode, or through one of its grids (for tubes); or through the source or emitter circuits (for solid-state amplifiers); the amount of modulating power is much less than would be required for high-level modulation. These methods are referred to as *low-level modulation*. As a general rule, tube-type amplifiers usually use high-level modulation, because of its greater efficiency. Solid-state power amplifiers usually are configured for low-level modulation and, in fact, often modulate several stages in phase. Any of these composite signals can then be applied to the input of a *"linear" RF power amplifier*, if further output power is needed.

Rf oscillators and RF power amplifiers are essentially the same types of circuits that you have already studied. Some of the peculiarities of RF circuits involve the critical nature of parts placement. At very high frequencies, even slight amounts of stray capacitance or lead inductance can provide unwanted signal coupling or attenuation. High frequency RF signals have the tendency to "ride on the surface" of conductors causing unexpected resistances on solid conductors (a phenomenon known as *skin effect*). For these, and other reasons, RF circuits must be designed with more attention toward the physical details of a circuit; much more so than for lower frequency circuits.

Most high-power RF amplifiers are designed for class-C operation in order to improve their efficiency. Class C amplifiers have a resonant LC circuit (called a *tank circuit*) between the amplifier and the load. A tank circuit can be thought of as a "flywheel" that smoothes out the short pulses produced by a class-C amplifier. The tank fills in the missing portions of the pulsed input signal. It produces a dampened sinusoidal waveform, at its resonant frequency, every time it is pulsed (just like the percussion synthesizer discussed in the previous chapter). Therefore, even though the class-C amplifier only pulses the tank circuit, the carrier frequency output is a clean sine wave.

After the modulated RF signal is output from the RF power amplifier, it is coupled to the transmission antenna for broadcast into the atmosphere. The other job of the tank circuit is to match the output impedance of the amplifier to the characteristic impedance of the transmission cable and the antenna.

Lower frequency radio waves (≤300 kHz), called "longwaves," have more of a tendency to travel along the curvature of the earth. Radio waves that propagate in this manner are called *groundwaves*. Midfrequency radio waves (from 300 kHz to 3 MHz), called *mediumwaves*, travel by a combination of groundwaves and *skywaves* (radio waves radiating upward, with a percentage of them bouncing off of the ionosphere layer of the upper atmosphere and returning to the earth). High-frequency (HF) radio waves (3 to 30 MHz), called *shortwaves*, are mostly propagated via skywaves; and they are commonly used for global communications. Very high frequency (VHF) and ultra high frequency (UHF) radio waves (30 to 300 MHz and 300 to 3000 MHz, respectively) behave more like a beam of light, traveling in a line-of-sight manner. For this reason, they are called *direct waves*.

## Radio receivers

The first stage of a radio receiver is, of course, the antenna. Receiver antennas are usually designed to be resonant at about the midpoint of the RF spectrum for the band being received. For lower frequency BCB reception (broadcast band; commonly and errantly referred to as AM), even a 1/4-wave antenna is usually impractical to use (because of the long wavelength). Therefore, many mediumwave BCB antennas are "loaded" (reactive components are added) to make the antenna resonant at lower frequencies. This loading does somewhat decrease efficiency, however.

Most modern BCB antennas are actually part of the "front end" tuned circuit (a tunable tank circuit). In a *loopstick antenna*, the antenna wire is wound on a ferrite rod, and placed in parallel with the tuning capacitor. The primary advantages of this system are that the antenna is very compact, and its *null* (direction of minimal signal reception) is very directional. This directional characteristic of the reception can be used to "null-out" an unwanted station on the same frequency, while allowing the desired station to be clearly received. The end of the ferrite rod must point toward the offending transmitting station for maximum signal attenuation.

Virtually all FM receivers need an external antenna. Because of the higher frequency of the FM band, a 1/2-wave FM dipole antenna is about 4.8 feet in length; a 1/4-wave whip antenna is only about 29 inches.

After the radio signal is received by the antenna, this extremely small RF signal voltage needs to be amplified by an *RF amplifier stage* (some smaller, inexpensive radios omit this stage). Referring to Fig. 16-3, the RF amplifier stage contains a tunable capacitor (or a varactor) to allow it to amplify only the desired station to which it is tuned.

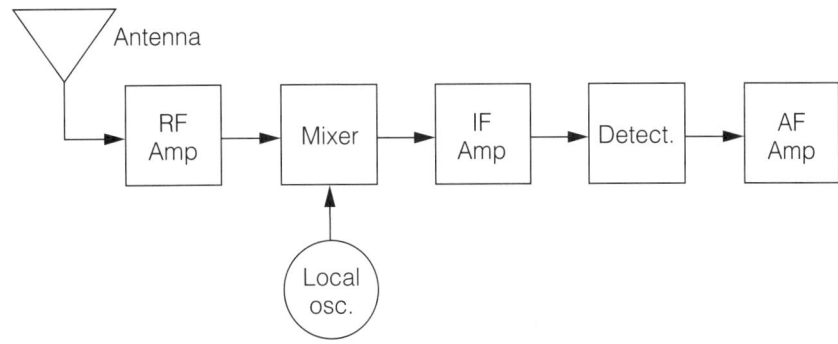

■ **16-3** *Block diagram of a superheterodyne receiver.*

The amplified RF signal is then applied to a mixer stage. The mixer stage (technically, a modulator) modulates the amplified RF with a signal from the local oscillator. The frequency of this local oscillator is controlled by the radio's tuner to continuously produce a lower "mixing product" of (usually) 455 kHz called the *intermediate frequency (IF)*. In other words, as you tune in a station on a radio, you are also changing the frequency of the local oscillator, causing it to "track" with the amplified RF signal, and to produce a constant 455-kHz IF.

This method of processing an RF signal provides higher gain than, and improved selectivity over, individually tuned RF amplifiers. By using a lower frequency, greater separation of adjacent stations might be afforded; thus resulting in greater selectivity. And, by concentrating the major part of the signal amplification process into one (IF) frequency, fewer components and stages are required; thus resulting in greater efficiency, simplicity, and savings. This system of mixing two signals in order to produce a third frequency is called *heterodyning*. Most modern radios use this process, and they are referred to as *superheterodynes*, or "superhets" for short. The processing of FM signals is identical to that described for AM, except that the i.f. frequency used in FM receivers is 10.7 MHz.

Continuing to refer to Figure 16-3, the intermediate frequency (IF) is then fed into one (or more) stages of i.f. amplifiers, and it is eventually applied to the "detector" stage. The *detector* is the stage that "extracts" the original program signal from the IF carrier.

## AM and FM detectors

Figure 16-4 illustrates an AM detection circuit. Notice that it is hardly more than a diode. In the early days of radio, diodes were "crystals" of galena or pyrite. Many of these early "crystal" radios were nothing more than a long, tunable antenna connected to a diode detector. Sensitive high-impedance headphones were used to listen to the program material.

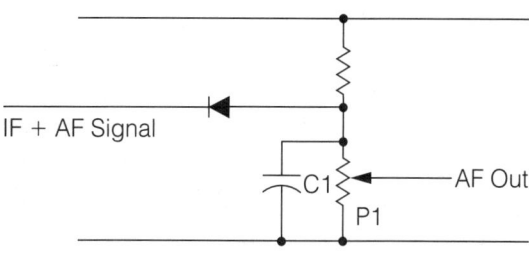

■ **16-4** *Basic AM detector circuit.*

To understand how a diode functions as an AM detector, refer back to Fig. 16-1. Notice that the program material is contained within the amplitude variations of the carrier wave. In actuality, it is carried on the top and the bottom of the carrier wave, with the bottom being a "mirror image" of the top. However, at any given instant, the audio component voltages (but not the RF voltages) of the upper and lower halves tend to cancel each other out; being 180 degrees out-of-phase with each other. Carefully examine where the positive and negative peaks of the two signals exist at several "instants." The RF signal alternates its peaks. The audio does not.

If a diode was used to "half-wave rectify" this signal, the result would be the full program signal (either the top, or the bottom) and half of the carrier wave, with no phase cancellation. In other words, it would be just like cutting the waveform horizontally, through the middle, and removing half of this energy to pass along to the audio stages. Referring back to Fig. 16-4, the half-wave rectified IF signal is applied across the RC network of C1 and P1.

The value of C1 is chosen to filter out the high-frequency IF carrier. This capacitance value has a low reactance to frequencies in the IF range, but it has a very high reactance to the lower audio frequencies. In this manner, the IF carrier is shunted to ground. The audio component, however, seeks the lower impedance path through the pot (P1), to the wiper, and on to the audio amplifier stages. P1 is the audio volume control.

FM detection is a little more complicated. FM detection circuits are called *discriminators* or autodetectors. Figure 16-5 is technically an autodetector. The primary and secondary of the input transformer, together with C1 and C2, are both designed to be resonant at the FM IF of 10.7 MHz. The frequency variations making up the program signal cause the resonant circuit to look more inductive (or more capacitive, depending on the direction of the frequency variation). This, in turn, creates unequal voltage levels across D1 and D2, in proportion to the original program signal. C3 and C4 filter out the unwanted IF (in a similar fashion to blocking capacitor in the AM detector), and the remaining program signal is applied to an audio amplifier for reproduction. The discriminator in a quality FM receiver is much more complicated than the simplified circuit of Fig. 16-5, but the operational principle is the same.

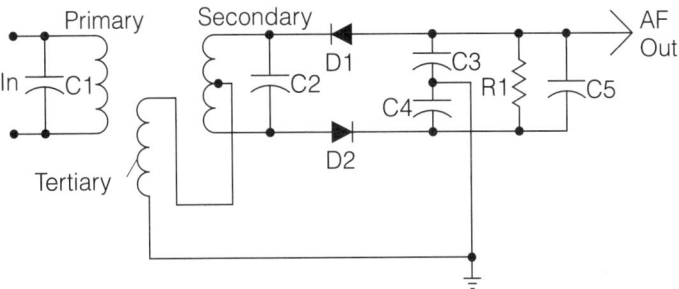

■ **16-5** *Basic FM discriminator.*

## Television

The majority of the principles that apply to radio also apply to television. The television RF signal (called the *composite video signal*) is transmitted from a broadcast station. It is received by a TV antenna, and the various components of the video signal (both AM and FM) are amplified and processed by superheterodyne action (in the same basic way as they are for radio). Of course, the composite video signal contains audio, video, and "chroma" (coloring) information. It is formatted with both FM (sound) and AM (picture) components. Even though the complexity of this signal is greatly increased, it only requires the addition of several resonant circuits to separate, or *split*, the various components: audio, video, synchronization, and colorburst signals.

The primary difference between radio and TV lies in the conversion of the video signal back into a visible picture. This conversion takes place in a special type of vacuum tube called a *cathode ray tube*, or "CRT."

A basic side-view diagram of a CRT is illustrated in Fig. 16-6. A CRT has many things in common with most other vacuum tubes: a *filament* to heat the cathode, a heated *cathode* which emits electrons, an *anode* to which electrons are attracted, and a *grid* to control the flow of electrons. In a loose way, you can consider these elements to be similar to a bipolar transistor's emitter, collector, and base, respectively.

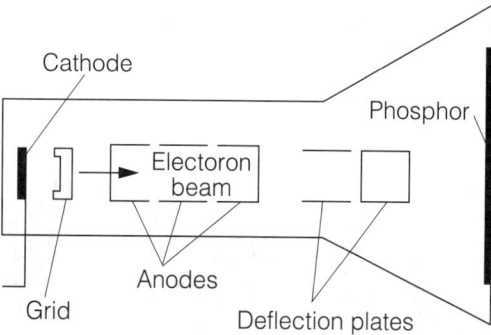

■ 16-6 *Basic diagram of a cathode ray tube.*

A unique aspect of a CRT is how the anodes are used as "electron accelerators." The electrons are attracted toward the anodes when they leave the cathode. But the ringed shape, and the voltage potentials (about 300 to 400 Vdc) of the beam-forming anodes, cause the electrons to "overshoot" them. This effect culminates into a narrow "beam." After this beam is formed, the electrons are further attracted toward the screen area by a very high-voltage (10 to 25 kVdc), positive potential applied toward the front of the screen by the second anode. The forward velocity of the electrons forces them to hit a phosphor coating on the front of the screen producing a tiny spot of light. The phosphor will absorb energy from the electron beam, and it will continue to glow for a short period, even after the beam is removed. This effect is known as *persistence*.

In order to light up the entire screen area, the tiny electron beam must be deflected in a linear pattern, called a *raster*, at a very high rate of speed. The movement of the beam is fast enough to allow it to cover the entire screen area, and to return to its starting point, before the beginning spot of phosphor totally looses its glow. In this manner, the whole screen surface can stay continually bright. This process is repeated at a 60-Hz rate.

The CRT illustration in Fig. 16-6 is actually the type of CRT used in oscilloscopes. CRTs used in TVs do not use deflection plates, as shown in the illustration. They incorporate *deflection coils* placed around the narrow part of the CRT, in an assembly called the "yoke." However, for illustration purposes, its much easier to show the action of deflection plates. The basic operational principle is the same for both.

For beam deflection purposes, two oscillators are utilized. The horizontal oscillator output is applied to the horizontal deflection plates. As the voltage varies on these plates, the electron beam will be attracted or repelled, depending on the voltage polarity, causing the electron beam to be deflected from right to left or left to right across the screen. The vertical oscillator output is applied to the vertical deflection plates, and this moves the beam up and down.

The horizontal and vertical oscillators are synchronized to cause the beam to start at the top of the screen and "draw" 263 horizontal lines of light from top to bottom; then the beam returns to the top, and draws another 262 lines "in-between" the original 263 lines. This whole process occurs 30 times per second, with the two rasters of 263 and 262 lines adding up to form a total raster of 525 lines. The total number of individual phosphor dots that must be excited for each full screen is about 315,000. This process is known as *interlaced scanning*.

By modulating the intelligence information onto the electron beam (causing it to vary in intensity, as it scans the screen) a black-and-white picture will be produced.

The conventional color CRT is made with three *electron guns* for the three primary colors; red, green, and blue. Each phosphor dot on the CRT screen is formed by separate red, blue, and green phosphors arranged in a triangle. The picture is thus scanned by a triangle of beams converging on each triangle of dots at the same rate as in a black-and-white picture. Individual intelligence information is provided to each gun, by the chroma board, telling it when (and how strongly) to fire. The various combinations of these guns' emissions form the resulting color picture on the screen.

# Symbols and equations

THE FOLLOWING SYMBOLS AND EQUATIONS ARE USED often in the electrical and electronic fields. In addition to equations discussed throughout the text, this section also contains many other equations for future reference. I have tried to illustrate the "case" of the symbols, and the abbreviations as they will appear most often; but many sources may use upper case letters, where this section lists them as lower case, and vice versa.

### Letter Symbols and Abbreviations

| Symbol | Meaning |
|---|---|
| A | Ampere; amp |
| $A$ | Length of the "adjacent" side; usually the longer; in a right triangle, in the same units as the other sides |
| AC(ac) | Alternating current |
| Ah | Ampere-hour |
| AM | Amplitude modulation |
| amp | Ampere, or amplifier |
| ant | Antenna |
| ASCII | American Standard Code for Information Interchange |
| assy | Assembly |
| aud | Audio |
| aux | Auxiliary |
| AWG | American Wire Gauge |
| $B$ | Susceptance (measured in seimens or mhos); reciprocal of reactance |
| $bw$ | Bandwidth |
| $C$ | Capacitance (measured in farads) |
| C | Collector |
| C | Coulomb |
| $D$ | Dissipation factor; reciprocal of storage factor $Q$ |
| DC(dc) | Direct current |
| DPDT | Double-pole, double-throw (switch) |
| DPST | Double-pole, single-throw (switch) |
| $d$ | Thickness of the dielectric material in a capacitor (measured in cm) |

| | |
|---|---|
| dB | Decibel (one-tenth of a bel); the logarithmic ratio between two levels of power |
| $E$(emf) | Electromotive force (measured in volts) |
| E | Emitter |
| EMT | Electrical metallic tubing |
| ERP | Effective radiated power |
| eV | Electronvolt |
| F | Farad (a measure of capacitance) |
| $F$ | Temperature (measured in degrees Fahrenheit) |
| FM | Frequency modulation |
| $f$ | Frequency (measured in hertz) |
| $G$ | Conductance (measured in seimens or mhos) |
| GHz | Gigahertz |
| H | Henry (a measure of inductance) |
| $H$ | Length of the hypotenuse of a right triangle, in the same units as the other (adjacent and opposite) sides |
| HF | High frequency |
| HT | High tension (high current and high voltage) |
| HV | High voltage |
| Hp | Horsepower |
| Hz | Hertz |
| $I$ | Electrical current (measured in amperes) |
| IC | Integrated circuit |
| ID | Inside diameter |
| IF | Intermediate frequency |
| IN(in) | Input |
| I/O | Input/output |
| IR | Infrared |
| $J$ | Energy, work, or quantity of heat (measured in joules) |
| $K$ | Coupling coefficient |
| $k$ | Dielectric constant |
| K | Temperature in kelvin |
| k | Kilo- (thousand) |
| kHz | Kilohertz |
| kV | Kilovolt |
| kW | Kilowatt |
| kWh | Kilowatt hour |
| $L$ | Inductance (measured in henries) |
| LW | Longwave |
| LF | Low frequency |
| LP | Low-pass |
| LSB | Lower sideband |

*Appendix A*

| | |
|---|---|
| LSI | Large-scale integration |
| $l$ | Length |
| MW | Mediumwave |
| M | Mega- (million) |
| MeV | Megaelectron volt |
| M$\Omega$ | Megohm (mega-ohm) |
| $M$ | Mutual inductance (measured in henries) |
| m | Milli- (one-thousandth) |
| mA | Milliampere |
| mH | Millihenry |
| mic | Microphone |
| mom | Momentary (switch) |
| ms | Millisecond |
| mV | Millivolt |
| mW | Milliwatt |
| N or n | General symbol for numbers |
| NC | No connection |
| NO | Normally open |
| n | Nano- (one-billionth) |
| neg | Negative |
| neut | Neutral |
| nF | Nanofarad |
| nom | Nominal |
| norm | Normal |
| ns | Nanosecond |
| $O$ | Length of the "opposite" side to the adjacent side (A) in a right triangle, in the same units as the other sides |
| osc | Oscillator |
| out | Output |
| $P$ | Power (measured in watts) |
| PC | Printed circuit |
| PCB | Printed circuit board |
| PCM | Pulse-code modulation |
| PWM | Pulse-width modulation |
| pF | Picofarad |
| pf | Power factor |
| pk | Peak |
| pos | Positive |
| pot | Potentiometer |
| preamp | Preamplifier |
| PS | Power supply |
| PU | Pickup |
| pri | Primary |

| | |
|---|---|
| pwr | Power |
| $Q$ | Quality factor; the ratio between reactance and resistance |
| $Q$ | Ratio between inductance and resistance, when specifying inductor quality |
| $Q$ | Quantity |
| $R$ | Resistance (measured in ohms) |
| RC | Resistive/capacitive |
| RF | Radio frequency |
| RFI | Radio frequency interference |
| RLY | Relay |
| RL | Resistive/inductive |
| rcv | Receive |
| rcvr | Receiver |
| rect | Rectifier |
| ref | Reference |
| res | Resistor |
| rms | Root mean square |
| rmt | Remote |
| rot | Rotate |
| rpm | Revolutions per minute |
| rps | Revolutions per second |
| rpt | Repeat |
| S | Siemens (conductance) |
| $S$ | Area of one plate of a capacitor (measured in square centimeters) |
| SB | Sideband |
| SNR | Signal-to-noise ratio |
| sec | Secondary |
| sel | Selector |
| sft | Shaft |
| sig | Signal |
| sol | Solenoid |
| spk | Speaker |
| spkr | Speaker |
| SPDT | Single-pole, double-throw (switch) |
| SPST | Single-pole, single-throw (switch) |
| sq | Square |
| stby | Standby |
| subassy | Subassembly |
| SW | Shortwave |
| sw | switch |
| $T, t$ | Temperature, or time |
| tel | Telephone |

| | |
|---|---|
| UHF | Ultrahigh frequency |
| UF | Ultrasonic frequency |
| V | Volt |
| V/A, VA | Voltampere |
| vid | Video |
| VLF | Very low frequency |
| VOM | Volt-ohmmeter |
| VTVM | Vacuum-tube voltmeter |
| W | Watt |
| wdg | Winding |
| WVdc | Working volts - dc |
| $\mu$ | Micro- (one-millionth) |
| $\mu A$ | Microampere |
| $\mu F$ | Microfarad |
| $\mu H$ | Microhenry |
| $\mu S$ | Microsecond |
| $X$ | Reactance (measured in ohms); the opposition to alternating current exhibited by reactive components |
| $Y$ | Admittance (measured in seimens or mhos); the reciprocal of impedance |
| $Z$ | Impedance (measured in ohms); the reactive and resistive opposition to the flow of alternating current |
| $\Theta$ | 90 - degrees |
| $\lambda$ | Wavelength (measured in meters) |
| $\pi$ | Pi; 3.1416 ... |
| $W_C$ | Cutoff frequency |
| $W_0$ | Midband frequency |

## Formulae

### Admittance:

$$Y = \frac{1}{\sqrt{R^2 + X^2}} \quad Y = \frac{1}{Z} \quad Y = \sqrt{G^2 + B^2}$$

### Average value:

$Average\ value = 0.637 (\text{peak value})$

$Average\ value = 0.900 (\text{rms value})$

### Capacitance:

Capacitors in parallel:

$$C_{\text{Total}} = C_1 + C_2 + C_3 + ... C_n$$

Capacitors in series:

$$C_{Total} = \frac{1}{\frac{1}{C_1} + \frac{1}{C_2} + \frac{1}{C_3} + \ldots \frac{1}{C_n}}$$

Two capacitors in series:

$$C_{Total} = \frac{C_1 C_2}{C_1 + C_2}$$

Capacitance value of a capacitor:

$$C = 0.0885 \frac{KS(N-1)}{d}$$

Quantity of charge stored:

$$Q = CE$$

where: $Q$ = charge (in coulombs)

$C$ = capacitance (in farads)

$E$ = voltage across capacitor (in volts)

Amount of stored energy:

$$W = \frac{CE^2}{2}$$

where: $W$ = energy (in joules)

$C$ = capacitance (in farads)

$E$ = voltage across capacitor (in volts)

### Conductance

$$G = \frac{1}{R} \quad G = \frac{I}{E}$$

$G_{Total} = G_1 + G_2 + G_3 + \ldots G_n$ (Resistors in parallel)

### Cosine

$$\cos\Theta = \frac{A}{H} \quad \cos\Theta = \sin(90 - \Theta)$$

$$\cos\Theta = \frac{1}{\sec\Theta}$$

## Cotangent

$$\cot\Theta = \frac{A}{0} \quad \cot\Theta = \tan(90 - \Theta)$$

$$\cot\Theta = \frac{1}{\tan\Theta}$$

## Decibel

$$dB = 10 \log \frac{P_1}{P_2}$$

$dB = 20 \log \frac{E_1}{E_2}$  Only if source and load impedance are equal

$dB = 20 \log \frac{I_1}{I_2}$  Only if source and load impedance are equal

$dB = 20 \log \frac{E_1 \sqrt{Z_2}}{E_2 \sqrt{Z_1}}$  Source and load impedances are unequal

$dB = 20 \log \frac{I_1 \sqrt{Z_1}}{I_2 \sqrt{Z_2}}$  Source and load impedances are unequal

## Figure of merit

$$Q = \frac{X}{R} \quad Q = \tan\Theta \quad Q = \frac{L}{R}$$

## Frequency

$$f = \frac{3.0 \times 10^5}{\lambda} \text{ (meters)} \quad f = \frac{9.84 \times 10^5}{\lambda} \text{ (feet)}$$

## Impedance

$$Z = \sqrt{R^2 + X^2} \quad Z = \sqrt{G^2 + B^2}$$

$$Z = \frac{R}{\cos\Theta} \quad Z = \frac{X}{\sin\Theta} \quad Z = \frac{E}{I}$$

$$Z = \frac{P}{I^2 \cos\Theta} \quad Z = \frac{E^2 \cos\Theta}{P}$$

*Formulae*

## Inductance

Inductors in series:
$$L_{Total} = L_1 + L_2 + L_3 + \ldots L_n$$

Inductors in parallel:
$$L_{Total} = \frac{1}{\frac{1}{L_1} + \frac{1}{L_2} + \frac{1}{L_3} + \ldots \frac{1}{L_n}}$$

Two inductors in parallel:
$$L_{Total} = \frac{L_1 L_2}{L_1 + L_2}$$

Coupled inductances in series with fields aiding:
$$L_{Total} = L_1 + L_2 + 2M$$

Coupled inductances in series with fields opposing:
$$L_{Total} = L_1 + L_2 - 2M$$

Coupled inductances in parallel with fields aiding:
$$L_{Total} = \frac{1}{\frac{1}{L_1 + M} + \frac{1}{L_2 + M}}$$

Coupled inductances in parallel with fields opposing:
$$L_{Total} = \frac{1}{\frac{1}{L_1 - M} + \frac{1}{L_2 - M}}$$

Mutual inductance of two coils with fields interacting:
$$M = \frac{L_A - L_O}{4}$$

where: $L_A$ is the total inductance of both coils with fields aiding

$L_O$ is the total inductance of both coils with fields opposing

Coupling coefficient of two rf coils inductively coupled so as to give transformer action:
$$K = \frac{M}{L_1 L_2}$$

## Meter formulas

$$\text{Ohms/volt} = \frac{1}{I} \text{ (meter sensitivity)}$$

where: $I$ = Full scale current in amperes

Meter resistance:

$$R_{\text{Meter}} = \frac{E_{\text{Full scale}}}{I_{\text{Full scale}}}$$

Current shunt:

$$R_{\text{Shunt}} = \frac{R_{\text{Meter}}}{N - 1}$$

where: $N$ is the new full-scale reading divided by the original full-scale reading (both in the same units)

Voltage multiplier:

$$R = \frac{\textit{full-scale reading required}}{\textit{full-scale current of meter}} - R_{\text{Meter}}$$

where: Reading is in volts, and current is in amperes

## Ohm's law for dc circuits

$$I = \frac{E}{R} \quad I = \sqrt{\frac{P}{R}} \quad I = \frac{P}{E} \quad R = \frac{E}{I}$$

$$R = \frac{P}{I^2} \quad R = \frac{E^2}{P} \quad E = IR \quad E = \frac{P}{I}$$

$$E = \sqrt{PR} \quad P = I^2R \quad P = EI \quad P = \frac{E^2}{R}$$

## Ohm's law for ac circuits

$$I = \frac{E}{Z} \quad I = \sqrt{\frac{P}{Z \cos\Theta}} \quad I = \frac{P}{E \cos\Theta}$$

$$Z = \frac{E}{I} \quad Z = \frac{P}{I^2 \cos\Theta} \quad Z = \frac{E^2 \cos\Theta}{P}$$

$$E = IZ \quad E = \frac{P}{I \cos\Theta} \quad E = \sqrt{\frac{PZ}{\cos\Theta}}$$

$$P = I^2 Z \cos\Theta \quad P = IE \cos\Theta \quad P = \frac{E^2 \cos\Theta}{Z}$$

**Peak value**

$$E_{Peak} = 1.414(rms\ value)\quad E_{Peak} = 1.57(average\ value)$$

**Peak-to-peak value**

$$E_{P-P} = 2.828(rms\ value)\quad E_{P-P} = 3.14(average\ value)$$

**Phase angle**

$$\Theta = arc\tan\frac{X}{R}$$

**Power factor**

$$pf = \cos\Theta\quad D = \cot\Theta\ (dissipation)$$

**Reactance**

$$X_L = 2\pi fL\quad X_C = \frac{1}{2\pi fC}$$

**Resistance**

Resistors in series:

$$R_{Total} = R_1 + R_2 + R_3 + ...\ R_n$$

Resistors in parallel:

$$R_{Total} = \frac{1}{\frac{1}{R_1}+\frac{1}{R_2}+\frac{1}{R_3}+...\frac{1}{R_n}}$$

Two resistors in parallel:

$$R_{Total} = \frac{R_1 R_2}{R_1 + R_2}$$

**Resonance**

$$f_{Res} = \frac{1}{2\pi\sqrt{LC}}\quad L = \frac{1}{4\pi^2 f^2 C}\quad C = \frac{1}{4\pi^2 f^2 L}$$

**Right triangle**

$$\sin\Theta = \frac{O}{H}\quad \cos\Theta = \frac{A}{H}\quad \tan\Theta = \frac{O}{A}$$

$$\sec\Theta = \frac{H}{A}\quad \cot\Theta = \frac{A}{O}$$

## Root mean square (Sinusoidal waveshapes only)

$$rms = 0.707(\text{peak value}) \quad rms = 1.111(\text{average value})$$

## Secant

$$\sec\Theta = \frac{H}{A} \quad \sec\Theta = \frac{1}{\cos\Theta}$$

$$\sec\Theta = \text{cosecant}(90 - \Theta)$$

## Sine

$$\sin\Theta = \frac{O}{H} \quad \sin\Theta = \frac{1}{\text{cosecant}\,\Theta}$$

$$\sin\Theta = \cos(90 - \Theta)$$

## Susceptance

$$B = \frac{X}{R^2 + X^2} \quad B = \frac{1}{X}$$

$$B_{\text{Total}} = B_1 + B_2 + B_3 + \ldots B_n$$

## Tangent

$$\tan\Theta = \frac{O}{A} \quad \tan\Theta = \frac{1}{\cot\Theta} \quad \tan\Theta = \cot(90 - \Theta)$$

## Transistors, bipolar

$$I_c = I_b(\beta_{dc}) \quad I_e \approx I_c \quad E_e = E_b - 0.7\ \text{Vdc (silicon)}$$

$$E_e = E_b - 0.3\ \text{Vdc (germanium)} \quad Z_b = R_e(\beta_{dc})$$

$$I_e = \frac{E_e}{R_e} \quad A_p = \frac{P_{\text{output}}}{P_{\text{input}}}$$

## Transistors, field-effect

$$G_{fs} = \frac{\Delta I_D}{\Delta E_{GS}}$$

where: $G_{fs}$ is the transconductance value

$\Delta E_{GS}$ is a change in gate to source voltage

$\Delta I_D$ is a subsequent change in drain current

## Temperature

$$°C = (0.556°F) - 17.8 \quad °F = (1.8°C) + 32 \quad °K = °C + 273$$

**Transformer ratio**

$$\frac{N_p}{N_s} = \frac{E_p}{E_s} = \frac{I_s}{I_p} = \sqrt{\frac{Z_p}{Z_s}}$$

**Units of energy**

$$watt\text{-}hours = PT$$

where: $P$ = power (in watts)
$T$ = time (in hours)

**Wavelength**

$$\lambda = \frac{300{,}000}{f} \text{ (in meters)}$$

# Sources for electronic materials

THE SOURCES LISTED IN THIS SECTION ARE CLASSIFIED under the following subheadings:

- [ ] Data book sources
- [ ] Electronic kit suppliers
- [ ] Full-line suppliers
- [ ] Surplus dealers
- [ ] Tools and hardware suppliers
- [ ] Electronic textbook publishers
- [ ] Electronic test equipment dealers
- [ ] Electronic periodicals

Please keep in mind that there is much overlapping within these categorizations. For example, few surplus dealers sell "only" surplus items, and many are quite diversified in their product lines. The goal here is to provide some means of organization, based on "primary" commercial efforts.

## Data book sources

(Note: Thomson Consumer Electronics is the source for the SK line of electronic components. MCM and Jameco offer multiple data book lines.)

NTE Electronics
44 Farrand ST.
Bloomfield, NJ 07003
800-631-1250

MCM Electronics
650 Congress Park Dr.
Centerville, OH 45459-4072
800-543-4330

Thomson Consumer Electronics
Accessories & Components
 Business
2000 Clements Bridge Rd.
Deptford, NJ 08096-2088

Jameco Electronic Components
1355 Shoreway Rd.
Belmont, CA 94002-4100
800-831-4242

## Electronic kit suppliers

(Note: SEAL Electronics offers many of the projects contained within this book in kit form.)

Ramsey Electronics Inc.
793 Canning Parkway
Victor, NY 14564
716-924-4560

SEAL Electronics
P.O. Box 268
Weeksbury, KY 41667
606-452-4135

Mark Five Electronics, Inc.
8019 E. Slauson Ave.
Montebello, CA 90640
800-423-3483

Active Kits
345 Queen Street W.
Toronto, Ontario M5V 2A4
800-465-5487

## Full-line suppliers

(Note: These are large retailers and distributors. They can supply every electronic need from data books to tools.)

Kelvin Electronics
10 Hub Drive
Melville, NY 11747
800-535-8469

MCM Electronics
650 Congress Park Drive
Centerville, OH 45459-4072
800-543-4330

Jameco Electronics
1355 Shoreway Rd.
Belmont, CA 94002-4100
800-831-4242

Allied Electronics
7410 Pebble Drive
Fort Worth, TX 76118
800-433-5700

Dalbani
4225 Northwest 72nd Ave.
Miami, Florida 33166
800-325-2264

Mouser Electronics
2401 Hwy. 287 N.
Mansfield, TX 76063-4827
800-346-6873

Parts Express
340 E. First St.
Dayton, OH 45402-1257
800-338-0531

Digi-Key Corp.
701 Brooks Ave. South
P.O. Box 677
Thief River Falls, MN 56701-0677
800-344-4539

## Surplus dealers

Fair Radio Sales
1016 E. Eureka St.
P.O. Box 1105
Lima, OH 45802
419-227-6573

Southpaw Electronics
P.O. Box 886
New Hyde Park, NY 11040-0311
800-851-8870

International Microelectronics
P.O. Box 170415
Arlington, TX 76003
800-999-0463

B.G. Micro, Inc.
P.O. Box 280298
Dallas, TX 75228
800-276-2206

Brigar Electronics
7-9 Alice St.
Binghamton, NY 13904
607-723-3111

Herbach and Rademan
P.O. Box 122
18 Canal St.
Bristol, PA 19007-0122
215-788-5583

C and H Sales
2176 E. Colorado Blvd.
Pasadena, CA 91107
800-325-9465

DC Electronics
P.O. Box 3203
Scottsdale, AZ 85271-3203
800-467-7736 or 800-423-0070

All Electronics Corp.
14928 Oxnard St.
P.O. Box 567
Van Nuys, CA 91408-0567
800-826-5432

Marlin P. Jones & Assoc., Inc.
P.O. Box 12685
Lake Park, FL 33403-0685
407-848-8236

Hosfelt Electronics, Inc.
2700 Sunset Blvd.
Steubenville, OH 43952-1158
800-524-6464

## Tools and hardware suppliers

Techni-Tool
5 Apollo Rd.
Box 368
Plymouth Meeting, PA 19462-0368
610-941-2400

Contact East, Inc.
335 Willow St.
North Andover, MA 01845-5995
800-225-5334

Jensen Tools Inc.
7815 S. 46th St.
Phoenix, AZ 85044-5399
800-426-1194

Time Motion Tools
12778 Brookprinter Place
Poway, CA 92064-06810
800-779-8170

Harbor Freight Tools
3491 Mission Oaks Blvd.
Camarillo, CA 93011-6010
800-423-2567

## Electronic textbook publishers

(Note: I highly recommend joining the electronics book club offered through TAB/McGraw-Hill. They are, by far, the largest and most comprehensive in their field.)

TAB/McGraw-Hill
Blue Ridge Summit, PA 17294-0840
800-822-8158

For book club information, write to:
Electronics Book Club
P.O. Box 508
Columbus, OH 43216-0508

## Electronic test equipment dealers

Tucker Electronics
P.O. Box 551419
Dallas, TX 75355-1419
800-527-4642

RAG Electronics, Inc.
2450 Turquiose Circle
Newbury Park, CA 91320-1200
800-380-3457

Global Specialties
70 Fulton Terrace
New Haven, CT 06512
800-572-1028

C & S Sales
1245 Rosewood Ave.
Deerfield, IL 60015
800-292-7711

## Electronic periodicals

Electronics Now
Subscription Service
P.O. Box 51866
Boulder, Colorado 80321-1866
800-999-7139

Popular Electronics
Subscription Dept.
P.O. Box 338
Mount Morris, IL 61054-9932
800-827-0383

Nuts & Volts Magazine
430 Princeland Court
Corona, CA 91719
800-783-4624

Depending on your personal interests, you may want to contact the Edmund Scientific Co. for their latest catalog. Although they do not cater specifically to the electronics industry, they offer many items that fall into a "gray" area. Their catalog is fascinating, and it is entertaining.

Edmund Scientific Co.
101 East Gloucester Pike
Barrington, NJ 08007-1380
609-573-6260

# Index

## A
abbreviations, 321-325
active vs. passive devices, 128
addressing, computers, 284
admittance, 325
air dielectric capacitors, 31
alligator clip leads, 12
alpha-numeric displays, LEDs, 161
alternating current (ac), 41-42, **41**, 57-69
ammeters, 17-18
amp-hour (Ah) rating, batteries, 241
ampere as measure of current, 40, 112
amplifiers, 134, 170-171, **171**, 273
  audio amplifiers (*see* audio amplifiers)
  power amplifiers (*see* power amplifiers)
  proportional amplifier, 298, **298**
amplitude, 39, 59-61
amplitude modulation (AM), 309, 310-311, **310**
Amprobe, 17-18
analog signals and computers, 261, 285
AND logic gate, 263-264, **264**
anode and cathode, diodes, 92, **92**
antennas, 312, 314
apparent power, 68
apparent resistance, 121
arithmetic logic unit (ALU), 280, 281
assembly language, 284
astable multivibrators, 165, 266
attenuation, 127, 187
audio amplifiers, 179-210, 313
audio taper, 29
automobile emergency flasher, 238, **238**
avalanche voltage, zener diodes, 151-152
average value, **59**, 60, 325

## B
bandpass filter, 257, **257**, 295, 296, **296**, 297
bandstop filter (*see* notch filters)
Bardeen, John, 128
base, transistors, 129-134
batteries, 119, 241-246
battery charger circuit, 243-246, **246**
beat frequency, 225, 310-311
beta multiplier, 171
beta value, 130-131, 174
biasing transistors, 137, 179-181
bidirectional transient suppression diodes, 158
binary coded decimal (BCD) counter, 269
binary number system, 261-262
bipolar transistors, 331
bistable multivibrators, 266, 267
bits and bytes, 281
blown-fuse alarm, 176-177, **176**
books, periodicals, and manuals, 4-7
bootstrapping, 194
bounce, 271
bounceless switch, 276, **276**
Brattian, Walter, 128
breadboards, 165
breakdown devices, 157
breakdown voltages, diodes, 92, 94
bridges, diode bridges, 33
broadcast signals, 310
buffers, 224, 273
buses, 281
bypass capacitors, 135

## C
capacitance and capacitors, 30-31, 107-126, 325, 326
  bypass capacitors, 135
  capacitive reactance, 288-289
  coupling, 134-135
  parallel configuration, 290, 326
  series configuration, 290, 326
  shunting, 135
  voltage-controlled capacitor (*see* varactor)
capacitive reactance, 288-289
carrier frequency, 310
cathode and anode, diodes, 92, **92**
cathode ray tubes (CRTs), 318-320, **318**
CD-ROM, 283
central processing unit (CPU), 280
charge coupled devices (CCD), 164
chirping circuit, 169-170, **169**
chokes or coils, 32-33, 64, 70
circuits, 36-50, **36**, **37**, **40**, **46**, **49**, **50**, **52**, **53**, 164-165
class-A audio amplifier circuit, 181-182, **181**
class-AB audio amplifier circuit, 182-184, **183**
cleaning supplies, 12
clocks, 266, 271, 273, **273**, 274, **274**
CMOS, 263
coefficient of coupling, inductors, 328
coils (*see* chokes or coils)
coincidential point, SCRs, 212
cold joints, soldering flaw, 73, 75
collector, transistors, 129-134

color codes for resistors, 26-28, **27**
common mode rejection, 192
common-base transistors configuration, 140-141,
common-collector transistors configuration, 138-140, **139**
common-emitter transistors, 135-138, **136**, 141
compilers, 285
complex ac waveforms, 300
computers, 279-285, **280**
conditional logic, 265-266, **265**
conductance and conductors, 42, 90, 326
constant current source circuit, 189-190, **190**
constant-K filters, 295
continuity, 37, 40, 81
cores, inductance and inductors, 64
cosine, 326
cotangent, 327
coulomb as measure of current, 39, 112
counter electromotive force (cemf), 65, 70, 88
counters, 267-269, **268**, **270**, 271, 273
coupling, 134-135
covalent bonding, 90
crossover distortion, 182
crystal-based timebase oscillator, 276-277, **277**
current, 36-38, **36**, 39-40, 43-54, 127, 133
current lag, inductance and inductors, 67
current limit mode, 77
current-regulating circuit, LED application, 175, **175**
current-transformer ammeter (Amprobe), 17-18
cycles per second (cps), 41, 57, **58**

## D
dampening, 304
Darlington pair, 171
data books, 4-6
dc circuits, Ohm's law, 329
decibels, 186-187, 327
decimals, 261
deflection coils, CRT, 319
depletion area, 129-130
desoldering (*see* soldering/desoldering)
detectors, AM and FM, 316-317, **316**, **317**

Illustrations are in **boldface**.

diacs, 157-158, 215, 216
dielectric absorption, 109
dielectric constant, 126
dielectrics, capacitance and capacitors, 107-108, 126
differential amplifiers, 191-193
differentiator circuit, 300, **300**
digital audio amplifiers, 185
digital decade counter, 269, **270**
digital electronics, 261-278
digital multimeter (DMM), 14
digital pulser or clock circuit, 273, **273**
digital voltmeter (DVM), 14-15, **14**, 102, 149-150
digitizing, 285
diodes, 33-34, **33**, 91-101, **91**, 129, 159-164
direct current (dc), 41-42, **43**, **44**, 69-70
direct waves, 314
discharge time constant, 118
discriminators, FM, 317, **317**
dissipation of power, 54-55, 60-61
distortion, 182, 188
doping, 91
doubler, voltage doubler circuit, 175-176, **175**
drain, field-effect transistors (FET) 227
drivers, 312
drop in voltage, 95, 131-132
dual inline package (DIP), 247
duty cycles, 167, 184, 278
dynamic range, 187-188

**E**

earth ground, 80
effective (rms) power, 68
effective values, 61
efficiency ratios, 127
electrical potential, 39
electromagnetic fields, 63, **63**, 64
electromotive force, 39, 65
electron flow, 40
electron guns, CRT, 320
electrons and atomic-level electrical movement, 36, 89-91
emitter, transistors, 129-134
energy, units, 332
EPROMs, 272, 283
equivalent circuits, 47-49
etching circuit boards, 200-201, **201**, **204**
exclusive OR and exclusive NOR logic gate, **263**, **265**
execute cycle, 282
exponential curves, 112

**F**

50-watt audio amplifier, 198-220, **198**, **204**, **205**, **206**, **208**
fall time, 271
fanout, 265
farad as unit of capacitance, 112
Faraday, Michael, 125
feedback, 137, 193
fetch cycle, 282
field effect transistors (FET), 227-240, **229**, 331

figure of merit, 327
filter capacitors, 114-119, **114**, **116**, 135
filters
 bandpass filter, 257, **257**, 295, **296**, 297
 constant-K filters, 295
 high-pass filters, 257-259, **258**, 295-297, **296**
 high-Q filter, 304
 LC filters, 295
 low-pass filters, 257-259, **258**, 295-297, **295**
 notch filter, 256-257, **256**, 295, 297, 298
 passive filters, 294-298, **295**, **296**, **297**
 rolloff, 257
 Sallen-Key filters, 257-259, **258**
firmware, 283
flashing light circuits, 165-169, **166**, **169**, 238, **238**
flip-flops, **263**, 266, 267, 273
flux, flux lines, 64
frequency, 41, 58-59, 327
frequency counters, 16
frequency generators, 167
frequency modulation (FM), 309, 311, **311**
frequency response, 188, 292, 294, **294**
full-wave bridge rectifiers, 96-98, **96**, 102-105, **103**, 116-119, **116**
full-wave rectifiers, 96, **96**, 98-101, **98**, **99**, **100**
function generators, 16, **17**, 303-304, **304**
fundamental frequency, 299

**G**

gain, 127, 133, 139
gate, field-effect transistors (FET), 227
grounding, 80, 233
groundwaves, 314

**H**

half-wave rectifiers, 94-96, **94**, 114-116, **114**
hardware, 282-283
harmonic distortion, 188
harmonics, 299
headphone amplifier circuit, 170-171, **171**
heatsinks, 34, 75, 78, 146
henry as unit of inductance, 65
hertz as measure of frequency, 41, 58
heterodyning, 315-316
HEXFETs, 238
high-pass filters, **257**-**259**, 295-297, **296**
high-Q filter, 304
holding current, zener diodes, 152
holes, transistors, 130
hum reducer circuit, 256-257, **256**
hybrid modules, 248

**I**

impedance, 134-135, 188-189, 291, 327

impedance matching, 141-142, **141**
inductance and inductors, 32-33, 63-69, 290, 328
inductive reactance, 88, 287-288, 289
input impedance, 189
input/output (I/O), 280, 284, 285
instantaneous voltage, 60
insulating sleeves, 147
insulators, 90
integrated circuits (IC), 22-23, 34-35, **35**, 247-260
integrator circuit, 299-300, **299**
interference, 192
interlace, CRT, 319
intermediate frequency (IF), 315
intermodulation distortion, 188
inventory of parts and materials, 18-24
inverter, power MOSFET inverter circuit, 239-240, **239**
isolation transformers, 13, **13**, 220-222, **221**

**J**

JFETs, 224, 227, **228**, 229, 231, **231**, 232

**K**

knee area, 225

**L**

lab or workroom setup, 7-14
lab power supply, 12-14, 76-88, **80**, 101-105, **104**, 122-125, **122**, **123**, 142-150, **145**, **146**, 252-254, **253**, **254**
lasers, 159-160, 162-163
latch circuit, SCR, 222-223, **222**
latches, 272
LC filters, 295
leakage current, 118, 136
leveling effect, 148
light-controlled sound circuit, 236-238, **237**
light-emitting diodes (LED), 22, 34, 159-160, 160-161, 175, **175**
line operated power supplies, 119
line-level inputs/outputs, 194-195
linear power amplifiers, 313
linear taper, 29
liquid crystal display (LCD), 163-164
load, 42, 118, 148, 179-181
load time constant, 118
locking terminal solder lug, 78
logarithmic taper, 29
logarithms, logarithmic scale, 186
logic gates, 262-266, **263**
logic probes, 16, 275, **275**
logic pulsers, 16, 274, **274**
low-pass filters, 257-259, **258**, 295-297, **295**

**M**

machine cycle, 282
machine language, 284
magazines, periodicals, 6-7
memory, 272, 273, 280, 283-284
meters, 329
mho as measure of conductance, 42

microphone noise-reduction circuit, 255-256, **256**
microprocessors, 280
mixing, 309
modulation, 225, 309, 313
modulo eight counter, 267-269, **268**
monostable multivibrators, 266
MOSFETs, 227, **228**, 229, 232, 233-236, **235**
motor controller, PWM, 277-278, **277**
"mouse" circuit, squeak or chirp circuit, 169-170, **169**
multivibrators, 165, 266-271
mutual inductance, 328

## N
N-type materials, 91
NAND logic gate, **263**, 264-265
neon tubes, **215**, 216
noise, 158-159
NOR logic gate, **263**, 265
notch filter, 256-257, **256**, 295, **297**, 298
NPN and PNP transistors, 129, 185

## O
1-watt audio amplifier, 260, **260**
Ohm's law, 40, 43-54, **43**, 62-63, 72, 117, 121, 329
ohms as measure of resistance, 41
one-shots, 266
operational amplifiers, 248-250, **249**
optocouplers, 161-162
optoisolators, 161-162
OR logic gate, **263**, 264
oscillators, 266, 273, 312
  crystal-based timebase oscillator, 276-277, **277**
  sine-wave oscillator, 304
  square-wave, 300-302, **301**
  UJT, 223-225, **224**
oscilloscopes, 15-16, **15**

## P
P-type materials, 91
parallel circuits, 46-50, **46**, **49**
parallel-in/parallel-out operation, 271-272
parallel-resonant circuit, 293, **293**
parts and materials inventory, 18-24
peak forward surge current, 117
peak reverse voltage (PRV), 94
peak value, 59, **59**, 62-63, 95, 120, 330
peak-inverse voltage (PIV), 94
peak-to-peak value, 59, **59**, 62-63, 119, 330
pentavalent impurities, 91
percussion synthesizer, 304-308, **305**, **306**, **307**
periodicals, magazines, 6-7
persistence (CRT), 319
phase modulation (PM), 309
phase relationships, 67-68, **67**, 87, **87**, 215, 330
photodiodes, 162
photoeyes, 161-162
photons, 159
photoresistive cells, 162

phototransistors, 162
photovoltaic cells, 159
PIN diodes, 157
pinch-off region, 229
polarity of voltage, 39, 87
ports, 284
potentiometers, 22, 26, 28-30, **30**
power, 42, 54-55, 68, 127, 139, 186
power amplifiers, 186, 189-193, **190**, 312, 313
power factor, 69, 330
power logs, 187
power supplies (see also batteries), 16, **17**
  50-watt audio amplifier power supply, 205-207, **206**
  battery supplies, 119
  dual polarity power supplies, 100-101
  lab power supply (see also power supply)
  line operated power supplies, 119
  noise, noise suppression, 158-159
  quad power supply, 254-255, **255**
  raw dc power supplies, 119-122, 123
  transients, transient suppression, 158-159
  zener-regulated power supplies, 153-155, **153**, 173-174, **174**
preamplifiers, 194, **194**, **195**
primaries and secondaries, transformers, 70-71, **70**
printed circuit boards, 74, 200-201, **201**, **205**
programming computers, 284-285
programs, 282
proportional amplifier, 298, **298**
protons, 89
pulse code modulation (PCM), 309
pulse width modulation (PWM), 184, 277-278, **277**, 309
pulses, 16
push-pull amplifiers, 182

## Q
quad power supply, 254-255, **255**
quantum physics, 159-160
quiescent transistor operation, 180

## R
radio, 309-317
random access memory (RAM), 272, 283
rasters, 319
raw dc power supplies, 119-122, 123
reactance, 88, 134-135, 287-289, 330
read-only memory (ROM), 272, 280, 281, 283
recharger circuit for batteries, 243-246, **246**
recovery time, diodes, 158
rectification and rectifiers, 89-105
  full-wave bridge rectifiers, 96-98, **96**, 102-105, **103**, 116-119, **116**
  full-wave rectifiers, 96, **96**, 98-101, **98**, **99**, **100**
  half-wave rectifiers, 94-96, **94**, 114-116, **114**

silicon-controlled rectifiers (SCR), 211-214, **212**
voltage doubler circuit, 175-176, **175**
redundant mathematics, 281-282
referencing, 101, 174
reflected impedance, 291
registers, 281
relay driver, trip-point, 259-260, **259**
resistance and resistors, 26-28, **27**, 36-38, 41, 121, 330
  direct current (dc), 69-70
  Ohm's law, 43-54
  parallel configuration, 326, 330
  photoresistive cells, 162
  series configuration, 330
resistive-capacitive (RC) circuit, 110, **110**
resolution, 285
resonance, 291-294, **292**, **293**, **294**, 312, 330
resonant frequency, 292
RF signals, transmission and reception, 311-314
rheostats (see also potentiometers), 26, 28-30, 213-214, **213**
right triangles, 330
ripple, 98, 105, 119, 135
rise time, 271
rms audio amplifier, 12-watt, 196-198, **196**
rolloff, 257
root mean square (rms), 59, 61, 62-63, 68, 120, 331

## S
safety, 7-8, 77, 84-85
Sallen-Key filters, 257-259, **258**
salvaging/recycling parts, 18-23
scanning, CRT, 319
schematics, 47
Schottky diodes, 157
secant, 331
semiconductors (see also diodes), 91
sensitivity, 189
sequential flashing light circuit, 167-169, **169**
serial-in/serial-out operation, 272
series circuits, 50-52, **50**, **52**
series-parallel circuits, 52-54, **53**
series-pass transistors, 144
series-resonant circuit, 291-292, **292**
seven-segment decoder/driver, 269, **270**, 271
shells and rings, atomic level, 89-91
shift registers, 271-272, 273
shock hazards (see safety)
Shockley, William, 128
short-circuit secondary current, 117
shorting, 137-138, 148, 149
shortwaves, 314
shunting, 135
sidebands, 311
siemen as measure of conductance, 42
signal diodes, 22
signal generators, 16, **17**
signal-to-noise ratio (SNR), 189
silicon hot-carrier diodes, 157
silicon-controlled rectifiers (SCR), 211-214, **212**, 222-223, **222**

sine, 331
sine waves, 57, **58**, 330
sine-wave oscillator, 304
single-junction devices (*see* diodes)
siren circuit, 302-303, **303**
skywaves, 314
slew rate, op amps, 248-249
slugs, transformer or coil slugs, 33
software, 282-283
solar cells, 159
soldering/desoldering, 72-76
  controller for soldering iron, 216-219, **217**, 219-220, **219**
  tools, irons, 9-12, **9**, **11**
solderless breadboards, 165, **166**
solid-state devices, 89-91
sound, 54
sound generator, light-controlled, 236-238, **237**
source time constant, 118
source, field-effect transistors (FET), 227
speaker protection circuit, 207-210, **208**
square waveforms, 299
square-wave generator, 167, 303-304, **304**
square-wave oscillators, 300-302, **301**
static-electricity damage, 232-233
step-up or step-down transformers, 71
strain relief, 78
substrate, field-effect transistors (FET), 230
superheterodyne receiver, 315-316, **315**
suppliers and sources, 333-336
surge current, 117
surplus-supply parts, 23-24
susceptance, 331
sweep generators, 16
switches, 37-38, 276, **276**
switching diodes, 22
symbols and abbreviations, 321-325

## T

12-watt rms audio amplifier, 196-198, **196**
tangent, 331
tank circuits, 313
taper, 29

television, 317-320
temperature, 331
temperature coefficients, 108-109, 136
test equipment, 14-18
thermal joint compound, 147
thermal runaway, 136
thyristors, 211, **215**
time constants, 66-67, 112-113, 118
time delay relay (TDR), 171-173, **172**
timebase, crystal oscillator, 276-277, **277**
tinning a soldering iron, 74
tone control circuit, 194-195, **195**
tone generator, 302, **302**
tools, 8-12
traces, 74
transconductance, FETs, 230-231
transducers, 312
transformers, 32-33, **32**, 70-72, **70**, **72**, 332
  isolation transformers, 13, **13**, 220-222, **221**
transients, 158-159
transistors, 34, **34**, 127-150, **128**, **130**, 215-216
transorbers, 158
triacs, 214-215, **215**
triangle-wave generator, 303-304, **304**
triangular waveforms, 299
triggering, 157, 215
trim pot (*see* potentiometers; rheostats)
trimmer capacitors, 31
trip-point relay driver, 259-260, **260**
trivalent impurities, 91
true power, 68
truth tables for logic gates, **264**
TTL, 263
tunnel diodes, 157
turns ratio, transformers, 71
twin-tee notch filter, 256-257, **256**, 304

## U

unidirectional surge clamping diodes, 158
unidirectional transient suppression diodes, 158
unijunction transistors (UJT), 215-216, 223-225, **224**

## V

vacuum tube voltmeters (VTVM), 14, 102
valence shell, atomic level, 90
varactor diodes, 156, **156**
varistor diodes, 158
VLSI chips, 247, 280
volatile vs. nonvolatile memory, 272, 283
volt-amp rating (VA), transformers, 71-72, 120
volt-ohm-milliammeter (VOM), 14, 102-103
voltage, 37-39, 65
  drop, 131-132
  gain, 127
  leveling effect, 148
  loading effect, 148
  Ohm's law, 43-54
  referencing, 101
  ripple, 98, 105, 119, 135
  zero reference line, 98
voltage and current decibel logs, 187
voltage divider, 137
voltage doubler circuit, 175-176, **175**
voltage regulators, 144-145, **145**, 153-155, **153**, 173-174, **174**, 251-252

## W

water-flow analogy of electrical current, 36-37, **36**
watts, 42, 186, 332
waveforms, waveform analysis, 15-16, 57, 101, 299-300
wavelengths, 332
window comparator, 259-260, **259**
workbenches, 8
working Vdc, capacitors, 116-117

## X

XNOR and XOR logic gates, **263**, 265

## Z

zener diodes, 151-155, **152**
  regulator for power supplies, 153-155, **153**, 173-174, **174**
zero reference line, 59, **59**, 98, 182